D0913161

TECHNIQUES OF CHEMISTRY

(Arnold Weissberger, *Editor*)

VOLUME V

TECHNIQUE OF ELECTROORGANIC SYNTHESIS

PART III

SCALE-UP AND ENGINEERING ASPECTS

TECHNIQUES OF CHEMISTRY

VOLUME V, PART III

TECHNIQUE OF ELECTROORGANIC SYNTHESIS
Scale-up and
Engineering Aspects

Edited by

N. L. WEINBERG

Electrosynthesis Company
E. Amherst, New York

B. V. Tilak

Hooker Research Center
Grand Island, New York

1807 1982

A WILEY-INTERSCIENCE PUBLICATION

JOHN WILEY & SONS

New York   Chichester   Brisbane   Toronto   Singapore

Library of Congress Cataloging in Publication Data:

Main entry under title:

Technique of electroorganic synthesis.

(Techniques of chemistry; v. 5)
    Part 3 also has subtitle: scale-up and engineering aspects/edited by N. L.
Weinberg, B. V. Tilak.
    "A Wiley-Interscience publication."
    Includes bibliographical references and index.
    1. Electrochemistry. 2. Chemistry, Organic—Synthesis. I. Weinberg,
Normal L. II. Tilak, B. V.

QD61.T4 vol. 5 [QD273]   542s [547'.2] 73-18447
ISBN 0-471-93271-X (v. 1)   AACR2
ISBN 0-471-06359-2 (r. 3)

Printed in the United States of America

10 9 8 7 6 5 4 3 2 1

AUTHORS OF PART III

RICHARD ALKIRE

Department of Chemical Engineering, University of Illinois, Urbana, Illinois

THEODORE BECK

Electrochemical Technology Corporation, Seattle, Washington

C. R. CAMPBELL

Monsanto Chemical Intermediates Company, Pensacola, Florida

L. CARLSSON

Swedish National Development Company, Akersberga, Sweden

C. Y. CHENG

Department of Chemical Engineering, Clarkson College of Technology, Potsdam, New York

W. V. CHILDS

Phillips Petroleum Company, Bartlesville, Oklahoma

D-T CHIN

Department of Chemical Engineering, Clarkson College of Technology, Potsdam, New York

D. E. DANLY

Monsanto Chemical Intermediates Company, Pensacola, Florida

D. DEGNER

BASF AG, Main Laboratory, Ludwigshafen, West Germany

HANS FEESS

Institut Für Chemische Technologie der TH Darmstadt, Federal Republic of Germany

H. HOLMBERG

Foundation for Industrial Organic Electrochemistry, Chemical Center, University of Lund, Lund, Sweden

B. JOHANSSON

Foundation for Industrial Organic Electrochemistry, Chemical Center, University of Lund, Lund, Sweden

K. B. KEATING

E. I. DuPont DeNemours and Company, Wilmington, Delaware

A.NILSSON

Foundation for Industrial Organic Electrochemistry, Chemical Center, University of Lund, Lund, Sweden

ROBERT RUGGERI

Electrochemical Technology Corporation, Seattle, Washington

S. SARANGAPANI

Union Carbide, Cleveland, Ohio

MARK STADTHERR

Department of Chemical Engineering, University of Illinois, Urbana, Illinois

V. D. SUTLIC

E. I. DuPont DeNemours and Company, Wilmington, Delaware

B. V. TILAK

Hooker Research Center, Long Road, Grand Island, New York

H. V. K. UDUPA

Central Electrochemical Research Institute, Karaikudi, India

K. S. UDUPA

Central Electrochemical Research Institute, Karaikudi, India

NORMAN L. WEINBERG

The Electrosynthesis Company, East Amherst, New York

HARTMUT WENDT

Institut für Chemische Technologie der TH Darmstadt, Federal Republic of Germany

INTRODUCTION TO THE SERIES

Techniques of Chemistry is the successor to the Technique of Organic Chemistry Series and its companion—Technique of Inorganic Chemistry. Because many of the methods are employed in all branches of chemical science, the division into techniques for organic and inorganic chemistry has become increasingly artificial. Accordingly, the new series reflects the wider application of techniques, and the component volumes for the most part provide complete treatments of the methods covered. Volumes in which limited areas of application are discussed can easily be recognized by their titles.

Like its predecessors, the series is devoted to a comprehensive presentation of the respective techniques. The authors give the theoretical background for an understanding of the various methods and operations and describe the techniques and tools, their modifications, their merits and limitations, and their handling. It is hoped that the series will contribute to a better understanding and a more rational and effective application of the respective techniques.

Authors and editors hope that readers will find the volumes in this series useful and will communicate to them any criticisms and suggestions for improvements.

ARNOLD WEISSBERGER

Research Laboratories
Eastman Kodak Company
Rochester, New York

PREFACE

What information does the industrial chemist or chemical engineer require to carry a successful bench-scale electroorganic process into the scale-up mode? If we assume nontechnical concerns, such as product marketability, profitability, growth, and competition, have been adequately answered, will the candidate process stand up to the following questions?

1. *Is the electrolytic route the best synthetic technique?* If the product is amenable to chemical catalytic hydrogenation or oxidation, the electrochemical route may not compete (see Chapters V and IX).

2. *How do the best chemical routes compare with the proposed electrochemical method?* Comparison of the economics, feedstock cost and availability, and present versus future energy requirements should be developed and potential environmental issues should be addressed.

3. *What scale-up problems could exist and what must be done to overcome them to achieve commercial viability?* Costly cell components and cell design, high-energy-consuming separation procedures, and electrode fouling have been the demise of some otherwise promising electrochemical processes.

More than 5 years have elapsed since the publication of Part I and Part II of *Technique of Electroorganic Synthesis* [1, 2]. The introduction to those parts contains a discussion of why so few industrial electroorganic processes exist today. After all, many thousands of small-scale processes have been described and catalogued [1-3] since the inception of this field in about 1801. Several reasons are advanced, including, "that commercial-size scale-up of an electrochemical process is difficult, more difficult generally than a chemical process. The reaction variables in electrochemical processes are complex and still poorly understood." Whereas chemical engineering is a well developed discipline, electrochemical engineering is still a relatively new field. The past two decades have witnessed significant advances in modeling of electrochemical systems, but therein lies the "difficult" task. We still must rely heavily on direct scale-up experience.

Certainly we owe a great deal of thanks to the significant advances made in the associated fields such as energy storage and conversion, electroplating, and chloralkali production. The direct technological benefits to electroorganic synthesis are many, including stable, more highly electrocatalytic materials, porous separators, membranes and novel cell designs.

It is the purpose of the present volume to provide a starting point and to set a stage for handling the questions and difficulties posed above and, by so doing, enable more rapid scale-up of worthy candidate processes. Whereas Parts I and II detail the results of small laboratory-scale studies and provide some insight into scale-up [4], Part III is devoted entirely to scale-up, engineering principles, the chemical and electrochemical variables and their interactions, economic considerations, comparison of energy requirements of chemical and electrochemical processes, and examples of direct experience.

Advances, properties, and the choice of membranes are not covered here in detail. Several excellent reviews on membranes exist [4-10]. Membrane technology is a rapidly evolving field not only for cell compartmentization, but with additional promise for more economical product separations in some cases. Several chapters (Chaps. III, V, VII, and VIII) do relate the use of membrane materials. Important issues to be considered with divided cells include

1. Costs associated with the separator or membrane.

2. Higher cell voltages (and energy costs).

3. More complex cell design than cells without membranes or separators.

The preferred configuration is an undivided cell if the electrochemical process will allow it. The Monsanto adiponitrile process (see Chapt. VI) started out with membrane cells, but through further cell design studies the membranes were completely eliminated.

The basic principles involved in scale-up and process engineering of electrochemical cells for electroorganic processes are described in Chapter I (see also Chapter IX). While some of the examples provided in Chapter I are nonelectroorganic, the fundamental concepts can still be readily employed to develop the needed scale-up parameters.

Experiences encountered during scale-up are detailed in Chapters II, III, V, VI, and VII. Chapter II by Feess and Wendt is devoted to novel synthetic techniques using two-phase-electrolyte electrolysis. Problems generally encountered during scale-up and solutions in developing modulized cells are described in Chapter III. Chapter VI on "Experience in the Scale-Up of the Monsanto Adiponitrile Process," by Danly and Campbell describes the scale-up of the most important ongoing electroorganic process today. Of particular interest is the evolution from a relatively complex membrane cell to an undivided cell design. Childs discusses the successfully piloted "Phillips Electrofluorination Process" (in Chap. VII)—a process peculiar in its cell design and product separation technique. Degner's contribution (Chapter V) provides the type of problems encountered during scale-up and solutions exemplified with case studies and prospects of application of electroorganic syntheses in chemical industry. While the chapter of Udupa and Udupa (Chap. VIII) describes the "Use of Rotating Electrodes for Small-Scale Electroorganic Processes," it should be of added interest for electrosyntheses of small-tonnage, high value, or value-added organic chemicals.

Of key importance to cell design is the choice of electrode and separator materials. Great advances have been made on both fronts. Thus the chapter on "Electrode Materials" describes some of the recent developments in the last decade in this still too poorly understood variable. Our present knowledge in this area is primarily the result of the significant research and developmental efforts in the associated fields of fuel cells, batteries, and especially the chloralkali industry. Factors involved in a comparative energy assessment of chemical and electrochemical routes for large-tonnage organic chemicals are exemplified in Chapter IX by Beck *et al.* Keating and Sutlic, in Chapter X, describe the type of scenario required for estimating the "inside battery limit" costs, which are essential for an economic appraisal of the electrochemical route(s) that could provide an incentive for corporate management to invest in or abandon a given process over other alternate routes.

Each of the chapters in this book could no doubt be expanded to a full book or more in itself. Neither space nor time allow such committment. It is our hope that this book serves the stated objectives and provides the reader with the basic tools needed to scale up electroorganic processes.

NORMAN L. WEINBERG
B. V. TILAK

East Amherst, New York
Grand Island, New York

March 1982

REFERENCES

1. *Technique of Electroorganic Synthesis,* Part I, N. L. Weinberg, Ed. Techniques of Chemistry Series, Vol. 5, A. Weissberger, Ed., Wiley, New York, 1974.
2. *Technique of Electroorganic Synthesis,* Part II, N. L. Weinberg, Ed. Techniques of Chemistry Series, Vol. 5, A. Weissberger, Ed., Wiley, New York, 1975.
3. S. Swann, Jr. and R. Alkire, *Bibliography of Electro-organic Synthesis,* 1801–1975, The Electrochemical Society Inc., New Jersey, 1980.
4. F. Goodridge and C. J. H. King, in *Technique of Electroorganic Synthesis,* Part I, N. L. Weinberg, Ed., Techniques of Chemistry Series, Vol. 5, A. Weissberger, Ed., Wiley, New York, 1974, Chap. II.

5. K. A. Mauritz and A. J. Hopfinger, *Mod. Aspects Electrochem.*, Vol. 14, in press, 1982.

6. S. U. Falk and A. J. Salkind, *Alkaline Storage Batteries*, Wiley, New York, 1969, pp. 240-275.

7. C. Jackson, B. A. Cooke, and B. J. Woodhall, *Industrial Electrochemical Processes*, A. T. Kuhn, Ed., Elsevier, New York, 1971.

8. B. V. Tilak, P. W. T. Lu, J. E. Coleman, and S. Srinivasan, *Comprehensive Treatise of Electrochemistry*, J. O. M. Bockris, B. E. Conway, and E. A. Yeager, Eds., Vol. 2, Plenum Press, New York, 1981, p. 1.

9. "Ion Exchange: Transport and Interfacial Properties," Abstracts #596-623, Extend Abstracts Vol. 80-2, The Electrochemical Society Inc., 1980.

10. M. M. Baizer, Ed., *Organic Electrochemistry*, Dekker, New York, 1973.

CONTENTS

Chapter I

ELECTROCHEMICAL ENGINEERING PRINCIPLES AS RELATED TO ELECTROORGANIC PROCESSES

D-T. Chin and C. Y. Cheng

Department of Chemical Engineering,
Clarkson College of Technology,
Potsdam, New York

1   INTRODUCTION

Electrochemical engineering is a branch of chemical engineering science that is concerned with the technological application of electrochemistry to large-scale industrial processes. The relationship between electrochemical engineering and electrochemistry is analogous to that between chemical engineering and chemistry. The function of the electrochemical engineer is to transfer the electrochemical technology into a profit-making process. In doing so, he or she must rely on the principles of thermodynamics, electrode kinetics, transport processes, and system analysis to scale-up, design, construct, and operate efficient electrochemical cells and auxiliary equipments. In addition, the electrochemical engineer must have good economic insight so that he or she takes into consideration the availability and cost of electrical energy, as well as the marketing and distribution of products to the consumers. The objective of this chapter is to identify the scope of electrochemical engineering as applied to the electroorganic processes.

The field of electroorganic chemistry has been in existence since 1801 [1]; however, the history of large-scale production of electroorganic chemicals is relatively short. One of the first electroorganic plants was built in 1937 by the Atlas Power Company [2] for the manufacture of sorbitol and mannitol by cathodic reduction of glucose. In 1964 Nalco started a plant in Freeport, Texas, to produce tetraalkyl lead by an electrochemical process [3]. In the following year, Monsanto began the electrochemical production of adiponitrile in Decatur, Alabama [3]. A variety of electroorganic processes have been investigated on the pilot-plant scale. These include cathodic reduction of salicylic acid to salicylaldehyde [4, 5], phthalic acid to dihydrophthalic acid [6, 7], N,N-dimethylaminoethyltetrachlorophthalimide to N,N-dimethylaminoethyltetrachloroisoindodine [8], benzoic acid to benzyl alcohol [9], anodic oxidation of propylene to propylene oxide [10, 11], hydrogen cyanide to cyanogen bromide [12], and starch to dialdehyde [13], and the Kolbe synthesis of dimethyl sebacate from monomethyl adipate [14]. Today the growing problems of energy, feedstocks, and the environmental considerations have provided new opportunities for the development of electroorganic processes [15].

Industrial electrochemical processes are more complex than laboratory processes; they must (1) be able to efficiently utilize electrical energy for the production of chemicals or vice versa; (2) be of a large scale and preferably continuous; (3) be able to operate at high current densities to reduce the capital investment of electrochemical cells; and (4) be able to generate a satisfactory capital return on the investment. The art of electrochemical engineering is to transfer a laboratory development into a large-scale industrial process that fulfills the above requirements. In the past, the approach to electrochemical cell and plant design was largely based on empirical experience. Recent advances in chemical engineering and the availability of large computers have made it possible to design more efficient and more complex electrochemical cells. Modern electrochemical engineering is based on five branches of fundamental science: thermodynamics, electrode kinetics, material and energy balances, transport processes, and optimization.

Thermodynamics can be used to predict whether the electrochemical reaction is spontaneous as written and to estimate the open-circuit voltage of electrochemical cells. The information on electrode kinetics and transport processes can be used to determine the operating cell current densities and voltages. Material balance is based on stoichiometry and is used to determine the production rate, current efficiency, and yield. Energy balance takes into account the ohmic and overpotential losses in an electrochemical cell and is used to determine energy efficiency and thermal management during the plant operation. The economics of the industrial process requires that the cell be operated at the highest possible current densities in order to minimize the capital costs associated with the cell. In some cases, the operating current density may approach the limit imposed by mass transfer of an ionic species to the electrode surface. In this situation, an understanding of the convective transport problem becomes essential in the design of electrochemical cells. Optimization is based on economical incentives and helps to determine the optimal cell design and operating conditions that would generate a maximum return on the capital investment. Also, the problem of current distribution plays an important role in the design of efficient electrochemical cells. Current distribution is affected by the cell geometry, reaction kinetics, transport process, and ohmic resistance of the electrolyte. This information can lead to a set of scale-up laws that help to determine the elements of cell construction.

In the following sections, we briefly examine various principles of electrochemical engineering. It should be noted that this chapter is not intended to give an exhaustive introduction of these subjects. Rather, it identifies the scope of electrochemical engineering, the role of electrochemical engineers in the process development, and the "language" that might help improve communications among electrochemical engineers.

2   THERMODYNAMIC AND KINETIC BACKGROUND

Open-Circuit Cell Voltage

Let us consider a typical electrochemical cell as shown in Fig. 1.1. The cell is composed of two solid electrodes immersed in an electrolyte; the anode and the cathode chambers are separated by a porous cell separator. The electrodes are connected to an external DC power supply. Two identical reference electrodes, one located in the anodic chamber and the other located in the cathodic chamber, are used to measure the electrode potential of the anode and the cathode. According to the convention used here, the cell voltage is the cathode-to-anode potential difference, and the electrode potential is the potential difference between the solid electrode and a reference electrode. These are indicated by the appropriate $+/-$ terminals of the DC voltmeters shown in Fig. 1.1.

The thermodynamics of electrochemical cells permits the calculation of the open-circuit cell voltage, E_0, when the switch in the circuit is open. We use the concept of the Gibbs free energy change [16] as a starting point of our discussion. It is known that the sign of the Gibbs free energy change serves as a criterion for the spontaneity of chemical reactions. If the sign is positive, the chemical reaction as written is not spontaneous. Conversely, if the sign is negative, the reaction as written is spontaneous. If the Gibbs free energy change is zero, then the reaction is at equilibrium.

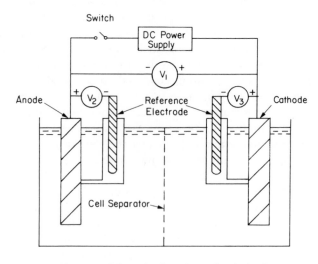

Fig. 1.1.   Schematic of an electrochemical cell.

For a reversible electrochemical reaction,

$$a\,A + b\,B = c\,C + d\,D \tag{1.1}$$

the Gibbs free energy change at constant temperature and pressure is equal to the sum of the partial molar free energy or the chemical potential μ_j, of all species involved in the reaction:

$$\Delta G = \sum_j \nu_j \mu_j = c\mu_c + d\mu_d - a\mu_a - b\mu_b \tag{1.2}$$

where ν_j is the stoichiometric coefficient of the j-species in the reaction. The chemical potential of the j-species is related to the chemical potential at reference state μ_j^0 by

$$\mu_j = \mu_j^0 + R_g T \ln a_j \tag{1.3}$$

where a_j is the activity of the j-species, R_g is the universal gas constant, and T is the absolute temperature of the system. Substituting (1.3) into Eq (1.2), one obtains

$$\Delta G = \Delta G^0 + R_g T \sum_j \nu_j \ln a_j \tag{1.4}$$

Here

$$\Delta G^0 = \sum_j \nu_j \mu_j^0 \tag{1.5}$$

is the Gibbs free energy change at the reference state. At 298°K and 1 atm, ΔG^0 is called the standard Gibbs free energy change and μ_j^0 is called the standard free energy of formation of the j-species whose value has been tabulated in the literature [17] for most common organic and inorganic chemicals. The negative value of the Gibbs free energy change is a measure of the maximum useful work obtainable from a system at constant temperature and pressure. In an electrochemical cell, this maximum useful work is equal to nFE_0; thus the open-circuit cell voltage of a reversible electrochemical cell is related to the Gibbs free energy change by

$$\Delta G = -nFE_0 \tag{1.6}$$

where F is the Faraday constant and n is the number of electrons transferred in the cell reaction. Combining (1.4) to (1.6), one obtains.

$$E_0 = E^0 - \frac{R_g T}{nF} \sum_j \nu_j \ln a_j \tag{1.7}$$

Here

$$E^0 = -\frac{\Delta G^0}{nF} = -\frac{1}{nF} \sum_j \nu_j \mu_j^0 \tag{1.8}$$

is the open-circuit cell voltage at the reference state; at 298°K and 1 atm, it is called the standard cell voltage. Equation (1.7) is called the Nernst equation and can be used to estimate the open-circuit cell voltage of an isothermal electrochemical cell from the activities or the concentrations of the reacting species.

In thermodynamics, the Gibbs free energy change is defined as

$$\Delta G = \Delta H - T \Delta S \tag{1.9}$$

where ΔH is the enthalpy change, or the heat of reaction of the electrochemical reaction occurring in the cell, and ΔS is the corresponding entropy change. Using the Gibbs-Helmoltz thermodynamic identity [18],

$$\Delta S = -\left(\frac{\partial \Delta G}{\partial T} \right)_p \tag{1.10}$$

Equation (1.9) can be rearranged to

$$\Delta H = -nF\left[E_0 - T\left(\frac{\partial E_0}{\partial T} \right)_p \right] \tag{1.11}$$

where the subscript p denotes the differentiation at constant pressure. Hence the enthalpy change of the cell reaction at constant temperature and pressure can be calculated from the open-circuit cell voltage and its temperature coefficient. The calculation of ΔG, E_0, and ΔH is necessary for the material and energy balances in the cell design. Also, for a reversible process at constant temperature and pressure, the heat absorbed by the system is equal to $T \Delta S$

$$Q = T \Delta S \tag{1.12}$$

Substituting (1.10) into (1.12) and making use of (1.6) gives

$$Q = nFT\left(\frac{\partial E_0}{\partial T} \right)_p \tag{1.13}$$

Thus the product of temperature and the temperature coefficient of the open-circuit cell voltage can be used to estimate the minimum heating effect during the cell operation.

EXAMPLE 1

The cathodic and the anodic reactions for the electrochlorination of ethylene are given below.

Cathode

$$Cl_2 + 2e \rightarrow 2Cl^-$$

Anode

$$C_2H_4 + 2Cl^- \rightarrow CH_2ClCH_2Cl + 2e$$

The overall cell reaction is

$$Cl_2(g) + C_2H_4(g) \rightarrow CH_2ClCH_2Cl(l) \qquad (1.14)$$

The thermodynamic data at the standard state (298°K and 1 atm) are given below [19].

Species and State	Standard Heat of Formation H_f^0 [J/(kg)(mole)]	Standard Free Energy of Formation μ^0 [J/(kg)(mole)]
$C_2H_4(g)$	52.34×10^6	68.17×10^6
$Cl_2(g)$	0	0
$CH_2ClCH_2Cl(l)$	-129.81×10^6	-73.91×10^6

Calculate: (a) the standard cell voltage; (b) from the thermodynamic point of view, the heat to be removed from the cell in order to maintain an isothermal condition at 298°K; and (c) the temperature coefficient of the standard cell voltage.

Solution

Basis: 1.0 kg.mole C_2H_4 oxidized in the operating cell at 298°K and 1 atm.

$$\Delta G^0 = \sum_j v_j \mu_j = (-73.91 \times 10^6) - (68.17 \times 10^6) = -142.08 \times 10^6 \, J \quad (a)$$

$$E^0 = -\frac{\Delta G^0}{nF} = \frac{-(-142.08 \times 10^6)}{2 \times 9.65 \times 10^7} = 0.736 \, V$$

E^0 is positive. Thus the reaction is spontaneous and the system is an electrogenerative process. When the reactions are thermodynamically favorable, no external energy supply is needed and the current is generated.

$$\Delta H^0 = \sum_j v_j H_{fj}^0 = (-129.81 \times 10^6) - (52.34 \times 10^6) \qquad (b)$$

$$Q = T \Delta S = \Delta H - \Delta G = -182.15 \times 10^6 - (-142.08 \times 10^6)$$
$$= -40.07 \times 10^6 \, J$$

Since Q is negative, the cell reaction is exothermic and the heat has to be removed to maintain an isothermal condition.

$$\frac{\partial E^0}{\partial T} \text{ (at 298°K and 1 atm)} = \frac{-40.07 \times 10^6}{2 \times 9.65 \times 10^7 \times 298} = -0.0007 \, V/°K \quad (c)$$

Thus at 1.0 atm pressure and for a small temperature deviation from 298°K the value of E^0 can be estimated from the following correlation:

$$E_T^0 \sim E_{298}^0 + \left(\frac{\partial E^0}{\partial T}\right)_{298°K, \, 1 \, atm} (T - 298) \sim 0.737 - 0.0007 \, (T - 298)$$

where E_T^0 is the value of E^0 at a temperature other than 298°K.

Polarization

In the foregoing section, we regard the electrochemical cell shown in Fig. 1 as a reversible system because there is no flow of electric current and no net reaction taking place inside the cell. In this case, the use of thermodynamics permits one to calculate the open-circuit cell voltage, E_0. Now if one switches on the external power supply and allows the flow of electric current through the cell, the cell voltage is no longer equal to E_0, because the current causes a composition change and the cell becomes irreversible. In reality, the value of cell voltage becomes more negative in order to support the flow of electric current; the greater the current, the more negative the cell voltage. This irreversible phenomenon is the basis of electrode kinetics.

The degree of deviation of the cell voltage from its thermodynamic open-circuit value is called polarization. According to elementary electrode kinetics [20], there are five components responsible for the polarization of an electrochemical cell: the surface overpotential at the cathode, $\eta_{s,c}$; the concentration overpotential at the cathode, $\eta_{conc,c}$; the surface overpotential at the anode, $\eta_{s,a}$; and the total ohmic potential drop in the cell, IR_{int}. Thus the cell voltage when there is current flowing through the cell can be related to the open-circuit cell voltage E_0 by the relationship

$$E = E_0 + (\eta_{s,c} + \eta_{conc,c}) - (\eta_{s,a} + \eta_{conc,a}) - IR_{int}$$

$$= E_0 + \eta_{total,c} - \eta_{total,a} - IR_{int} \tag{1.15}$$

where I is the total current through the cell and R_{int} is the sum of all internal cell resistance, including that of electrolyte, cell separator, and electrode bodies. The sum of the surface and the concentration overpotentials of a particular electrode is called the total overpotential of that electrode. The total overpotential, η_{total}, can be obtained by means of a reference electrode as shown in Fig. 1. The potential difference between a given electrode and a reference electrode has been defined as electrode potential. The value of the electrode potential in an open-circuit cell is called the equilibrium electrode potential (or the rest potential). When the cell is closed, the electrode potential departs from its equilibrium value, and the amount of deviation is a measure of total overpotential of that electrode. It should be noted that the total overpotential of the anode is always positive and that of the cathode is always negative. Thus the cell voltage, E in (1.15) is always more negative than its thermodynamic reversible value, E_0.

Surface Overpotential

The surface overpotential is associated with the electron-transfer reaction at the electrode surface. A typical single electrode reaction involves adsorption of reactants and desorption of products, together with surface diffusion and electron transfer steps. The slowest step is the rate-determining step and has the maximum activation energy barrier. This barrier must be overcome by the surface overpotential for the reaction to proceed at an appreciable rate. Since the reaction rate at a single electrode is determined by the current density at the electrode surface, it is obvious that the surface overpotential is a function of current density, that is, it increases with increasing current density. Let us consider a simple electrode reaction.

$$O + ne = R \tag{1.16}$$

where O is the oxidized species and R is the reduced species of the single electrode reaction. For this type of reaction, the relationship between the surface overpotential, η_s, and the current density at the electrode surface can often be expressed as [20]

$$i = i_0 \left[\exp\left(\frac{\eta_s}{\beta_1} \right) - \exp\left(\frac{\eta_s}{\beta_2} \right) \right] \tag{1.17}$$

This equation is known as the Butler-Volmer equation of electrode kinetics [21]. The parameter i_0 is called the exchange current density and represents the rate of forward and backward reactions at equilibrium. The value of exchange current density depends on the nature of the electrode surface, the temperature, and the concentrations of O and R species in the electrolyte. For a given electrode system and temperature, it can be related to the concentrations of R and O species by

$$i_0 = i^0 \, C_R^{\omega_1} \, C_0^{\omega_2} \tag{1.18}$$

where i^0 is the standard exchange current density at unit concentrations of R and O, and ω_1 and ω_2 are constants associated with the reaction orders with respect to R and O species, respectively. At high anodic or high cathodic overpotential, (1.17) can be reduced to

$$i = i_0 \exp\left(\frac{\eta_s}{\beta_1} \right) \qquad \text{(for high anodic overpotential)} \tag{1.19}$$

$$-i = i_0 \exp\left(- \frac{\eta_s}{\beta_2} \right) \qquad \text{(for high cathodic overpotential)} \tag{1.20}$$

Equations (1.19) and (1.20) are called the Tafel equations, according to which a plot of η_s *versus* ln $|i|$ should result in a straight line. The slopes of the straight lines are equal to β_1 and β_2. Thus β_1 and β_2 are called the Tafel slopes. These parameters are of fundamental importance in electrode kinetics; they have been used to interpret the mechanism of electrode reaction. Generally speaking, the surface overpotential decreases with increasing i_0 and increases with increasing values of β_1 and β_2.

Concentration Overpotential

The concentration overpotential of an electrode arises from a difference in the concentrations of ionic species between the region of electrolyte adjacent to the electrode surface and that in the bulk of the cell. At open-circuit, the cell can be considered to be at equilibrium, and the electrolyte concentration is uniform throughout the cell. However, the passage of electric current through the cell causes a net reaction at the electrode surface, and this requires the flow of ionic species to or from the electrode surface to sustain the rate of the charge transfer reaction. Since the reactant is consumed at the electrode, its concentration at the electrode surface becomes smaller than that in the bulk. Conversely, the product is produced at the electrode; its concentration at the electrode surface becomes greater than that in the bulk. All ionic species have a natural tendency to move from the area of high concentration to the area of low concentration by a diffusion process in order to equalize the concentrations in both areas. To maintain such a concentration gradient, one obviously needs external energy, and the concentration overpotential can be regarded as the energy needed to maintain the concentration differences. Thus the concentration overpotential, like the surface overpotential, is also associated with the electrode systems that are perturbed from the equilibrium state.

Let us consider the single electrode reaction shown in (1.16). According to the Nernst equation (1.18), the reversible single electrode potential at open-circuit may be defined in terms of the bulk concentrations of the reacting species:

$$\phi_0 = \phi^0 - \frac{R_g T}{nF} \ln \frac{C_{R,\infty}}{C_{O,\infty}} \tag{1.21}$$

When the electrode is perturbed by the flow of electric current, the surface concentrations of the ionic species become different from those in the bulk. Under these conditions the electrode senses an apparent "reversible" single electrode potential that has a value different from that given in (1.21). This apparent "reversible" potential can be expressed in terms of the surface concentrations:

$$\phi_{0,\text{app}} = \phi^0 - \frac{R_g T}{nF} \ln \frac{C_{R,s}}{C_{O,s}} \tag{1.22}$$

where the subscript s denotes the surface property. The concentration overpotential is defined as the difference between $\phi_{0,\text{app}}$ and ϕ_0:

$$\eta_{\text{conc}} = \phi_{0,\text{app}} - \phi_0 = \frac{R_g T}{nF} \ln \frac{C_{O,s} C_{R,\infty}}{C_{O,\infty} C_{R,s}} \tag{1.23}$$

If the reaction given by (1.16) is taking place at the cathode, the oxidized species O is consumed, and the reduced species R is generated at the electrode. This results in the following inequalities:

$$\frac{C_{O,s}}{C_{O,\infty}} < 1; \quad \frac{C_{R,\infty}}{C_{R,s}} < 1 \tag{1.24}$$

Consequently, from (1.23), the concentration overpotential of a cathodic process is negative. Consideration of an anodic process would yield a positive concentration overpotential at the anode.

Often in the literature, the concentration overpotential is expressed in terms of a limiting current density at the electrode. The maximum reaction rate an electrode can sustain is generally limited by how fast a reactant can be supplied to the electrode surface. At this maximum rate, the reactant is consumed for the charge-transfer reaction as fast as it arrives at the electrode surface, and the surface concentration of the reactant becomes zero. In these circumstances, the electrode reaction is said to be limited by mass transfer, and the current density at the electrode is called the limiting current density. It has been shown [21] that for the cathodic process, (1.23) may be reduced to

$$\eta_{\text{conc},c} = \frac{R_g T}{nF} \ln \left(1 - \frac{i_c}{i_{\text{lim},c}} \right) \tag{1.25}$$

where i_{lim} is the limiting current density, and the subscript c denotes that the process is taking place at the cathode. For the anodic process, the concentration overpotential $\eta_{\text{conc},a}$ is related to the anodic limiting current density by [21]

$$\eta_{\text{conc},a} = \frac{R_g T}{nF} \ln \left(1 - \frac{i_a}{i_{\text{lim},a}} \right) \tag{1.26}$$

Equations (1.22) and (1.23) describe the relationship between the current density and the concentration overpotential of an electrode process. In many systems, where $|i_{\text{lim}}| \gg |i|$ (i.e., the rate of the electrode reaction is very small compared to the maximum rate permitted by the mass transfer limitations), the concentration overpotential becomes very small and can be dropped from the analysis. However, a number of industrial processes require that the cell be operated at the highest possible current densities, and the concentration overpotential becomes very large. A common method to reduce the concentration overpotential is to increase the bulk electrolyte concentration and to use forced convection to increase the limiting current density. The effect of fluid flow and mass transfer on the rate of electrochemical reaction is a special area of electro-chemical engineering and is treated in more detail in the section concerning the transport processes.

3   MATERIAL AND ENERGY BALANCES

The material and energy balances are based on the concept of the conservation of mass and energy. They have been extensively used by chemical engineers to calculate the capacity and the operating characteristics of modern chemical plants. An understanding of the material and energy balances is essential to the design of electrochemical cells.

The material and energy balance equations are obtained by choosing a control volume and equating the changes of mass or energy in the control volume to the amount of mass or energy entering, leaving, and produced within the system. The control volume can be a quantity of mass or a region of space such as a tank, or an electrochemical cell as shown schematically in Fig. 1.2. The "inlet" and the "outgoing" streams in the figure contain all chemical species and energy entering and leaving the cell. "Generation" and the "accumulation" represent the net production and the physical buildup of the materials and energy within the cell. According to the principle of conservation of mass and energy, one may write the following equation:

(rate of accumulation   (rate of mass or   (rate of mass or
of mass or energy   = energy flow into − energy flow out
within the cell)   cell)   of cell)

(rate of generation
+ of mass or energy   (1.27)
within the cell)

Equation (1.27) can be written for the material balance of each chemical species involved in the process. The summation of all individual balance equations gives an overall material balance of the cell. These equations can be used to determine the production rate, yield, feed rate, and so forth. The generation rate of a chemical species, j, due to an electrochemical reaction is related to the total cell current by

$$rate\ of\ \text{generation of j-species} = \frac{I\nu_j\epsilon_f}{nF} \tag{1.28}$$

where ν_j is the stoichiometric coefficient of species j and n is the number of electrons transferred in the electrochemical reaction. The quantity ϵ_f is called the faradaic efficiency (or current efficiency), which is defined as the ratio of the amount of species j produced in the cell to that theoretically expected from Faraday's law. This value is usually obtained through experimental studies. One unique feature of the electrochemical cell is that the material balance can be separately performed for the anodic and the cathodic reactions.

Equation (1.27) can be used to deduce a generalized energy balance equation of a steady-state flow system [22]:

$$\Delta H + \Delta K + \Delta \Pi = Q - W \tag{1.29}$$

where H, K, and Π are the enthalpy, the kinetic energy, and the potential energy associated with the materials flowing in and out of the system, Q is the heat absorbed by the system, and W is the work done by the system. For most electrochemical processes, the changes in the kinetic and the potential energy are often very small, and (1.29) reduces to

$$\Delta H = Q - W \tag{1.30}$$

An important application of the energy balance is to determine the overall energy efficiency of the electrochemical process. For electrolysis, the overall efficiency can be expressed as

$$\epsilon_0 = - \frac{\Delta H}{NFE} \epsilon_f \tag{1.31}$$

where ΔH is the heat of reaction, which is the amount of energy theoretically needed for the conversion of chemicals, E is the closed-circuit cell voltage,

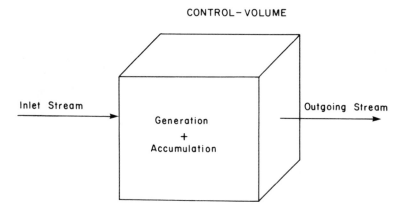

Fig. 1.2. Control volume and materials and energy balances.

and ϵ_f is the faradaic efficiency. Also during the cell operation, the amount of heat absorbed by the system is equal to

$$Q = T \Delta S + nF(E - E_0) \tag{1.32}$$

We use an example to illustrate the usefulness of the material and energy balance in the design of electrochemical systems. The following example is taken from the area of electrochemical energy conversion. Although specific applications of engineering principles are largely guided by the process characteristics, the strategy of mathematical modeling and the method of numerical computations described below should also be valid for the analysis of electroorganic synthesis plants.

EXAMPLE 2

A 20-mw hydrogen/chlorine fuel cell plant has been considered for the storage of off-peak electric energy [23]. The plant will be designed to operate at 10-hr charge and 10-hr discharge cycles. A simplifed flow sheet is shown in Fig. 1.3. The heart of the plant is a cell stack consisting of a number of solid-polymer electrolyte (SPE) batteries. During the charge cycle, the HCl pump transfers concentrated hydrochloric acid to the chlorine-electrode compartments of the cell stack. Where the following reaction occurs:

$$\text{HCl (aq)} \xrightarrow[\text{discharge}]{\text{charge}} \tfrac{1}{2}\,H_2(g) + \tfrac{1}{2}\,Cl_2(l) \tag{1.33}$$

The Cl_2-electrode and the HCl/Cl_2 storage are pressurized at 2760 kPa (400 psi); at this pressure, the Cl_2 produced by the electrolysis in the liquid form. The liquid Cl_2 is carried away by the recirculating electrolyte to the HCl/Cl_2 storage, where it is stored as a separate liquid phase. The H_2 gas produced by the electrolysis is passed through a H_2 clean-up device and is

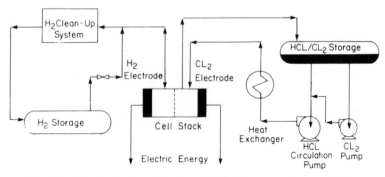

Fig. 1.3.   Flow sheet of an electrochemical H_2/Cl_2 energy storage system.

stored at 298°K in the H_2 storage. The SPE electrochemical cell is capable of compressing the H_2 gas to 4140 kPa (600 psi) without the need of a gas compressor. In the present plant, the H_2-storage pressure is 690 kPa (100 psi) at the beginning of charge; at the end of the 10-hr charge cycle, the H_2 pressure increases to 4140 kPa (600 psi). During discharge, the HCl circulation pump and the Cl_2 pump transfer the aqueous HCl and liquid Cl_2 into the cell stack; the H_2 gas also passes through the expansion valve to the H_2 electrode of the cell stack. During this period the reverse of (1.33) occurs.

Equation (1.33) has a large entropy change; at 298°K the value of $T \Delta S$ amounts to 36.24×10^6 J/(kg)(mole HCl). This positive $T \Delta S$ results in the cooling of the electrolyte during charge and a large heating effect during discharge. The function of the heat exchanger in Fig. 1.3 is to limit the electrolyte temperature to a maximum (T_{max}) and a minimum (T_{min}) operating temperature. When the electrolyte temperature reaches T_{max}, heat is removed from the heat exchanger to maintain the temperature at T_{max}. Conversely, when the electrolyte temperature decreases to T_{min}, heat is added from the heat exchanger to maintain the temperature at T_{min}. Practical design considerations limit T_{max} at 368°K and T_{min} at 288°K. Also, the electrolyte concentration is limited to 35 wt% HCl at the beginning of charge and 5 wt% HCl at the end of charge. The SPE cell performance data have been determined experimentally [24]. They are summarized below [25, 26].

For charge

$$E = -[1.28 - 0.96(\chi - 0.1) - 0.0017(T - 298)]$$
$$+ 4.3 \times 10^{-5} T(\ln P - 4.62) + 9.87 \times 10^{-6} P$$
$$- [0.78 - 0.006(T - 298)] \times 10^{-4} i \qquad (1.34)$$

For discharge

$$E = 1.28 - 0.96(\chi - 0.1) - 0.0017(T - 298)$$
$$+ 4.3 \times 10^{-5} T(\ln P - 4.62) + 9.87 \times 10 \times 10^{-6} P$$
$$- (4.54 - 0.011 T)[1 + 1.07 \exp(-0.0087 P)] \times 10^{-4} i \qquad (1.35)$$

where E is the closed-circuit cell voltage, χ is the weight fraction of HCl in the electrolyte, i is the cell current density, T is temperature, and P is the pressure of the Cl_2 electrode in kilopascals. It is desired to determine the plant characteristics as a function of the operating cell current densities.

1. Calculate the changes of the electrolyte temperature during the charge and the discharge cycles.

2. Determine the cell voltage changes during the charge and the discharge cycles.

3. Assuming a 100% faradaic current efficiency, and neglecting all other

parasitic energy losses, determine the overall energy conversion efficiency in the form of

$$\epsilon_0 = \frac{\text{total output of electric energy during discharge}}{\text{total input of electric energy during charge}} \times 100 \qquad (1.36)$$

Solution

For this problem, the material and energy balances can be performed by choosing the entire energy storage plant as the control volume. In this way, the system can be regarded as a closed system, and according to the first law of thermodynamics

$$\Delta U = Q - W \qquad (1.37)$$

Also, the enthalpy change is related to the internal energy change by

$$\Delta H = \Delta U + \Delta(PV) \qquad (1.38)$$

In the present plant, the total volume is constant; thus combining (1.37) and (1.38), one obtains

$$\Delta H = Q - W - V \Delta P \qquad (1.39)$$

Let us consider a simplified control volume as shown in Fig. 1.4. During the charge period, the amount of work done by the system between time t_0 and t is equal to the amount of electric energy fed to the cell stack:

$$W = 3600 \int_{t_0}^{t} EI \, dt \qquad (1.40)$$

where the unit of time is an hour, and the coefficient 3600 is the conversion factor between joule and watt-hour. During this period, the electrolyte temperature changes from T_0 to T, and the amount of HCl decomposed is equal to $-\Delta n_{HCl}$ kg-moles. The total amount of heat absorbed by the system is equal to

$$Q = Q_{ex} + Q_{H_2} = Q_{ex} - 2.02 \frac{\Delta n_{HCl}}{2} \Big|_{T}^{298} C_{V,H_2} \, dT \qquad (1.41)$$

where Q_{ex} is the amount of heat input into the system through the heat exchanger and Q_{H_2} is the heat absorbed by hydrogen from the surroundings. Also, the hydrogen storage pressure changes from P_{H_0} to P_H. Since the pressure of the HCl/Cl_2 is kept at a constant value of 2760 kPa (400 psi), the contribution to the value of $V \Delta P$ is mainly from the hydrogen pressure change. Assuming that the hydrogen behaves as an ideal gas, it can be shown that

$$V \Delta p = - \frac{\Delta n_{HCl}}{2} R_g (298) \qquad (1.42)$$

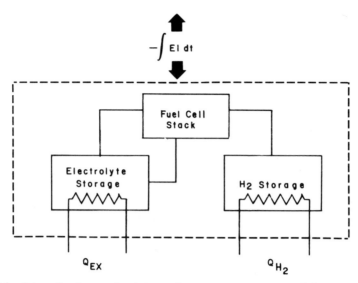

Fig. 1.4. Schematic of energy flow between the energy storage system and the surroundings.

The calculation of ΔH is more complex; the standard procedure for a nonisothermal process is shown in Fig. 1.5. Since the change of enthalpy is independent of the path of change, one may choose a hypothetical path consisting of three steps: (1) change of the electrolyte and hydrogen temperature from T_0 to 298°K; (2) at 298°K electrolysis of $-\Delta n_{HCl}$ kg-mole of HCl to $-\frac{1}{2}\Delta n_{HCl}$ kg-mole each of H_2 (g) and Cl_2 (l), and the change of electrolyte concentration from χ_0 to χ; (3) the change of electrolyte temperature from 298 to $T°$K. The corresponding enthalpy change in each step is termed ΔH_1, ΔH_2, and ΔH_3 as shown in Fig. 1.5, and

$$\Delta H = \Delta H_1 + \Delta H_2 + \Delta H_3 \tag{1.43}$$

The value of ΔH_1, ΔH_2, and ΔH_3 can be determined from the principles of thermodynamics; the details have been described elsewhere [25, 26] and are not repeated here. Substituting (1.40) to (1.43) into (1.39), one obtains

$$W_{HCl,0} \int_{T_0}^{298} C_{p,HCl}\, dT + W_{Cl_2,0} \int_{T_0}^{298} C_{p,Cl_2}\, dT$$

$$+ \Delta n_{HCl}\, H^0_{f,HCl}(\chi) + \frac{\chi_0 W_{HCl,0}}{36.5}\, [H^0_{f,HCl}(\chi) - H^0_{f,HCl}(\chi_0)]$$

$$- \frac{\Delta n_{HCl}}{2}\, H^0_{f,Cl_2} + \left(W_{Cl_2,0} - \frac{71}{2}\, \Delta n_{HCl} \right) \int_{298}^{T} C_{p,Cl_2}\, dT$$

$$- \frac{\Delta n_{HCl}}{2} R_g (298) + (W_{HCl,0} + 36.5 \, \Delta n_{HCl}) \int_{298}^{T} C_{p,HCl} \, dT$$

$$= Q_{ex} + 2.02 \frac{\Delta n_{HCl}}{2} \int_{T_0}^{298} C_{V,H_2} \, dT - 3600 \int_{t_0}^{t} EI \, dt$$

$$- \frac{\Delta n_{HCl}}{2} R_g (298); \, (Q_{ex} = 0 \text{ for } T_{min} < T < T_{max}) \tag{1.44}$$

Also from the material balances, one has

$$\Delta n_{HCl} = \mp \frac{3600}{F} \int_{t_0}^{t} I \, dt \tag{1.45}$$

$$W_{HCl} = W_{HCl,0} + 36.5 \, \Delta n_{HCl} \tag{1.46}$$

$$W_{Cl_2} = W_{Cl_2,0} - 71 \frac{\Delta n_{HCl}}{2} \tag{1.47}$$

$$\chi = \frac{W_{HCl,0} \chi_0 + 36.5 \, \Delta n_{HCl}}{W_{HCl,0} + 36.5 \, \Delta n_{HCl}} \tag{1.48}$$

The symbols in the foregoing equations are explained at the end of the chapter. The quantities with subscript 0 are the variables and the physical

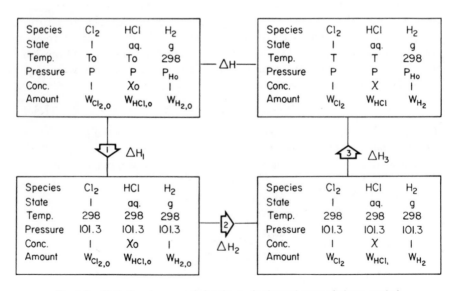

Fig. 1.5. Enthalpy-change path for the nonisothermal energy balance analysis.

and thermodynamic properties at time t_0; those without subscript 0 are the variables and the properties at the time t. The convention used here is that the cell voltage E and Δn_{HCl} have negative values during charge and positive values during discharge. The values of C_{p,H_2}, $C_{p,HCl}$, C_{p,Cl_2}, $H^0_{f,HCl}$, and ρ_{HCl} are functions of temperature and electrolyte concentrations; they have been summarized in the literature [25, 26].

Equations (1.44) to (1.48) together with the cell voltage given by Eq (1.34) and (1.35) and the thermodynamic properties reported in the literature [25, 26] can be used to calculate the changes in the electrolyte concentration, the electrolyte temperature, and the cell voltage during the charge and the discharge cycles. Numerical solutions have been obtained for a constant I of 2×10^7 A [25, 26].

RESULTS

1. Figure 1.6 shows the variation of electrolyte concentrations. The electrolyte concentration changes from 35 wt % at the beginning of charge to 5 wt % at the end of charge. In the discharge cycle, the concentration changes from 5 to 35 wt %. The curves are nearly linear, and the average concentration over the entire period is approximately 23 wt %. The change of electrolyte temperature is shown in Fig. 1.7 for various operating cell current densities. The electrolyte temperature at the beginning of charge is at T_{max} of 368°K. The temperature decreases during the charge period. During the discharge, the temperature increases rapidly. After the temperature reaches T_{max}, the heat exchanger removes heat from the electrolyte to keep the temperature at T_{max}.

2. The operating cell voltages during charge and discharge cycles are

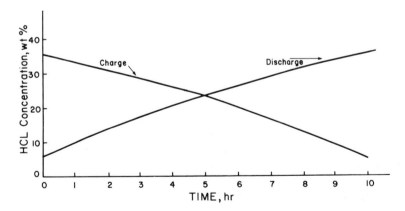

Fig. 1.6. Variation of HCl electrolyte concentration during charge and discharge.

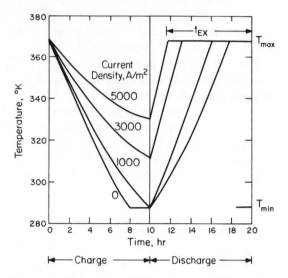

Fig. 1.7. Electrolyte temperature change during charge and discharge cycles for $T_{max} = 368°K$, $\chi_i = 0.35$, $P = 2760$ kPa, $T_{min} = 288°K$.

given in Fig. 1.8 for various cell operating current densitites. It is seen that the cell voltage increases during the charge cycle and decreases during the discharge period. Also there is a discontinuity in E at the end of charge and the beginning of discharge except for the curve with $i = 0$.

3.   The overall electric-to-electric efficiency can be calculated from Fig. 1.8 with

$$\epsilon_0 = - \frac{\left(\int_0^{10} EI\, dt \right)_{discharge}}{\left(\int_0^{10} EI\, dt \right)_{charge}} \times 100 \qquad (1.49)$$

The results are given in Fig. 1.9 as a function of cell operating current densities. It is seen that ϵ_0 decreases with increasing current density. Also given in the figure is the overall efficiency obtained with different initial concentrations of HCl at the beginning of the charge. For all the curves, the concentration of HCl at the end of charge is maintained at 5 wt %. The effect of χ_i on the overall electric-to-electric efficiency is discussed in the optimization section.

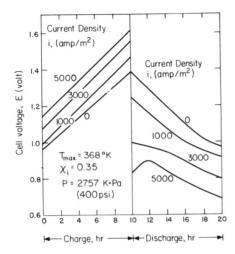

Fig. 1.8. Variation of cell voltages for various operating current densities at $T_{max} = 368°K$, $\chi_i = 0.35$, $P = 2760$ kPa, $T_{min} = 288°K$.

4   THE ELECTROCHEMICAL CELL DESIGN

Electrochemical cell design is based on the macroscopic material and energy balances wherein the entire cell or an entire electrode compartment is considered as the control volume. In analogy to chemical reactor analysis, one may classify electrochemical cells into three categories: (1) batch electrochemical cells; (2) continuous stirred tank electrochemical cells; and (3) plug flow electrochemical cells.

The batch electrochemical cell is one where the reactants are charged before electrolysis, and products are removed only after the completion of electrolysis. It is assumed that the electrolyte mixes well within the cell, and its composition and temperature change only with time. The continuous stirred tank electrochemical cell (CSTEC) is a steady-state flow cell where reactants are fed continuously and products are continuously removed from the cell. An ideal CSTEC has a complete mixing of electrolyte such that the composition of the electrolyte in the exit stream is the same as the electrolyte in the cell. The plug flow electrochemical cell, on the other hand, does not permit the mixing of electrolyte in the flow direction, even though the elements of electrolyte perpendicular to the flow path may freely mix with each other. Thus the composition of the electrolyte in a plug flow electrochemical cell changes as it flows through the cell. This classification calls for a certain degree of idealization. Many industrial electrochemical cells exhibit an immediate behavior. However, the idealization helps to simplify the mathematical formulation of design equations, and a reasonable approxima-

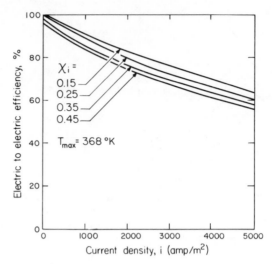

Fig. 1.9. Overall energy conversion efficiency as a function of i for various χ_i at $T_{max} = 368°K$, $P = 2760\,kPa$, $T_{min} = 288°K$.

tion to one or two of the idealized cells has been often used as the basis of cell design.

Batch Electrochemical Cell

The batch electrochemical cells are still commonly used in electroorganic synthesis even though the intermittency of operation makes it unsuitable for large-scale productions. The advantage of the batch cell is its ability to follow the rate of reaction as a function of electrolyte compositions, and the relationship between the faradaic efficiency and the cell current density can be obtained easily with a laboratory setup.

Let us consider the simple reaction (1.16) taking place at the cathode of a one-compartment batch electrochemical cell. Since there is no input or output, the material balance of (1.27) reduces to

$$\text{rate of accumulation} = \text{rate of generation} \tag{1.50}$$

For the oxidized species O, this translates to

$$V \frac{dC_O}{dt} = \frac{A_c i_c \epsilon_f}{nF} \tag{1.51}$$

where V = volume of the electrolyte in the cell
C_O = concentration of O-species

t = time

A_c = cathode area

i_c = cathode current density

ϵ_f = faradaic efficiency.

In chemical reactor analysis, one often introduces a conversion, X, defined as

$$X = 1 - \frac{C_O}{C_{Oi}} \qquad (1.52)$$

where C_{Oi} is the initial concentration of O-species in the cell. Substituting (1.52) into (1.51) and rearranging the resulting equation, one obtains

$$\frac{dX}{dt} = - \frac{\epsilon_f A_c}{nFC_{Oi}V} i_c \qquad (1.53)$$

If one assumes that the reaction is occurring at the Tafel potentials, and the initial concentration of the reduced species R is equal to zero, then the current density at the cathode, i_c, can be related to the cathode overpotential by combining (1.18) and (1.20)

$$i_c = - i^0 C_{Oi}{}^{\omega_1 + \omega_2} X^{\omega_1} (1 - X)^{\omega_2} \exp \left(- \frac{\eta_{s,c}}{\beta_2} \right) \qquad (1.54)$$

Equations (1.53) and (1.54) can be used to calculate the conversion X and the cathodic overpotential as a function of time if the kinetic parameters, i^0, ω_1, ω_2, and β_2 are known.

For constant-current operations, dX/dt is a constant, and (1.53) and (1.54) can be rearranged to give

$$X = \tau - i^* \qquad (1.55)$$

$$-\eta_{sc}{}^* = \omega_1 \ln X + \omega_2 \ln (1 - X) - \ln i^* \qquad (1.56)$$

where

$$\tau \text{ (dimensionless time)} = \frac{\epsilon_f A_c i^0 C_{Oi}{}^{\omega_1 + \omega_2} t}{nFC_{Oi}V} \qquad (1.57)$$

$$i^* \text{ (dimensionless current density)} = - \frac{i_c}{i^0 C_{Oi}{}^{\omega_1 + \omega_2}} \qquad (1.58)$$

$$\eta_{sc}{}^* \begin{array}{l} \text{(dimensionless cathodic} \\ \text{surface overpotential)} \end{array} = \frac{\eta_{sc}}{\beta_2} \qquad (1.59)$$

In electroorganic synthesis, one often operates the cell at a constant over-potential rather than at a constant cell current. For the constant potential control, one may substitute (1.54) into (1.53) and carry out the integration:

$$\int_0^X \frac{dX}{X^{\omega_1}(1-X)^{\omega_2}} = \tau \exp(-\eta_{sc}^*) \tag{1.60}$$

The dimensionless parameters τ, i^*, and η_{sc}^* are very useful in the design and the scale-up of electrochemical cells. We use the following example to illustrate their implications.

EXAMPLE 3

For the electrochemical reduction of quinoline to dihydroquinoline [27],

$$\tag{1.61}$$

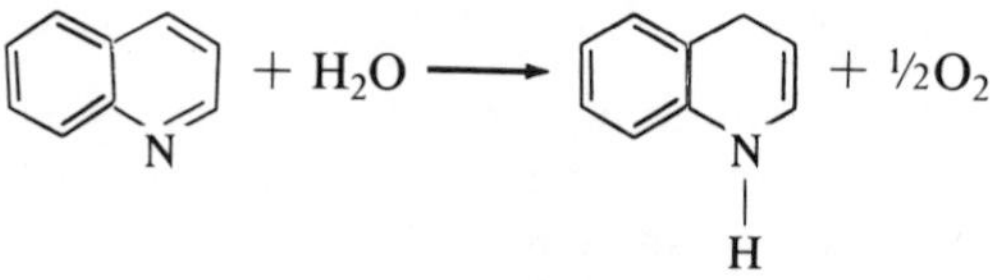

Quinoline   Dihydroquinoline

the corresponding anodic and cathodic reactions are given below.

Anodic Process

$$H_2O \rightarrow 2H^+ + \tfrac{1}{2}O_2 + 2e \tag{1.62}$$

Cathodic Process

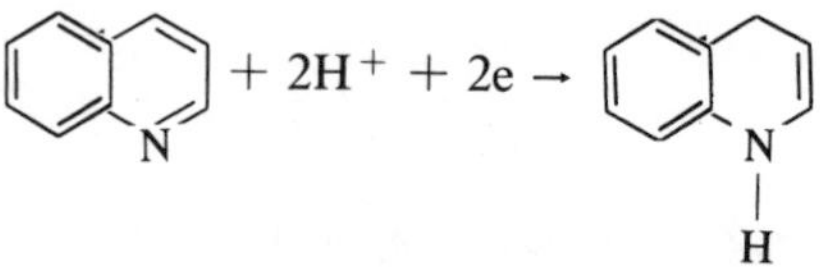

$$\tag{1.63}$$

The electrolysis is carried out in 25% H_2SO_4 containing 0.1 M quinoline with lead electrodes. The Tafel relation for the cathodic process at 25°C has been determined to be

$$i_c = -8.357 \times 10^{-7} C_O \exp\left(-\frac{\eta_{sc}}{0.025}\right) \tag{1.64}$$

where C_O refers to the concentration of quinoline. The reaction order with respect to dihydroquinoline is found to be zero. Assuming that the reaction is taking place in a batch electrochemical cell with 100% faradaic efficiency, determine the conversion as a function of time for (a) constant current operation and (b) constant cathodic overpotential operation.

Solution

For this problem, $\omega_1 = 0$, $\omega_2 = 1$ and (1.55) to (1.60) reduce to the following equations.

For constant current control

$$X = \tau i^* \tag{1.65}$$

$$\eta_{sc}^* = \ln(1 - X) - \ln i^* \tag{1.66}$$

For constant potential control

$$\ln \frac{1}{1 - X} = \tau \exp(-\eta_{sc}^*) \tag{1.67}$$

with

$$\tau = 4.33 \times 10^{-15} \frac{A_c t}{V} \tag{1.68}$$

$$i^* = -\frac{i_c}{8.357 \times 10^{-8}} \tag{1.69}$$

$$\eta_{sc}^* = \frac{\eta_{sc}}{0.025} \tag{1.70}$$

Equations (1.65) to (1.66) are plotted in Figs. 1.10 and 1.11 in the form of X versus τ and η_{sc}^* versus X, respectively, with the dimensionless cathodic current densities i^* as the parameters. Equation (1.67) is plotted in Fig. 1.12 in the form of X versus τ and the dimensionless cathode overpotential, η_{sc}^*, as the parameters. These curves can be used to determine the time of electrolysis and the cathode surface overpotential for a desire conversion of quinoline. It is interesting to note that with the constant-current-control mode, the cathode surface overpotential becomes very large at large values of τ when the conversion of quinoline approaches unity. Once the dimensionless time τ is determined, one has the flexibility to adjust the values of real electrolysis time t, the cathode area A_c, and the cell volume V. The ratio of the cathode area to the cell volume is an important design parameter. Often the electrolysis time can be shortened if one uses a cell with a large area-to-volume ratio.

The foregoing analyses are limited to the cathode overpotential and the conversion of the cathodic reaction. For the complete analysis of a batch electrochemical cell, a separate material balance must be performed for the anode to determine the anode overpotential. The voltage balance of (1.15) can then be used to determine the cell voltage as a function of time.

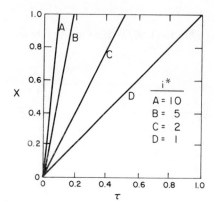

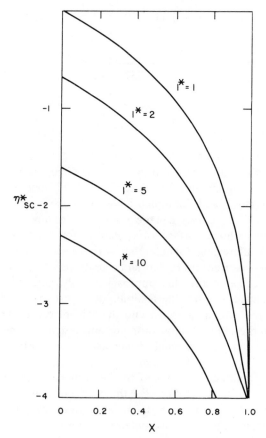

Fig. 1.10. X versus τ with i^* as the parameters for a batch cell under the constant-current-control condition.

Fig. 1.11. X versus η_{sc}^* for a batch cell under the constant-current-control condition.

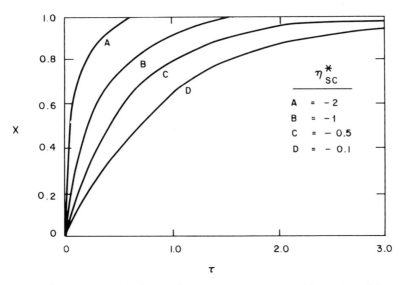

Fig. 1.12. X versus τ for a batch cell under the constant-potential-control condition.

Continuous Stirred Tank Electrochemical Cell

As discussed earlier an ideal CSTEC is a steady-state flow cell where there is a complete mixing of reactants and products. Let us consider that the reaction

$$O + K = R + L \qquad (1.71)$$

is taking place in a single compartment CSTEC as shown in Fig. 1.13. The cathodic and the anodic partial processes can be represented by (1.72) and (1.73).

Cathodic reaction

$$O + ne \rightarrow R \qquad (1.72)$$

Anodic reaction

$$K \rightarrow L + ne \qquad (1.73)$$

Under steady-state conditions, the volumetric flow rates of the entering and the outgoing streams are constant; they are represented by v as shown in the figure. The entering stream does not contain R and L species. The concentrations of species O and K in the entering stream are C_{Oi} and C_{Ki}, respectively; the steady-state concentrations of O and K in the cell are represented by C_O and C_K. The design of a CSTEC basically involves the following steps:

1. *Voltage balance.*

$$E = E_O + \eta_{total,c} - \eta_{total,a} - IR_{int} \tag{1.74}$$

2. *Charge balance.*

$$I = A_a i_a = -A_c i_c \tag{1.75}$$

3. *Material balance for species O and K.*

$$v(C_{Oi} - C_O) = - \frac{\epsilon_{fc} A_c i_c}{nF} \tag{1.76}$$

$$v(C_{Ki} - C_K) = \frac{\epsilon_{fa} A_a i_a}{nF} \tag{1.77}$$

4. *Electrode kinetic equations.* Assuming that the Tafel relation is valid for both the cathodic and anodic reactions, $\eta_s = \eta_{total}$, and

$$i_c = - i_c^0 C_R^{\omega_{1c}} C_O^{\omega_{2c}} \exp\left(- \frac{\eta_{sc}}{\beta_{1c}} \right) \tag{1.78}$$

$$i_a = i_a^0 C_K^{\omega_{1a}} C_L^{\omega_{2a}} \exp\left(\frac{\eta_{sa}}{\beta_{1a}} \right) \tag{1.79}$$

The notations used in the above equations are defined at the end of the chapter; the subscripts c and a denote the variables and the parameters

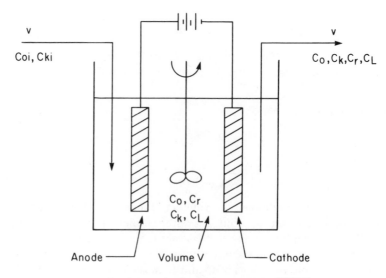

Fig. 1.13. Schematic of a single-component CSTEC.

associated with the cathode and the anode reactions, respectively. If one introduces a conversion defined as

$$X = 1 - \frac{C_O}{C_{Oi}} \tag{1.52}$$

then

$$C_K = C_{Ki} - C_{Oi} X \tag{1.80}$$

$$C_R = C_L = C_{Oi} X \tag{1.81}$$

Substituting (1.52) and (1.80) to (1.81) into (1.74) to (1.79), one can obtain the following design equations:

$$X = \tau_c \, i_c^* = \tau_a \, i_a^* \tag{1.82}$$

$$E = E_O - \beta_{2c} \ln \frac{i_c^*}{X^{\omega_{1c}}(1 - X)^{\omega_{2c}}}$$

$$- \beta_{1a} \ln \frac{i_a^*}{X^{\omega_{2a}}\left(\dfrac{C_{Ki}}{C_{Oi}} - X\right)^{\omega_{1a}}} - IR_{int} \tag{1.83}$$

where the dimensionless quantities are defined as

$$\tau_c \begin{array}{l} \text{(dimensionless cathodic} \\ \text{residence time)} \end{array} = \frac{\epsilon_{fc} i_c^0 A_c C_{Oi}^{\omega_{1c}+\omega_{2c}}}{nFC_{Oi}v} \tag{1.84}$$

$$\tau_a \begin{array}{l} \text{(dimensionless anodic} \\ \text{residence time)} \end{array} = \frac{\epsilon_{fa} i_a^0 A_a C_{Oi}^{\omega_{1a}+\omega_{2a}}}{nFC_{Oi}v} \tag{1.85}$$

$$i_c^* \begin{array}{l} \text{(dimensionless cathodic} \\ \text{current density)} \end{array} = \frac{I}{A_c i_c^0 C_{Oi}^{\omega_{1c}+\omega_{2c}}} \tag{1.86}$$

$$i_a^* \begin{array}{l} \text{(dimensionless anodic} \\ \text{current density)} \end{array} = \frac{I}{A_a i_a^0 C_{Oi}^{\omega_{1a}+\omega_{2a}}} \tag{1.87}$$

Equations (1.82) to (1.87) describe the general properties of a CSTEC; they can be used to determine the conversion and the cell voltage as a function of the feed rate and the operating cell current. It should be noted that these equations are developed for a single compartment CSTEC and a relatively simple electrochemical reaction in order to illustrate the design procedures.

For more complicated cases, such as the two-compartment CSTEC and the CSTEC with recycles, the reader is referred to Ref. 28 for detailed analyses.

Plug Flow Electrochemical Cell

The design of a plug flow electrochemical cell is more difficult than that of the CSTEC or the batch cell, since it calls for microscopic material and energy balances. Consider a limiting reactant, O, in an electrolyte stream flowing into a one-compartment parallel-plate plug flow electrochemical cell (Fig. 1.14), where an electrochemical reaction given by (1.71) is taking place. The concentration of O-species in the inlet stream is C_{Oi} and the volumentric flow rate of the electrolyte is v. The anode and the cathode have the same surface area, and the area per unit length along the flow direction is represented by s. Since the plug flow cell does not permit the dispersion and the diffusion of reactants and products along the y-direction, a material balance of O-species over a differential reactor element, dy, results in the equation

$$vC_{Oi} \frac{dX}{dy} = \frac{si_y}{nF} \tag{1.88}$$

where $i_y = -i_c = i_a$ is the absolute value of the local electrode current density along the flow direction. Equation 1.88 can be integrated to give

$$X = \frac{s}{nFC_{Oi}v} \int_0^1 i_y \tag{1.89}$$

Equations (1.88) or (1.89) may be used to determine the cell length for a given conversion in the cell. However, the integration is an involved task, for the value of i_y not only depends on y, the conversion X, and the local cathodic overpotential η_{sc}, but also on the local anodic overpotential η_{sa} and the local ohmic potential drop across the electrode gap. These quantities are tied together by the voltage balance, and the solution of the resulting differential equation often requires the use of large-scale digital computers. Only under limited circumstances can one obtain a closed-form solution for (1.88). One case, which is of a particular interest to electroorganic synthesis, is the potential controlled operation. Let us assume that species O is reduced in the plug flow cell under a cathodic potentiostatic control, and the cathode polarization can be described by a simple Tafel relation of the type:

$$i_y = -i_c = i_c^0 C_O \exp\left(-\frac{\eta_{sc}}{\beta_{2c}}\right) \tag{1.90}$$

Substituting this rate expression into (1.88) and carrying out the integration, one obtains

$$\ln \frac{1}{1-X} = Y \exp\left(-\frac{\eta_{sc}}{\beta_{2c}}\right) \tag{1.91}$$

where Y is defined as

$$Y(\text{dimensionless cell length}) = \frac{s i_c^{\,0} y}{nFv} \tag{1.92}$$

Equation 1.91 is useful to estimate the cell size. It is interesting to note that (1.91) has the same form as (1.60) for the potential controlled batch electrochemical cells. Thus the physical significance of Y is analogous to the reaction time τ of the batch cells.

We have thus far neglected the longitudinal dispersion in the plug flow system. In practical cells, considerable intermixing and backmixing can occur in the flowing stream. An excellent discussion of this phenomenon can be found in the monographs by Levenspiel [28] and Aris [29]. Also, a pressure drop is needed to force the electrolyte flowing through the long parallel electrode gaps, and the amount of energy consumption must be considered in the practical systems. One drawback of the plug flow cell is its relatively low electrode area. Recent use of the particulate electrodes and the flow-through

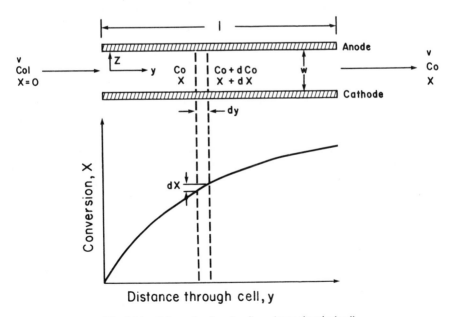

Fig. 1.14.   Schematic of a plug flow electrochemical cell.

porous electrodes [27, 30] has overcome some of the problems. More examples of the plug flow electrochemical cells can be found in a book by Pickett [31].

5 MASS TRANSFER

Transport Equations

We mentioned earlier that the flow of electric current causes a concentration change inside the cell, and this in turn results in a movement of ionic species in the electrolyte. For the systems with fast electrode kinetics (i.e., high exchange current density, small Tafel slopes, etc.) the concentration gradient is large, and the rate of reaction may be limited by the rate of transport of ionic species to or from the electrode surface. Under these circumstances, one must account for the concentration polarization in the cell analysis. As shown in (1.25) and (1.26), the concentration overpotential is related to the limiting current density at the electrode, and the information concerning the limiting current requires an understanding of the transport process of electrochemical cells.

The transport laws for a dilute electrolyte have been well established [20]. The flux of an ionic species j in the electrolyte can be attributed to the sum of (1) migration in the electric field, (2) molecular diffusion due to the concentration gradient, and (3) convection by the movement of the bulk of the electrolyte. In mathematical notations, this can be expressed as

$$\mathbf{N_j} = -Z_j U_j C_j F \nabla \Phi - D_j \nabla C + C_j \mathbf{v} \qquad (1.93)$$

<table>
<tr><td>Total
flux</td><td>Migrational flux</td><td>Diffusional
flux</td><td>Convective
flux</td></tr>
</table>

Here the flux $\mathbf{N_j}$ is the number of moles of j-species moving across a unit area per unit time [kg-mole/(m²) (sec)]. Since the movement is directional, $\mathbf{N_j}$ is a vector. The quantities, $\nabla \Phi$ and $\mathbf{v}$, are the potential gradient and the flow velocity of the electrolyte; Z_j and U_j are the charge number and the mobility of j-species. The electric current density in the electrolyte can be represented by the vectorial sum of the fluxes of all the charged species in the electrolyte:

$$\mathbf{i} = F \sum_j Z_j \mathbf{N_j} \qquad (1.94)$$

Also the condition of electric neutrality

$$\sum_j C_j Z_j = 0 \qquad (1.95)$$

should be true everywhere inside the electrolyte except in the diffuse double layer regime immediately adjacent to the electrode. This regime is of the order of several Angstroms and can be regarded as a part of the electrode surface. Now if we take a small differential element in the electrolyte as a control volume and make a material balance for the j-species, it can be shown that

$$\frac{\partial C_j}{\partial t} = - \vec{\nabla} \cdot \vec{N}_j + R_j \tag{1.96}$$

Here the $\partial C_j / \partial t$ term represents the accumulation of j-species inside the differential control volume, the $- \nabla \cdot N_j$ term represents the net flux of the j-species into the control volume, and R_j is the rate of generation due to a homegeneous bulk reaction in the control volume. In electrochemical cells, the term R_j is often equal to zero, and substituting (1.93) and (1.94) one obtains

$$\frac{\partial C_j}{\partial t} + \mathbf{v} \cdot \nabla C_j = Z_j U_j F \nabla \cdot (C_j \nabla \Phi) + D_j \nabla^2 C_j \tag{1.97}$$

Equation (1.97) describes the microscopic material balance of an ionic species in dilute electrolyte. The equation is difficult to work with, for the potential gradient, $\nabla \Phi$, is not only a function of cell geometry, but also depends on the local current density, the concentration, and the conductivity of the electrolyte. However, for the following two cases, the effect of $\nabla \Phi$ is ignored.

1. Binary electrolyte. In the binary electrolyte containing a single salt of the type $M_{\nu+} X_{\nu-}$, one can express the average concentration, and the effective diffusion coefficient as

$$c_\pm = \frac{C_M}{\nu_+} = \frac{C_X}{\nu_-} \tag{1.98}$$

$$D_\pm = \frac{Z_+ U_+ D_- - Z_- U_- D_+}{Z_+ U_+ - Z_- U_-} \tag{1.99}$$

Here the subscripts $+$ and $-$ denote the properties associated with the cation and the anion, respectively. It has been shown in Refs. 20 and 32 that by proper substitution of (1.98) and (1.99) into (1.97), the potential gradient term can be eliminated, and the equation becomes

$$\frac{\partial C_\pm}{\partial t} + \mathbf{v} \cdot \nabla C_\pm = D_\pm \nabla^2 C \tag{1.100}$$

2. Supporting electrolyte. A similar simplification of (1.97) applies when an inert supporting electrolyte is added to increase the conductivity of the solution and to reduce the potential gradient. For the ionic species whose concentration is low in the electrolyte, the migrational flux $-Z_j U_j C_j F \nabla \Phi$ becomes negligibly small as compared to the diffusional and the convective fluxes and (1.97) reduces to [20, 32]

$$\frac{\partial C_j}{\partial t} + \mathbf{v} \cdot \nabla C_j = D_j \nabla^2 C_j \qquad (1.101)$$

Equations (1.100) and (1.101) are called the equations of convective diffusion. Their form is similar to those of convective heat transfer and convective mass transfer in nonelectrolytic systems. Heat and mass transfer in nonelectrolytes have been extensively studied in the literature, and many of these results can be directly applied to the electrochemical systems. The solution of (1.100) and (1.101) requires the knowledge of fluid velocity, the cell geometry, and the boundary conditions at the electrode surface. The velocity profile is governed by a set of momentum equations. This is the area of fluid mechanics, and there are many excellent textbooks available to the readers dealing with this subject [33–35].

Nernst Diffusion Layer and Limiting Current Density

The solution of the convective diffusion equation gives the surface concentration gradient of a diffusing ion j, normal to the electrode surface. According to Fick's first diffusion law, this quantity can be related to the current density at the electrode by

$$\frac{\nu_j i}{nF} = D_j \left(\frac{\partial C_j}{\partial z} \right)_{z=0} \qquad (1.102)$$

where ν_j is the stoichiometric coefficient of species j in the electrode reaction and z is a spatial coordinate normal to the electrode surface and pointing toward the bulk of the electrolyte. Since the flow of electric current causes the charge-transfer reaction at the electrode surface, the concentration changes of reactants and products are always greatest in a thin electrolyte layer adjacent to the electrode surface. The concept of a diffusion layer was introduced by Nernst in 1904 [36]. He assumes that if the concentration change is a linear function of the distance from the electrode surface, and this change occurs entirely within a thin electrolyte layer of a thickness δ_N adjacent to the electrode as shown in Fig. 1.15, then

$$\left(\frac{\partial C_j}{\partial z} \right)_{z=0} = \frac{C_{j\infty} - C_{js}}{\delta_N} \qquad (1.103)$$

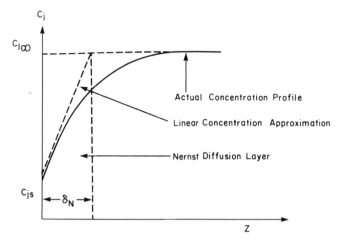

Fig. 1.15.   Concept of a Nernst diffusion layer.

where $C_{j\infty}$ is the bulk concentration of j-species. The quantity δ_N is called the thickness of the Nernst diffusion layer. Substituting (1.103) into (1.104), one has

$$i = \frac{nF}{\nu_j} \frac{D_j}{\delta_N} (C_{j\infty} - C_{js}) \qquad (1.104)$$

or

$$i = \frac{nF}{\nu_j} k_{jm} (C_{j\infty} - C_{js}) \qquad (1.105)$$

Here the quantity $k_{jm} = D_j/\delta_N$ is called the mass transfer coefficient of j-species; its value depends on the electrode geometry, the flow velocity, the diffusion coefficient of the j-species, and the kinematic viscosity of the electrolyte. The maximum current density allowed by mass transfer of the j-species for a given convective diffusion condition specified by k_{jm} occurs when the surface concentration becomes zero. Thus

$$i_{lim} = \frac{nFk_{jm}C_{j\infty}}{\nu_j} \qquad (1.106)$$

The concept of the Nernst diffusion layer assumes that there is a near stagnant electrolyte layer immediately adjacent to the electrode surface, and the transport of ionic species within the layer is due mainly to the diffusion process. The effect of convection on the transport process is to reduce the thickness of the Nernst diffusion layer; the thinner the thickness, the greater is the rate of mass transfer to the electrode surface. How much the flow velocity can exert an influence to the transport of ionic species is determined by a dimensionless quantity called the Schmidt number, Sc, which is defined as the ratio of the kinematic viscosity of the electrolyte, ν, to the diffusion coefficient, D_j,:

$$Sc = \frac{\nu}{D_j} \tag{1.106}$$

A small Schmidt number refers to the system having a large diffusion coefficient and low viscosity. Under this condition, the diffusional flux is greater than the convective flux, and the flow velocity would have little influence on the rate of mass transfer. A large Schmidt number means that the convective flux proceeds at a high rate, and the rate of transfer can be greatly enhanced by the convection of fluid. Most electrochemical systems have a large Schmidt number on the order of 10^3. This implies that the Nernst layer is confined to a very thin regime near the electrode and the value of δ_N is often on the order of 10^{-2} cm. This fact permits one to make reasonable simplifications in the solution of the convective diffusion equation.

EXAMPLE 4

A rotating hemispherical electrode (RHE) has been proposed as a complimentary tool to the rotating disk for investigating the kinetics of electrochemical reactions [37–39]. The RHE has the advantage that its current distribution is uniform below the limiting current potentials [40], and its shape change during the high-rate dissolution and deposition reactions has less effect on the flow pattern near the electrode. A typical RHE is shown schematically in Fig. 1.16. It is mounted on an inert support rod of an equal radius, a, and is rotating with a constant angular velocity, Ω, in a solution containing an excess of supporting electrolyte. Obtain an expression for the limiting current density at the electrode.

Solution

STEP 1. CONVECTIVE DIFFUSION EQUATION

For this problem, we use a set of spherical polar coordinates r, θ, ψ as shown in Fig. 1.16. The radial coordinate r is measured radially outward from the center of curvature of the spherical surface, θ is an angle measured from the axis of rotation, and ψ is the azimuth parallel to the direction of rotation. We assume that the hemispherical electrode is rotating at a sufficiently high velocity such that $\delta_N/a \ll 1$. This permits one to use a boundary layer approximation of the transport equation [33]. In addition, one may

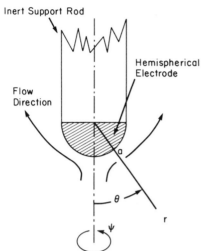

Inert Support Rod

Hemispherical
Electrode

Flow
Direction

a

θ

r

ψ

Fig. 1.16. Schematic of a rotating hemispherical electrode and the spatial coordinates.

take advantage of the fact that the system is axially symmetrical and there is no concentration gradient in the ψ direction. Thus (1.101) can be simplified to

$$v_r \frac{\partial C_j}{\partial r} + \frac{v_\theta}{a} \frac{\partial C_j}{\partial \theta} = D_j \frac{\partial C_j^2}{\partial r^2} \tag{1.107}$$

where v_r and v_θ are the radial and the latitudinal velocity components, respectively. The boundary conditions are

$$\text{at } r = a \qquad C_j = C_{js}$$
$$\text{at } r \to \infty \qquad C_j = C_{j\infty} \tag{1.108}$$

STEP 2. INFORMATION ON FLUID MECHANICS

The solution of (1.107) and (1.108) requires information concerning the velocity components v_r and v_θ near the electrode surface. For laminar flow, the momentum equations have been treated by Cochran [41] and Banks [42]. They expressed the velocity profiles in terms of a power series of r and θ:

$$v_r = (\nu\Omega)^{1/2} [(-0.51023\zeta^2 + \tfrac{1}{3}\zeta^3 + \cdots) + \theta^2 (0.52762\zeta^2$$
$$- \tfrac{1}{2}\zeta^3 + \cdots) + \cdots] \tag{1.109}$$

$$v_\theta = a\Omega [\theta (0.51023\zeta - \tfrac{1}{2}\zeta^2 + \cdots) + \theta^3 (-0.22129\zeta + \tfrac{1}{3}\zeta^2$$
$$+ \cdots) + \cdots] \tag{1.110}$$

where ζ is a dimensionless radial coordinate defined by

$$\zeta = \left(\frac{\Omega}{\nu}\right)^{1/2}(r - a) \tag{1.111}$$

Since an electrochemical mass transfer process is occurring within a very thin region of the order of δ_N near the electrode surface, the power series expansions of the type of (1.109) and (1.110) give a sufficiently accurate estimate of the mass transfer rate to the electrode.

STEP 3.   MATHEMATICAL SOLUTION OF THE PARTIAL DIFFERENTIAL EQUATION

An analytical solution of (1.107) and (1.108) has been obtained by Chin [37]. The details of the analysis are beyond the scope of this chapter and are not described here. Chin's results can be expressed as

$$\frac{C_j - C_{j\infty}}{C_{js} - C_{j\infty}} = 1 - 0.62045 \mathrm{Sc}^{1/3} \int_0^\zeta \exp\left(-0.17008 \mathrm{Sc}\,\zeta^3\right) d\zeta$$

$$- 0.12833 \mathrm{Sc}^{1/3} \theta^2 \zeta \exp\left(-0.17008 \mathrm{Sc}\,\zeta^3\right) \tag{1.112}$$

Using (1.103), the thickness of the Nernst diffusion layer δ_N is

$$\delta_N = -\frac{C_{js} - C_{j\infty}}{\left(\dfrac{\partial C_j}{\partial r}\right)_{r=a}} = \frac{1.61}{1 - 0.207\,\theta^2}\left(\frac{\nu}{\Omega}\right)^{1/2} \mathrm{Sc}^{-1/3} \tag{1.113}$$

and the average current density on the hemispherical electrode may be given as

$$i = -\frac{\left(\dfrac{nF}{\nu_j}\right) \displaystyle\int_0^{2\pi}\int_0^{\pi/2} a^2 \sin\theta\, D_j \left(\dfrac{\partial C_j}{\partial r}\right)_{r=a} d\theta\, d\psi}{\displaystyle\int_0^{2\pi}\int_0^{\pi/2} a^2 \sin\theta\, d\theta\, d\psi}$$

$$= 0.474\left(\frac{n}{\nu_j}\right) F(C_{j\infty} - C_{js}) D_j^{2/3} \nu^{1/6} \Omega^{1/2} \tag{1.114}$$

At the limiting current potentials, C_{js} is zero and (1.114) reduces to

$$i_{\lim} = 0.474\left(\frac{n}{\nu_j}\right) FD_j^{2/3} \nu^{-1/6} \Omega^{1/2} \tag{1.115}$$

Here ν_j is the stoichiometric coefficient of the j-species in the electrode reaction and ν is the kinematic viscosity of the electrolyte.

In mass transfer analysis, one often expresses the fluid flow condition and the mass transfer rate in dimensionless quantities called the Reynolds numbers (Re) and the Sherwood numbers (Sh):

$$Re = \frac{a\Omega^2}{\nu} \qquad (1.116)$$

$$Sh_{loc} \text{ (local Sherwood number)} = \frac{k_{jm}\, a}{D_j} = \frac{a}{\delta_N} \qquad (1.117)$$

$$Sh_{av} \text{ (average Sherwood number)} = \frac{k_{jm,av}\, a}{D_j} = \frac{a}{\delta_{N,av}} \qquad (1.118)$$

where $K_{jm,av}$ and $\delta_{N,av}$ are the average mass transfer coefficient and the average thickness of the Nernst diffusion layer over the entire electrode surface, respectively. For the hemispherical electrode surface one obtains the following dimensionless correlations:

$$Sh_{loc} = Re^{1/2}\, Sc^{1/3}\, (0.6205 - 0.1283\theta^2) \qquad (1.119)$$

$$Sh_{av} = 0.474\, Re^{1/2}\, Sc^{1/3} \qquad (1.120)$$

These dimensionless quantities are useful in the scale-up of electrochemical cells. We discuss their implications in a later section.

Use of Semiempirical Correlations for Mass Transfer Calculations

The rigorous solution of the convective diffusion equation is often difficult to obtain, and in many cases, an analytical solution may not exist. For two-dimensional problems, there are a number of simple semiempirical correlations reported in the literature, and they can be used to provide a quick and reasonably accurate estimate of mass transfer rate at high Schmidt numbers. One such correlation was introduced by Chilton and Colburn in 1934 [43]. The Chilton–Colburn relation is based on heat and mass transfer to a flat plate situated in a uniform flow stream as shown in Fig. 1.17. The coordinate y in the figure is the surface distance measured from the leading edge of the flow boundary layer, and the coordinate z is the perpendicular distance from the surface of the flat plate. For mass transfer to the plate at high Schmidt numbers, the thickness of the Nernst diffusion layer δ_N is much less than the fluid flow boundary layer. Under this condition, the Chilton–Colburn relation may be expressed as

$$Sh_y = Re_y\, Sc^{1/3} \left(\frac{f_y}{2} \right) \qquad (1.121)$$

where

$$\text{Sh}_y \begin{pmatrix} \text{local Sherwood number based on the} \\ \text{surface distance from the leading} \\ \text{edge of the boundary layer} \end{pmatrix} = \frac{k_{jm} y}{D_j} \qquad (1.122)$$

$$\text{Re}_y \text{ (local Reynolds number)} = \frac{U_\infty y}{\nu} \qquad (1.123)$$

$$f_y \begin{pmatrix} \text{local friction coefficient} \\ \text{at the plate surface} \end{pmatrix} = - \frac{\nu \left(\dfrac{\partial v_y}{\partial z} \right)_{z=0}}{\tfrac{1}{2} U_\infty^2} \qquad (1.124)$$

and v_y is the velocity component along the y-direction. Equation (1.121) is useful for both laminar and turbulent flows; it can be applied to two-dimensional transport problems, provided one has the information concerning the velocity gradient at the solid surface. Similar correlations have been proposed by many workers and are summarized in Table 1.1.

For the flow induced by the rotating hemispherical electrode described in Example 4, the leading edge of the flow boundary layer occurs at the pole of rotation. Thus the surface distance y is equal to $a\theta$. Also one may take $a\Omega \sin \theta$ as the characteristic velocity, U_∞, for every local point on the hemispherical surface. The quantities, Sh_y, Re_y, and f_y can then be expressed as [47]

$$\text{Sh}_y = \frac{k_{jm} a\theta}{D_j} \qquad (1.125)$$

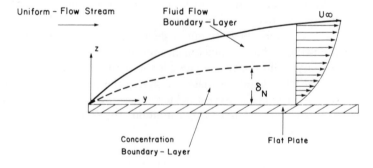

Fig. 1.17. Schematic diagram of flow and concentration boundary layers on a flat plate.

Table 1.1. Some Useful Semiempirical Correlations for Mass Transfer Calculations

Correlations	Investigators and References
$\mathrm{Sh}_y = \mathrm{Re}_y\, \mathrm{Sc}^{1/3}\, \dfrac{f_y}{2}$	Chilton, Colburn [43]
$\mathrm{Sh}_y = 0.0789\, \mathrm{Re}_y\, \mathrm{Sc}^{1/4}\, f_y^{1/2}$ (turbulent flow only)	Deissler [44]
$\mathrm{Sh}_y = \mathrm{Re}_y\, \mathrm{Sc}^{1/3}\, f_y^{1/2}$	Levich [32]
$\mathrm{Sh}_y = \mathrm{Re}_y\, \mathrm{Sc}^{1/4}\, f_y^{1/2}$ (turbulent flow only)	Levich [32]
$\mathrm{Sh}_y = 0.057\, \mathrm{Re}_y\, \mathrm{Sc}^{1/3}\, f_y^{1/2}$ (turbulent flow only)	Lin, Moulton, Putnam [45]
$\mathrm{Sh} = 4.86\, \mathrm{Re}_y\, \mathrm{Sc}^{1/3} \left(\dfrac{f_y}{2}\right) \dfrac{U_{\max}}{U_{\mathrm{av}}}$	Veith, Porter, Sherwood [46]

$$\mathrm{Re}_y = \frac{a^2 \Omega \theta \sin \theta}{\nu} \tag{1.126}$$

$$f_y = -\,\frac{\nu \left(\dfrac{\partial v_\psi}{\partial r}\right)_{r=a}}{\tfrac{1}{2}\, a^2 \Omega^2 \sin^2 \theta} \tag{1.127}$$

where v_ψ is the azimuthal velocity component of the flow stream. The quantity $(\partial v_\psi/\partial r)_{r=a}$ has been calculated by Manohar [48]; his results may be given as

$$-\left(\frac{\partial v_\psi}{\partial r}\right)_{r=a} = \frac{a\,\Omega^{3/2}}{\nu^{1/2}}\,(0.61592\theta - 0.18946\theta^3 - 0.05819 \sin^3 \theta) \tag{1.128}$$

Substituting (1.125) to (1.128) into (1.120) and rearranging, one obtains

$$\mathrm{Sh}_{\mathrm{loc}} = \frac{K_{jm}\, a}{D_j} = \mathrm{Re}^{1/2} \mathrm{Sc}^{1/3}\, \frac{0.61592\theta - 0.18946\theta^3 - 0.05819 \sin^3 \theta}{\sin \theta} \tag{1.129}$$

A comparison of (1.119) and (1.129) is shown in Fig. 1.18, where $\mathrm{Sh}_{\mathrm{loc}}/\mathrm{Re}^{1/2}\mathrm{Sc}^{1/3}$ is plotted as a function of θ. Also given in the figure are Newman's results [49] based on a numerical solution of (1.107). The agreement of

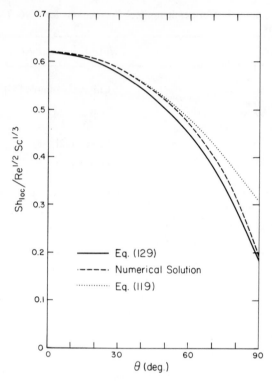

Fig. 1.18.   Local mass transfer rate to a rotating hemispherical electrode.

(1.129) with the results of two rigorous analyses is obvious. Integration of (1.129) gives

$$Sh_{av} = 0.433 \, Re^{1/2} Sc^{1/3} \tag{1.130}$$

which agrees with (1.120) to within 10%. Chin [50] has used the Chilton-Colburn relation to obtain a mass transfer correlation to the rotating hemispherical electrode under turbulent flow conditions:

$$Sh_{av} = 0.0198 \, Re^{4/5} Sc^{1/3} \tag{1.131}$$

Equations (1.120) and (1.131) have been experimentally verified by the mass transfer measurement in a ferricyanide–ferrocyanide redox system. The flow transition from laminar to turbulent flow is found to occur at Re = 1.5 × $10^4 \sim 4 \times 10^4$ [50].

6 CURRENT DISTRIBUTION

In the design of the idealized batch and continuous stirred-tank electrochemical cells, we tacitly assume above that there is a uniform current distribution at the electrodes to simplify the mathematical formulations. However, this seldom occurs in practical cells, and the degree of nonuniformity depends on the cell geometry, the electrolyte conductivity, the degree of polarization, and the flow velocity. Nonuniform current distribution in electrochemical cells can lead to serious problems, such as loss of the faradaic current efficiency, formation of undesirable dentrites during the charging of secondary batteries, and variation of the coating thickness in electroplating. To fabricate and operate the cells having uniform current distributions is a difficult task, for it requires a knowledge of potential theory, reaction kinetics, and convective transport within the cell. A normal procedure is to consider the anode and the cathode separately; the two current distributions are then tied together by a common boundary condition or a common electric field in the bulk electrolyte. The monograph by Newman [20] gives a detailed review of various current distribution problems.

To illustrate the principles of current distributions, let us consider that the reaction given by (1.16) is taking place at one electrode of an electrochemical cell. The governing equations for the calculation of current distribution are summarized below.

1. *Voltage balance.*

$$\phi - \phi_0 = \eta_s + \eta_{conc} + \eta_{ohm} \qquad (1.132)$$

Here η_{ohm} is the ohmic potential drop between the working and the reference electrodes; it has a negative value at the cathode and a positive value at the anode.

2. *Electrode kinetic equation for surface overpotential.*

$$i = i_0 \left(\frac{C_{R_s}^{\omega_1}}{C_{R\infty}} \right) \left(\frac{C_{O_s}^{\omega_2}}{C_{O\infty}} \right) \left[\exp\left(\frac{\eta_s}{\beta_1} \right) - \exp\left(-\frac{\eta_s}{\beta_2} \right) \right] \qquad (1.133)$$

Here i_0 is the exchange current density based on the bulk concentrations, which can be assumed to be constant under the steady-state conditions.

3. *Concentration overpotential.*

$$\eta_{conc} = \frac{R_g T}{nF} \ln \frac{C_{Os}}{C_{O\infty}} \frac{C_{R\infty}}{C_{R\infty}} \qquad (1.134)$$

4. *Steady-state convective diffusion equations for calculation of surface concentrations.*

$$\mathbf{v} \cdot \nabla C_O = D_O \nabla^2 C_O \tag{1.135a}$$

$$\mathbf{v} \cdot \nabla C_R = D_R \nabla^2 C_R \tag{1.135b}$$

It has been assumed that an excess supporting electrolyte is present in the cell, and the migrational fluxes of the O- and R-species can be neglected. The choice of spatial coordinate and the boundary conditions depends on the electrode geometry. In general, they can be expressed as follows:

In bulk electrolyte

$$C_O = C_{O\infty}, C_R = C_{R\infty} \tag{1.136a}$$

At the electrode surface

$$i = -nFD_0 \left(\frac{\partial C_0}{\partial z} \right)_{z=0}$$

$$= nFD_R \left(\frac{\partial C_R}{\partial z} \right)_{z=0} \tag{1.136b}$$

where i is the local current density and z is the coordinate normal to the electrode surface.

5. *Potential equation for calculation of electric field and ohmic potential drop in the electrolyte.*

$$\vec{\nabla}^2 \Phi = 0 \tag{1.137}$$

Again the choice of the spatial coordinate depends on the electrode and the cell geometry. The boundary conditions can be generally expressed as follows:

At reference electrode

$$\Phi = \Phi_r \tag{1.138a}$$

At the working electrode surface

$$i = -\kappa \left(\frac{\partial \Phi}{\partial z} \right)_{z=0} \tag{1.138b}$$

Here, κ is the conductivity of the electrolyte and is assumed to be constant in the cell. Equations (1.137) and (1.138) can be used to determine the electric potential at the surface of the working electrode, Φ_s, which is related to the ohmic potential drop by

$$\eta_{ohm} = \Phi_s - \Phi_r \tag{1.139}$$

It should be noted that the electric potential of the electrolyte adjacent to the reference electrode depends on the cell geometry and the nature of the counterelectrode. One may choose the counterelectrode as the reference point. In this case the value of Φ_r becomes a function of the local current distribution at the counterelectrode.

Equations (1.132) to (1.139) are tied together by the local current density and the local surface concentrations. They can be used to determine i as a function of the spatial coordinate on the electrode surface. The solution, even for a relatively simple geometry and flow conditions, is extremely involved and hence certain approximations are often made to simplify the mathematical analysis. Depending on the degree of simplifications, the current distribution may be classified into three categories.

1. *Primary current distribution.* If one neglects the surface overpotential and the concentration overpotential in (1.132), the solution then depends entirely on the electric potential field and the conductivity of the electrolyte in the cell. The current distribution thus obtained is called the primary current distribution. The solution of the Laplacian equation of (1.137) to (1.139) is similar to the problems of steady-state heat conduction and diffusion in solids. The book by Carslaw and Jaeger [51] contains many solutions of the Laplacian equations for various boundary conditions; it is a useful reference in the study of the primary current distributions.

2. *Secondary current distribution.* The secondary current distribution is obtained by neglecting the concentration overpotential in (1.132). Since there is no concentration polarization, the surface concentrations of the reacting species are then equal to the bulk concentrations, and the convective diffusion equations of (1.134) to (1.136) can be dropped from the analysis. The calculation may be further simplified by replacing the full Butler-Volmer equation of (1.133) with a linear polarization relation at small overpotentials and with the Tafel equation at high overpotentials. The secondary current distribution gives a good approximation to the systems having slow electrode kinetics and high mass transfer rates. It is generally more uniform than the primary current distribution.

3. *Tertiary current distribution.* The current distribution that fully accounts for the surface, the concentration, and the ohmic overpotentials as represented by (1.132) to (1.139) is called the tertiary current distribution. It describes the systems being operated at an appreciable fraction of the limiting current density, where the concentration variations near the electrode cannot be ignored. Equations (1.132) to (1.139) involve five variables; Newman [20] has suggested an iterative procedure for the numerical calculations using digital computers. The example presented here is taken from the area of electroplating. Nevertheless, the method of analysis is also valid for evaluating the current distribution patterns in electroorganic cells.

EXAMPLE 5

Continuous moving electrodes have long been used in the electrotinning of steel sheets and copper wires [52-54]. Let us consider a continuous semiinfinite flat sheet electrode moving with a constant velocity U_s through a stationary electrochemical cell as shown in Fig. 1.19. The sheet enters the cell through a watertight slot at one end and leaves the cell through a second slot at the opposite side. The motion of the solid surface induces a flow of adjacent electrolyte along the direction of the sheet movement. This creates a flow boundary layer that originates at the inlet slot and grows in thickness along the direction of the sheet movement. The counterelectrodes are located at a distance b on both sides of the moving electrode, and the cell walls are made of an insulation material. Assuming that the concentration overpotential and the surface overpotential at the countercurrent electrode are zero and that there is an excess supporting electrolyte present in the solution, obtain a current distribution at the moving electrode.

Solution

We consider that the electrolyte flow in the vicinity of the moving electrode is laminar in nature, and the effect of the cell wall on the flow boundary layer is negligible. The governing equations describing the electrode process at the continuous moving electrode are given below.

Voltage balance

$$\phi - \phi_0 = \eta_s + \eta_{conc} + \eta_{ohm} \tag{1.140}$$

Butler-Volmer equation of electrokinetics

$$i = i_0 \left(\frac{C_s^{\omega}}{C_{\infty}} \right) \left\{ \exp\left(\frac{\alpha n F}{R T} \eta_s \right) - \exp\left[-\frac{(1 - \alpha) n F}{R T} \eta_s \right] \right\} \tag{1.141}$$

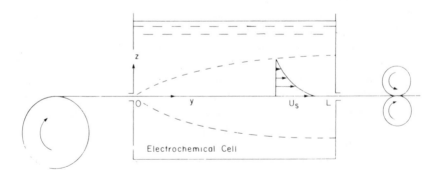

Fig. 1.19. Schematic diagram of a continuous moving sheet electrode.

Concentration overpotential

$$\eta_{conc} = \frac{RT}{nF} \ln \frac{C_s}{C_\infty} \tag{1.142}$$

Convective diffusion equations

$$v_y \frac{\partial C}{\partial y} + v_z \frac{\partial C}{\partial z} = D \frac{\partial^2 C}{\partial z^2} \tag{1.143}$$

$$C = C_\infty \quad \text{at} \quad y = 0 \quad \text{and} \quad z = \infty$$

$$\frac{\partial C}{\partial z} = \frac{i}{nFD} \quad \text{at} \quad z = 0 \tag{1.144}$$

Laplacian equation for potential distribution in the electrolyte

$$\frac{\partial^2 \Phi}{\partial y^2} + \frac{\partial^2 \Phi}{\partial z^2} = 0$$

$$\Phi = 0 \quad \text{at} \quad z = b \tag{1.145}$$

$$\frac{\partial \Phi}{\partial y} = 0 \quad \text{at} \quad y = 0 \quad \text{and} \quad y = L$$

$$\frac{\partial \Phi}{\partial z} = -\frac{i}{\kappa} \quad \text{at} \quad z = 0 \tag{1.146}$$

Here C and D are the concentration and the diffusion coefficient of the metal ion that is reduced at the moving electrode, and α is the transfer coefficient. Equations (1.143) to (1.145) can be integrated to give the surface concentration and the ohmic potential drop [55,56]:

$$C_s = C_\infty + \frac{1}{nF\sqrt{\pi U_s D}} \int_0^y i(t) \frac{dt}{\sqrt{y - t}} \tag{1.147}$$

$$\eta_{ohm} = \Phi_s = \frac{b}{L\kappa} \int_0^L i(t)\, dt + \sum_{m=1}^\infty \frac{z}{m\pi\kappa} \tanh \frac{m\pi b}{L} \cos \frac{m\pi y}{L}$$

$$\times \int_0^L i(t) \cos \frac{m\pi t}{L}\, dt \tag{1.148}$$

where t is a dummy variable. Equations (1.140) to (1.142), (1.147) and (1.148) must be solved simultaneously for five unknowns, that is, the surface overpotential η_s, the concentration overpotential η_{conc}, the ohmic potential drop η_{ohm}, the surface concentration C_s, and the current density at the electrode i. In numerical calculations it is useful to introduce the following parameters:

$$N\text{(dimensionless limiting}\atop\text{current density)} = \frac{n^2 F^2 D C_\infty}{RT\kappa \sqrt{\pi}} \left(\frac{U_s L^{1/2}}{\nu} \right) Sc^{1/2} \qquad (1.149)$$

$$\xi\text{(dimensionless exchange}\atop\text{current density)} = \frac{nFL}{RT\kappa} i_0 \qquad (1.150)$$

The dimensionless limiting current density N represents the ratio of ohmic potential drop to the concentration overpotential at the electrode. A large value of N implies that the ohmic resistance tends to be the controlling factor for the current distribution. For small values of N, the concentration overpotential is large and the mass transfer tends to be the rate-limiting step of the overall process. The dimensionless exchange current density ξ, on the other hand, represents the ratio of the ohmic potential drop to the activation overpotential. When both N and ξ approach infinity, one obtains the geometrically dependent primary current distribution.

Viswanathan and Chin [56] have obtained the current distribution for a simple first-order cathodic reaction with $\alpha = \omega = \frac{1}{2}$. Their results are given in Fig. 1.20 for various values of N. Also shown in the figure is a limiting current distribution designated by $i_{av}/i_{lim,av} = 1$. For this geometry the primary current distribution is uniform; the secondary current distribution, which is obtained by letting $N \to \infty$, is also uniform. For small values of N, the concentration overpotential is large and the process tends to be mass transfer limited, so that the tertiary current distribution becomes nonuniform. The results suggest that the uniform current distribution can be achieved at the continuous moving electrode when the values of N are large.

7   DIMENSIONAL ANALYSIS AND SCALE-UP SIMILARITIES

We have thus far introduced many dimensionless variables in the discussion of cell design, mass transfer, and current distribution problems. These dimensionless variables provide a set of similarities between a laboratory cell and its corresponding full-size industrial cells. In the scale-up of an electrochemical cell, the following similarities should be considered.

1. *Geometric similarity.*   The geometric similarity between two different size cells is achieved by choosing a set of dimensionless lengths and letting these quantities of one cell be equal to the corresponding quantities of the second cell. For example, in a filter-press electrochemical cell where the parallel electrodes have a dimension of l_x, l_y, and l_z, one may keep the cells geometrically similar on scale-up by specifying constant values of l_x/l_z and l_y/l_z. The geometric similarity is generally limited to the shape and area of the electrode. The thickness of the cell separator and the gap-space between

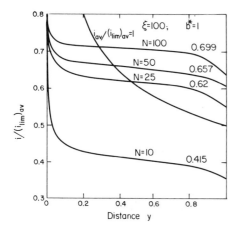

Fig. 1.20. Current distribution on a continuous moving electrode at $\xi = 100$ for various values of N.

the anode and the cathode are often kept at the same values during the scale-up of electrochemical cells because of the cell voltage considerations.

2. *Kinematic similarity.* The kinematic similarity refers to the driving forces concerning the motion of the electrolyte and the transport of ionic species. In two geometrically similar cells, a kinematic similarity exists if

$$\frac{\text{inertia force in cell I}}{\text{inertia force in cell II}} = \frac{\text{viscous force in cell I}}{\text{viscous force in cell II}} \tag{1.151}$$

For example, in a cell with forced convections, the Reynolds number, Re, is the ratio of inertia force to viscous force. The Schmidt number, Sc, is the ratio of viscous force to diffusion force, and the Sherwood number is the ratio of the convective flux to the diffusional flux. These values should be the same if the two cells are kinematically similar.

3. *Chemical similarity.* The chemical similarity refers to the same stoichiometric ratios of reactants in the feed and the same conversion in the product stream. For the continuous stirred tank electrochemical cell, this may be defined by

$$\left(\frac{\epsilon_f I}{nFC_{oi}V} \right)_{\text{cell I}} = \left(\frac{\epsilon_f I}{nFC_{oi}V} \right)_{\text{cell II}} \tag{1.152}$$

as seen from (1.81). For the batch cell being operated at constant current conditions, the chemical similarity may be expressed as

$$\left(\frac{\epsilon_f I t}{nFC_{oi}V} \right)_{\text{cell I}} = \left(\frac{\epsilon_f I t}{nFC_{oi}V} \right)_{\text{cell II}} \tag{1.153}$$

For the plug flow cells operated at constant potentials, the dimensionless cell length must be equal for two chemically similar cells.

$$\left(\frac{s i_c^0 y}{nFv} \right)_{\text{cell I}} = \left(\frac{s i_c^0 y}{nFv} \right)_{\text{cell II}} \tag{1.154}$$

4. *Electrical similarity.* The electrical similarity is the most important factor in the scale-up of electrochemical cells. It is defined as the situation where the geometrically, kinematically, and chemically similar cells bear the same cell voltage and similar current distributions inside the cells. Two geometrically similar cells would have an identical current distribution if they had the same values of the dimensionless limiting current density, N, and the dimensionless exchange current density, ξ, as indicated by (1.149) and (1.150). To achieve the same cell voltage for two different size cells being operated at the same current density often requires that the anode–cathode spacing and the thickness and the conductivity of the cell separator be identical in the two cells.

The aim of scale-up is to determine the required cell size for a given electrochemical process from the rate equations. The mathematical model of many important engineering problems cannot be solved theoretically. Problems of this type are especially common in fluid flow, heat transfer, and mass transfer processes. Under these circumstances, the dimensional analysis is a useful tool to provide information concerning the process variables [57–59]. The method is based on the fact that if there exists a theoretical equation correlating the variables of the process, that equation must be dimensionally homogeneous. Because of this requirement, it is possible to group many process variables into a smaller number of dimensionless groups and simplify the task of fitting experimental data with an empirical design equation. The method is useful in the scale-up of laboratory data to predict the performance of full-scale equipment. The dimensional analysis does not give a complete mathematical correlation for the process variables and hence the determination of the final equation form must rely on the empirical efforts. Also, in performing the dimensional analysis, one should have a good knowledge of the physical principles of the process and know what variables are important in the problem. We use the following example to illustrate the procedure of dimensional analysis.

EXAMPLE 6

The parallel-plate electrodes are of a common configuration in industrial electrochemical cells. One such system is shown in Fig. 1.21. The cell consists of a set of planar anode and cathode separated by a short distance d. The electrolyte is pumped from a solution reservoir and flows through the electrode gap as shown in the figure. The electrodes have a length L along the

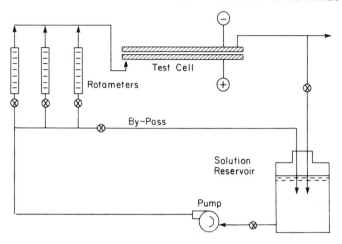

Fig. 1.21. Schematic diagram of a parallel-plate electrochemical cell.

flow direction, and a width w perpendicular to the flow. Assuming that the velocity profile in the cell is fully developed, obtain a dimensionless correlation for the rate of mass transfer to the planar electrodes.

Solution

In the parallel-plate cells, an important variable, which determines the convective conditions, is the equivalent diameter d_e, defined as

$$d_e = 4 \, \frac{\text{cross sectional area}}{\text{periphery perpendicular to flow}} = \frac{2dw}{d+w} \qquad (1.155)$$

We assume that the cell is sufficiently wide so that the mass transfer rate varies only along the direction of fluid flow. The other variables that would affect the average mass transfer coefficient, k_{jm}, are the electrode length L, the average flow velocity U_{av}, the kinematic viscosity of the electrolyte ν, and the diffusion coefficient of the diffusing species D_j. The dimensions of these variables are given in Table 1.2. If there exists a theoretical equation for the problem, it can be expressed as

$$k_{jm,av} = \psi(U_{av}, L, d_e, D_j, \nu) \qquad (1.156)$$

Equation (1.156) must be dimensionally homogeneous, and any term in function ψ must conform to the dimensional formula

$$(k_{jm,av}) = (U_{av})^a \, (L)^b \, (d_e)^c \, (D_j)^e \, (\nu)^f \qquad (1.157)$$

Substitution of the dimensions from Table 1.2 gives

$$\frac{l}{t} = \left(\frac{l}{t}\right)^a l^b\, l^c \left(\frac{l^2}{t}\right)^e \left(\frac{l^2}{t}\right)^f \tag{1.158}$$

Equating the exponents of the dimensions, l and t, on both sides of (1.158), one obtains

$$a + b + c + 2e + 2f = 1$$

$$a + e + f = 1 \tag{1.159}$$

Equation (1.159) can be rearranged to yield

$$a = 1 - e - f$$

$$b = -c - e - f \tag{1.160}$$

Thus (1.157) becomes

$$(k_{jm,av}) = (U_{av})^{1-e-f}(L)^{-c-e-f}(d_e)^c\,(D_j)^e\,(v)^f \tag{1.161}$$

Rearranging (1.161) gives

$$\left(\frac{k_{jm,av}\, d_e}{D_j}\right) = \left(\frac{U_{av}\, d_e}{v}\right)^{1-e-f} \left(\frac{v}{D_j}\right)^{1-e} \left(\frac{d_e}{L}\right)^{c+e+f} \tag{1.162}$$

The dimensions of each of the four bracketed groups in the above equation are zero. Any functions of these dimensionless groups will be dimensionally homogeneous, and the function will be a dimensionless one. Thus we may express the final results in the form:

$$Sh_{av} = f\left(Re,\ Sc,\ \frac{d_e}{L}\right) \tag{1.163}$$

Table 1.2. Process Variables and Their Dimensions for Mass Transfer in a Parallel-Plate Electrochemical Cell

Variables	Symbol	Dimensions[a]
Average mass transfer coefficient	$k_{jm,av}$	l/t
Average flow velocity	U_{av}	l/t
Equivalent diameter	d_e	l
Cell length	L	l
Kinematic viscosity	v	l^2/t
Diffusion coefficient	D_j	l^2/t

[a] The notation l refers to the dimension of length, and t is the dimension of time.

where

$$Sh_{av} = \frac{k_{jm,av}d_e}{D_j}$$

$$Re = \frac{U_{av}d_e}{\nu}$$

$$Sc = \frac{\nu}{D_j} \tag{1.164}$$

The function f must be found experimentally if a theoretical solution is not available. The study of mass transfer to the parallel-plate electrochemical cells has been reported extensively in the literature. For laminar flow between two flat plates, the Leveque analysis [60, 61] gives

$$Sh_{av} = 1.85\left(Re\ Sc\ \frac{d_e}{L}\right)^{1/3} \quad \text{for } Re < 2100 \tag{1.165}$$

For turbulent flow, Van Shaw and Hanratty [62, 63] have carried out experiments with circular tubes. Their results can be expressed in the present notations as

$$Sh_{av} = 0.276\ Re^{0.58}\ Sc^{1/3}\left(\frac{d_e}{L}\right)^{1/3} \quad \text{for } Re > 4100 \tag{1.166}$$

Figure 1.22 shows a comparison between (1.165) and (1.166) and the experimental data obtained by the limiting current measurement for the reduction of ferricyande ion in a rectangular flow cell. The agreement is obvious, and (1.166), though originally obtained for circular tubes, is shown to be valid also for turbulent flow between two flat electrodes.

8 OPTIMIZATION AND COST ANALYSIS

Optimization in the chemical process industry infers the selection of equipment and operating conditions for the production of a given material such that the profit will be a maximum. This could be interpreted as meaning the maximum output for a given capital outlay or the minimum investment for a specified production rate. The former is a mathematical problem of determining the appropriate values of a set of variables to maximize a dependent variable, whereas the latter is the problem of locating a minimum value. In terms of profit, both types of problem are maximization problems, and the solution is generally accomplished by an economic balance between capital and operating costs.

In the electrochemical industry, the selection of equipment and operating

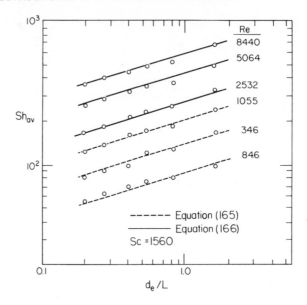

Fig. 1.22. Dependence of average Sherwood number on electrode length.

conditions involves a balance between the cost of power usage and the capital cost of electrolysis systems. A high operating current density for a given production rate can save the capital investment of an electrolytic cell; however, the electrical power consumption will increase with current density in order to sustain the internal polarization losses. Let us consider an electrochemical process where the capital investment of electrolytic cells is estimated to be $C_{\$f}$ dollars per square meter of electrode area. The cells have a useful life of L_f years. If one assumes a straight-line depreciation and a zero residue value for the cell, the annual capital cost would be

$$\text{annual capital cost (\$/year)} = \frac{C_{\$f}A}{L_f} = \frac{C_{\$f}I}{L_f i} \tag{1.167}$$

where A is the total electrode area of the cells, I is the total current, and i is the operating current density. If a species j is produced by the electrolysis with a 100% faradaic efficiency, the annual production rate can be calculated from the Faraday law as

$$\text{annual production (kg/year)} = \frac{\nu_j I M_j (3600 \times 24 \times 365)}{nF} \tag{1.168}$$

where n is the number of electrons transferred in the electrochemical reaction, and ν_j is the stoichiometric coefficient of the j-species. For many electrochemical processes, the cell voltage can be approximately expressed as a linear function of the operating current density i:

$$E = E_0 - \frac{\delta}{\kappa_{av}} i \qquad (1.169)$$

where E_0 is the open-circuit cell voltage, δ is the anode-to-cathode distance, and κ_{av} is the average conductivity of the cell. The annual cost of electricity is given by

$$\text{annual electricity cost (\$/year)} = C_{\$e}\left(-\frac{EI}{1000}\right)(24 \times 365) \qquad (1.170)$$

where $C_{\$e}$ is the cost of electrical energy in \$/kWh. Neglecting the costs of raw materials, labor, overhead charges, interest, and other parasitic energy consumptions, the total production cost for this simple problem can be expressed as

$$Y_P \text{ (production cost in \$/kg)} = \frac{\text{annual capital cost}}{\text{annual production}}$$

$$+ \frac{\text{annual electricity cost}}{\text{annual production}} \qquad (1.171)$$

Substituting (1.167) to (1.170) into (1.171), one obtains

$$Y_P = \frac{nF}{3600\,\nu_j M_j}\left[\frac{1}{8760}\frac{C_{\$f}}{L_f i} + \frac{C_{\$e}}{1000}\left(-E_0 + \frac{\delta i}{\kappa_{av}}\right)\right] \qquad (1.172)$$

A common problem facing the electrochemical engineers during the equipment-design phase is to determine an optimal operating current density and the cell size in order to yield a minimum production cost. Mathematically this means that the following condition must be satisfied:

$$\frac{dY_P}{di} = 0 \qquad (1.173)$$

Differentiating (1.172) with respect to i, one obtains

$$\frac{dY_P}{di} = \frac{nF}{3600\,\nu_j M_j}\left(-\frac{C_{\$f}}{8760\,L_f i^2} + \frac{C_{\$e}\delta}{1000\,\kappa_{av}}\right) = 0 \qquad (1.174)$$

Equation (1.174) is an algebraic function and can be solved for i:

$$i_{opt} = 0.338\left(\frac{C_{\$f}\kappa_{av}}{L_f C_{\$e}\delta}\right)^{1/2} \qquad (1.175)$$

An analysis similar to that of (1.175) has been used by Ibl and Schelch [64, 65] for the production of chlorine and electrolytic copper. A more elaborated analysis of the cost optimization of electrochemical cells has been treated by Beck [66]. Although this specific problem is concerned with the analysis of an electrochemical energy storage plant, the same cost estimate procedures and the method of optimization can also be used for the cost analysis of a large-scale electroorganic plant.

EXAMPLE 7

For the electrochemical hydrogen–chlorine energy storage plant described in Example 2, the results of material and energy balances indicate that the overall energy conversion efficiency is affected by (1) operating cell current density i; (2) the maximum electrolyte temperature, T_{max}; and (3) the initial concentration of hydrochloric acid, χ_i, at the beginning of charge. It can be shown that the volume requirement for the HCl/Cl_2 storage increases as the electrolytic concentration decreases; on the other hand, a higher energy conversion efficiency is obtained at lower values of χ_i (Fig. 1.9). The maximum electrolyte temperature affects the capital investment of heat exchangers, as well as the overall energy conversion efficiency. These variables have a great influence on the capital investment of the energy storage plant. Perform an analysis to determine the optimal values of these process variables.

Solution

To simplify the analysis, 1.0 kWh of energy output during the discharge period is used as the basis of calculation. However, numerical computations are made for a 20-MW plant being operated at 10-hr charge and 10-hr discharge cycles. As is described in Example 2, the minimum electrolyte temperature, T_{min}, and the final electrolyte concentration, χ_f, at the end of the 10-hr charge period are kept at 288°K and 5 wt% HCl, respectively, because of the environmental and the technoeconomic considerations.

STEP 1.   MODELING OF OBJECTIVE FUNCTION

According to the flowsheet shown in Fig. 1.3, we minimize the capital cost consisting of the following plant components: (1) cell stack, (2) H_2 storage, (3) HCl/Cl_2 storage, (4) heat exchanger, (5) piping, and (6) pumps. The costs of installation, chemicals, the equipment for power conditioning, H_2 cleanup, sewerage, water distribution, safety precautions, and so on are not directly related to the variables to be optimized; therefore, they are not considered in this analysis for the simplicity of mathematical formulations. The present task is to minimize the following function for a 20-MW/200MWh energy storage plant:

$$Y_P\,(i,\,T_{max},\,\chi_i) = \frac{\text{cost of (cell stack } + \; H_2 \text{ storage } + \; HCl/Cl_2 \text{ storage } + \text{ heat exchanger } + \text{ piping } + \text{ pumps)}}{\text{net energy output during the discharge period}}$$

$$= \frac{1000 \sum\limits_{j=1}^{6} C_{\$j} F_j^{\alpha_j}}{\left(\int\limits_{10}^{20} EI\,dt \right) \text{discharge}} \quad (\$/\text{kWh}) \qquad (1.176)$$

Here $C_{\$j}$ and F_j are the cost coefficient and the design capacity of equipment j; α_j is an exponent associated with the scale-up of the equipment [25, 26]; and E and I are the voltage and the total current from the cell stack. Equation (1.176) is subject to the following constraints:

$$0 \le i \le 5000 \text{ A/m}^2$$

$$318°\text{K} \le T_{\max} \le 368°\text{K}$$

$$0.05 < \chi_i \le 0.5 \qquad (1.177)$$

These values are chosen because of the availability of the cell performance data, the temperature limitation for the plant materials to withstand the corrosion of the aqueous HCl/Cl_2 mixture, and the stability of hydrochloric acid at high temperature and concentrations. The other conditions associated with the objective function are

$\chi_f = 0.05$

$T_{\min} = 288°\text{K}$

Pressure of HCl/Cl_2 storage and Cl_2 electrode, $P = 2760$ KPa

H_2-storage pressure at the beginning of charge, $P_1 = 690$ KPa

H_2-storage pressure at the end of charge, $P_2 = 4100$ KPa   (1.178)

STEP 2.   COST ESTIMATE

We summarize the cost estimate for each of the capital items listed in (1.176). The details of the analysis are given in Refs. 25 and 26. All the prices given below are based on a constant 1977 U.S. dollar.

$$\text{Cost of cell stack} = C_{\$f} A_F = C_{\$f} \frac{I}{i} \qquad (1.179)$$

since the electrode is made of $RuO_x/graphite$ and Nafion is used as the cell separator, $C_{\$f}$ is $162/m^2$.

$$\text{Cost of } H_2 \text{ storage} = C_{\$H_2} V_{H_2}^{0.53} \qquad (1.180)$$

For carbon steel storage tank, $C_{\$H_2}$ is $28,000/m^{1.6}$.

$$\text{Cost of } HCl/Cl_2 \text{ storage} = C_{\$acid} V_{acid}^{0.48} \qquad (1.181)$$

where the cost coefficient, $C_{\$acid}$, is equal to $\$21,300/m^{1.4}$ [69]. The design variable V_{acid} is the total volume of hydrochloric acid and liquid chlorine required for storage. It is a function of initial electrolyte concentration χ_i.

$$\text{Cost of heat exchanger} = C_{\$ex} A_{ex}^{0.94} \tag{1.182}$$

where A_{ex} is the heat exchange area and $C_{\$ex}$ has a value of $\$5000/m^{1.9}$. The function of the heat exchanger is to remove the waste heat during discharge and to limit the electrolyte and cell temperature within the operating limits. The heat exchanger area, A_{ex}, depends on T_{max} and the amount of heat, $-Q_{ex}$, to be removed; it is a function of the operating current density, i, the maximum electrolyte temperature, T_{max}, and the initial electrolyte concentration, χ_i, at the beginning of charge.

$$\text{Cost of pipe} = M \cdot C_{\$pipe} \cdot L \tag{1.183}$$

Here M is the number of parallel pipelines; the cost coefficient, $C_{\$pipe}$, is equal to $\$290/m$ [67] for the 0.0394-m i.d. glass-lined steel pipes, and L is the pipeline length of 300 m.

$$\text{Cost of pumps} = MC_{\$pump} W_p^{0.98} + 47,600M \tag{1.184}$$

where $C_\$ = \$1.7(J/S)^{0.8}$ is the cost coefficient of tantalum pumps for the recirculation of electolyte, and W_p is the brake power of each pump whose value depends on the rate of recirculation. The Cl_2 pump is a PTFE-lined metering pump with a double Teflon diaphram; it is estimated that each Cl_2 pump costs $\$47,600$.

STEP 3.   NUMERICAL CALCULATIONS

Hsueh et al. [25, 26] have used a digital computer to determine the minimum of (1.176) to (1.184). Their numerical calculation takes into account the changes of electrolyte temperature, cell voltage, and the physical properties of electrolyte during the charge and discharge cycles. The results are summarized in Table 1.3. The optimal point is located at $\chi_i = 0.3$, $T_{max} = 343°K$, and $i = 1900$ A/m². The total input electric energy during the charge cycle is 268 MWh, and the net output electric energy during the discharge cycle is 200 MWh; this yields an overall electric-to-electric efficiency of $\epsilon_0 = 74.6\%$. The total electrode area required for the cell stack to operate at the optimal condition is 10,500 m². The design volumes for the H_2 storage and the HCl/Cl_2 storage are 2680 and 1070 m³, respectively. The waste heat to be removed during the discharge cycle is 2.4×10^{11} J $(5.7 \times 10^7$ kcal); this requires a heat exchange area of 281 m². Also, a total of seven pipelines and seven sets of Cl_2 pumps and HCl pumps is needed to recirculate the electrolyte at a rate of 1880 m³/hr (8280 gal/min); the brake power required for each HCl pump is 2400 J/sec (4 hp). The minimum cost for the six capital items included in (1.176) is $\$31.9/kWh$; the second and

TABLE 1.3.   Results of the Optimization Calculation for Example 6[a]

Items	Capital Cost ($/kWh)	Percentage	Design Capacity
Cell Stack	8.5	26.6	$1.05 \times 10^4 m^2$ cell area
H₂ Storage	9.5	29.8	$2.68 \times 10^3 m^3$ storage volume
HCl/Cl₂ storage	3.1	9.7	$1.07 \times 10^3 m^3$ storage volume
Heat exchanger	6.0	18.8	$2.81 \times 10^2 m^2$ heat exchanger area
Pump	1.8	5.6	2.4×10^3 J/sec pump brake power
Piping	3.0	9.5	7 parallel pipelines
Total cost	31.9 $/kWh	100.0	

[a] Optimal operating condition: $I = 1900$ A/m², $T_{max} = 343°$K (70°C), $\chi_i = 0.3$, $\epsilon_0 = 74.6\%$.

third columns in Table 1.3 give the cost breakdown for each item. In Fig. 1.23 the total capital cost and the capital cost of each item are plotted as a function of the operating current density at the optimal $T_{max} = 343°$K and $\chi_i = 0.3$. It can be seen that the capital costs of each item except the cell stack increase as the operating current density increases. Only the cost of the cell stack decreases as the operating current density increases. The influence of the current density on the cost of H₂ storage and HCl/Cl₂ storage is due to a decrease in the electric-to-electric efficiency as mentioned earlier. The influence of current density on the cost of heat exchanger, pump, and piping is due to a combined effect of decreasing the electric-to-electric efficiency and increasing the design flow rate of the electrolyte. The major cost items in the objective function are the cell stack, the H₂ storage, and the heat exchanger, which constitute 75% of the total capital investment.

9   INDUSTRIAL CELLS FOR ELECTROORGANIC PROCESSES

General Considerations

Cell Configuration

A variety of cell configurations have been developed: flat plate, packed bed, fluidized bed, and pumped slurry-bed electrode. The cells are suitable for continuous flow-through operation; there is no need to periodically

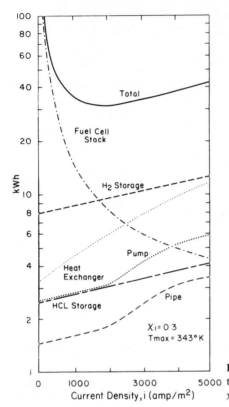

Fig. 1.23. Capital investment cost as a function of i at the optimal $T_{max} = 343°K$ and $\chi_i = 0.3$.

change and withdraw ingredients from the cells. In electroorganic synthesis the capital cost can be a major portion of the total production costs. Since capital costs are directly proportional to the electrode surface, it is a basic criterion to maximize the ratio of electrode surface area to cell volume in the design of the electrochemical cells. The simplicity of construction and ease of maintenance are other cost-saving factors. The specific surface area for the filter-press (plate-and-frame) cell is 10 to 100 times as great as that of the plate electrodes in tank cells. The former construction is less complex, more compact, and easier to maintain; it has been often considered as a first choice in many electrochemical processes.

Electrical Energy

The basic design goal of a chemical process is to produce a product of high quality with a minimum energy consumption and capital costs. To maximize the performance of an electrochemical cell, the parameters affecting the process economics must be optimized; these include the faradaic current efficiency, cell voltage, and current density. The cell voltage is composed of thermodynamic reversible cell voltage, electrode overpotentials, and the ohmic losses through the cell. It increases with operating current densities and generally decreases with increasing operating temperature. As the current density increases, the number of the cells is reduced so that the capital cost is reduced. However, there is a mass transfer controlled limiting current density that depends on the electrolyte concentration, nature of electrode surface, temperature, and fluid velocity. The components contributing to ohmic losses are the electrolytes, cell membranes, resistance of electrodes, and external bussings. These losses can be reduced by decreasing the spacing between separator and electrodes. The connection of individual cells in series or in parallel is determined by a cost analysis of electric power utilizations. This cost analysis shows how the cost of power supplies varies with the current and voltage. As the operating current and voltage of individual cells are known, the method of connection can be deduced from the results of the power cost analysis.

Fluid Distribution

An electrolyte stream must be distributed uniformly within a cell to eliminate stagnant zones. Sufficient fluid velocity past the electrode must be attained to provide a sufficiently high mass transfer rate. The situation can be achieved by enough agitation. The design of fluid distribution through the cells depends on current density, heat transfer, the nature of products, and the pressure drop across the cell.

Materials of Constructions

The electrode materials must have adequate electrical and thermal conductivity, good mechanical properties, corrosion resistance, and especially low capital cost. The membrane is used to minimize the mixing of anolyte with catholyte to reduce undesired side reactions. The basic properties of membrane are high mechanical strength, good ionic selectivity, and chemical stability over a wide range of temperature and electrolyte concentrations.

Operating Conditions

The operating temperature and pressure depend on the cell design and the properties of the cell components. The material and energy balances should be performed to determine if there is a need of heat exchangers for the thermal management. The best way to determine the operating conditions is to optimize the process variables.

Types of Industrial Cells

Filter-Press Cells or Plate-and-Frame Cells

The most common type of cells for large-scale electroorganic synthesis is the filter-press or plate-and-frame type cell. This design consists of plate electrodes separated by insulating gaskets or frames. The electrodes may be supported on nonconducting plates. A diaphram may be employed between a pair of gaskets to provide hydrodynamically separated anode and cathode compartments. A number of individual cells are combined to form a filter-press-like enclosure by end plates and tie bolts. The flow of electrolyte and electric current to the cells may be arranged in series as well as in parallel.

One type of plate-and-frame cell used in the electroorganic synthesis is the Foreman/Veatch cell [68]. As shown in Fig. 1.24, it is composed of a stack of alternating plate electrodes and insulating frames [69]. Electrolytes flow serially from compartment to compartment through slots in the electrodes, while electrical current is supplied by parallel connections to alternating electrodes. If cooling is required, a heat-exchanger plate may be inserted into the filter press. This design is suitable for both oxidation and reduction reactions, but is limited to the systems in which a diaphram is not needed. The preferred range of current densities is 300 to 5000 A/m^2.

A 2000-A plate-and-frame cell has been described by Mantel [70]. This cell is used for the production of periodic acid. Cation-selective membranes are used to separate anode and cathode compartments. Both the electrolyte flow and the electrical connection are in parallel.

A plate-and-frame cell with diaphrams has been used by Monsanto [71] for the production of adiponitrile from acrylonitrile. Figure 1.25 shows the Monsanto cell. It consists of a lead cathode and a lead alloy anode attached to the opposite sides of a polypropylene plate. An ion-exchange membrane mounted between a pair of gasketed polypropylene frames is used to separate the anode and cathode compartments of a single cell. The design employs a flow rate of 1 to 2 m/sec through the cathode compartment; this flow rate is in the turbulent flow regime to increase the mass transfer rate. The electrolyte flow through the cells is in parallel, while the flow electrical current is in series through the bank of cells.

Capillary Gap Cells

The capillary gap cell, originally developed by Beck and Guthke [72], is bipolar and has the electrodes stacked on top of each other as shown in Fig. 1.26 [71]. The cell is composed of a base plate and a stack of round electrodes with a central bore. The electrodes are isolated from each other by spacers 0.3 to 1.0 mm thick and are held in position by compressing the cover and base plate, which act as the current collectors. The electrolyte inlet is in the bottom plate. The outlet is in the upper part of the outer casing. The capillary gap cell has been used for the preparation of different dicarboxylic acid esters, especially sebacic and suberic acids. It is also used for the anodic methoxylation of furan to 2,5 dimethoxy-2,5 dihydrofuran.

Tank Cells with Plate Electrodes

The use of plate electrodes immersed in a lined, rectangular tank represents a direct extension of laboratory beaker cells. The tank may be fitted

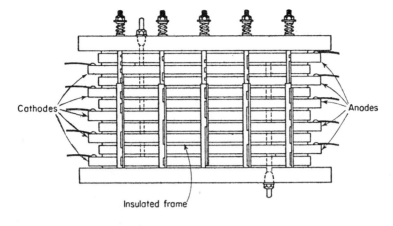

Cathodes

Anodes

Insulated frame

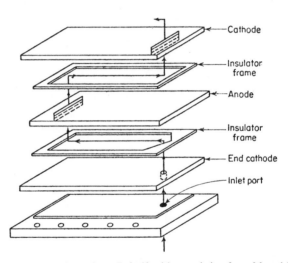

Cathode

Insulator frame

Anode

Insulator frame

End cathode

Inlet port

Fig. 1.24.   Foreman/Veatch cell. Reprinted from Ref. 69 with permission from Marcel Dekker, Inc.

with a cover if collection of gaseous products is desired. A 1000-A tank-type Kryschenko cell as shown in Fig. 1.27 has been described for the production of isobutyric acid [74]. The cell is made from sheet vinyl using welded joints. The anodes suspended from the tank cover are constructed of lead dioxide electroplated on a nickel grid. The cathode is composed of planar copper coils positioned between the rows of anodes.

An elaborate variation of the tank cell has been proposed by LeDuc [10] for electrochemical production of olefin oxides. Special features of the cell in-

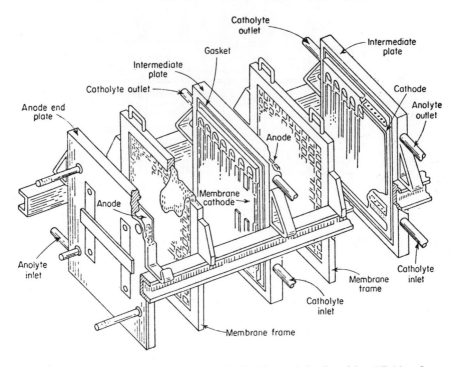

Fig. 1.25. Monsanto cell. Reprinted from Ref. 69 with permission from Marcel Dekker, Inc.

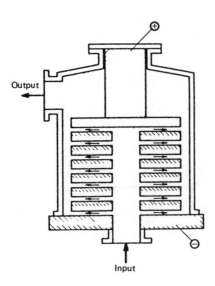

Fig. 1.26. An improved capillary gap cell. Reprinted from Ref. 73 with permission from AIChE.

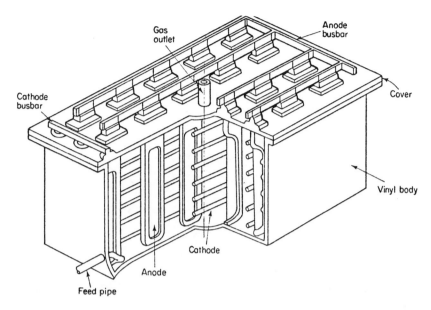

Fig. 1.27. Kryschenko cell for synthesis of isobutyric acid. Reprinted from Ref. 69 with permission from Marcel Dekker, Inc.

clude (1) hollow porous anode plates of graphite through which olefin gas is introduced (2) flattened turbular cathodes formed from steel wire screen and (3) a diaphram of asbestos fiber deposited on the outside surfaces of the cathodes tubes (Fig. 1.28). The cathodes and anodes are spaced alternately in the tank with a gap of about 10 mm. Chambers in the base and the sides of the tank distribute the incoming olefin feeds and withdraw cathode products. A dilute brine solution is fed into the tank that serves as the anode compartment. A typical assembly contains 16 anode plates, 15 intermediate cathodes, and 2 half cathodes at the end. It is capable of passing a current density of 11,000 A/m² of apparent electrode area.

Rotating Electrode Cell

The majority of electroorganic process are carried out in filter-press cells, which have low surface area per unit volume and low yield. When the reactions are limited by mass transfer, the capacity of the cell can be increased by flow of electrolyte over the electrode surface or by rotation of the working electrode. Figure 1.29 shows a rotating electrode cell for the oxidation of glucose to calcium gluconate [75]. These electrodes are attached to a mechanical rotating assembly and can be easily removed for cleaning. The rotating electrodes may be either cylinders or a number of disks suitably interplaced. The

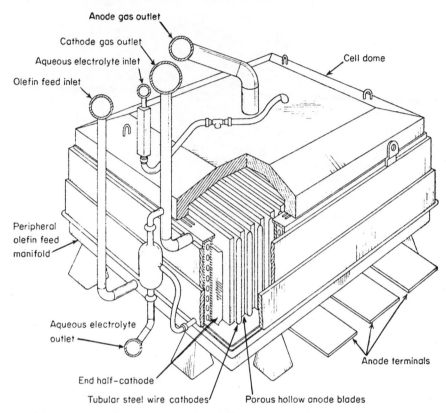

Fig. 1.28.   Le-Duc Cell for olefin oxidation. Reprinted from Ref. 69 with permission from Marcel Dekker, Inc.

rotating cylinder electrodes have uniform current distribution. However, the disk electrodes are lighter in weight and have more surface area per unit cell volume.

The Swiss-Roll Cell

The essential feature of the Swiss-roll cell [76] is a flexible sandwich of electrodes and separators as shown in Fig. 1.30. This two-dimensional structure is rolled up around an axis to give a compact, but large, electrode structure. The roll is normally enclosed in a container with the provision for electric connections to the electrodes. Since the construction is very simple, its cost is small. The scale-up can be readily achieved by winding more layers of electrodes onto the roll. Electrodes that may be employed in the form of sheets, nets, or expanded metal. The separators may be inert plastic cloths, ion ex-

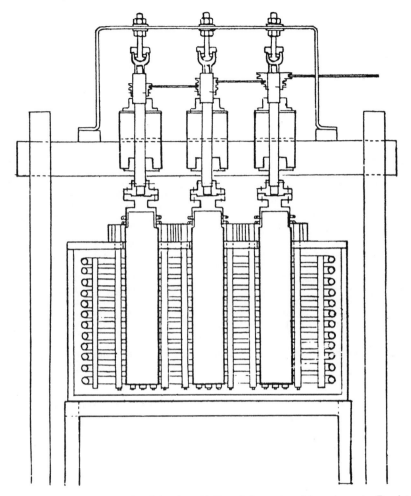

Fig. 1.29. Rotating electrode cell for the oxidation of glucose to calcium gluconate. Reprinted from Ref. 75 with permission from AIChE.

change membranes, or porous plastic fabric sheets. The Swiss roll cell may be regarded as a narrow-gap parallel-plate electrochemical cell. It has been used for the oxidation of diacetone-L-sorbose (DAS) to diacetone-L-katogulonic acid (DAG) [77].

Particulate Electrode Cells

A number of cell designs have been proposed employing electrodes in the form of small particles rather than cylindrical or flat plates. The advantage

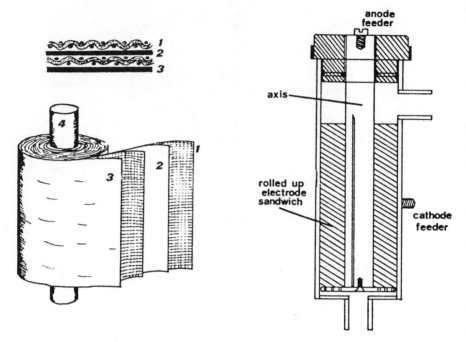

Fig. 1.30.   Swiss roll cell construction. Reprinted from Ref. 77 with permission from AIChE.

of this design is that it provides a high ratio of electrode surface to cell volume. The particle electrodes can be classified into the packed bed and the fluidized bed electrochemical cells.

PACKED BED ELECTROCHEMICAL CELLS

A cell with stationary particulate electrode is called a packed bed electrochemical cell. It is typically a cylindrical vessel, filled with metal shots. The wall of the vessel serves as the counterelectrode, while the metal shots act as the working electrode.

An example of such a cell is that developed by Nalco Chemicals [78] for the production of tetralkyl lead. The cell is composed of a steel pipe 50 mm in diameter by 75 cm long filled with lead pellets. A lining of woven polypropylene screen is placed to insulate the lead from steel. The pipe is cathodic. A lead rod is inserted into the center of the particle bed to serve as the anodic current collector. The electrolyte consisting of Grignard reagent (RMgCl) and excess alkyl halide is circulated through the cell. The electrolyte reacts with the lead anode to produce tetralkyl lead. Figure 1.31 shows the diagram of a Nalco cell used in Freeport, Texas. The capacity of the cell is approximately 2 million kg/year. Fleet and Gupta [79] have developed an electrochemical

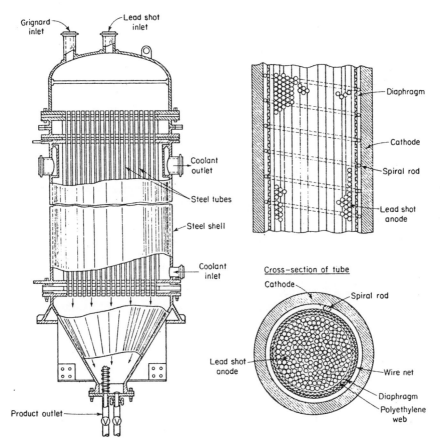

Fig. 1.31. Nalco cell for the production of tetramethyl lead. Reprinted from Ref. 69 with permission from Dekker Inc.

cell using packed carbon fibers as the working electrode. The cell has a high mass transfer rate and uniform potential over the entire electrode surface.

The bipolar particulate cell [80] is a packed bed consisting of a mixture of conducting and nonconducting particles. Each conducting particle can theoretically function as a bipolar electrode. This bipolar unit gives closely packed anodes and cathodes and can handle dilute electrolytes. It has been used for the synthesis of hypobromite and the preparation of propylene oxide.

FLUIDIZED BED ELECTROCHEMICAL CELLS

Several cell designs have been proposed in which a slurry or fluidized suspension of conductive particles is employed [81]. The particles act as a discontinuous extension of the electrode surface. A fluidized bed electrode has been proposed by Backhurst et al. [82, 83] for electroorganic reactions. In this design, one or both electrodes contain a bed of small metallic particles. The bed is fluidized to provide 10 to 20% bed expansion by upward flow of electrolytes as shown in Fig. 1.32. The flow of electrical current to the beds is applied through a screen current conductor. The fluidized bed electrochemical cell has been used in the cathodic reduction of m-nitrobenzene sulfuric acid to metanilic acids, with a current intensities up to 175 A/dm^3 of fluidized bed [82].

EXAMPLE 8

A flow-through porous platinum electrode has been considered for the electrooxidation of 9,10-diphenylanthracene (DPA) in a laboratory setup. Dry acetonitrile containing 0.2 N tetraethylammonium perchlorate is used as the electrolyte. The reaction consists of two successive electrochemical steps:

$$DPA \rightarrow DPA^+ + e \qquad (1.185a)$$

$$DPA^+ \rightarrow DPA^{2+} + e \qquad (1.185b)$$

The 9,10-diphenylanthracene dication (DPA^{2+}) is not a stable species; it rapidly reacts with traces of nucleophilic species (Nu), such as water, in the electrolyte [84]:

$$DPA^{2+} + Nu \rightarrow product \qquad (1.185c)$$

At moderately high overpotentials, the rate of reaction is limited by mass transfer of DPA to the electrode surface. The porous electrode under consideration has a specific surface area(s) of 1.53×10^4 m^2/m^3 and a porosity of 0.73. It is installed in a circular cylindrical cell having a cross-sectional area (A) of 1.76×10^{-2} m^3. The electrolyte containing 0.01 M DPA is introduced into the cell at a rate of 5.23×10^{-6} m^3/sec. It flows through the porous anode bed and DPA is oxidized to the product. The flow of electrolyte in the cell is parallel to the flow of electric current. The physical properties of the electrolyte are given below [85].

Diffusion coefficient of DPA (D_j) = 1.64×10^{-8} m^2/sec
Viscosity of electrolyte (μ) = 2.36×10^{-4} kg/(m)(sec)
Density of electrolyte (ρ) = 0.79×10^3 kg/m^3
Conductivity of electrolyte (κ) = 2.06 ($\Omega \cdot$m)$^{-1}$

Assuming a mass transfer limitation in the first step, determine the following:

1. The thickness of the flow-through porous electrode, l, required for a 99.9% conversion of DPA to the product.

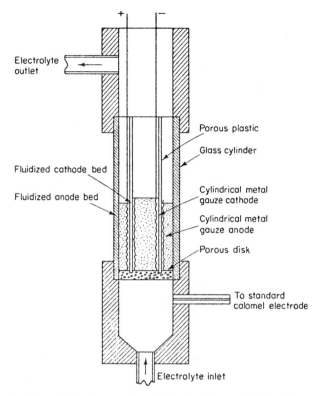

Fig. 1.32. Backhurst fluidized-electrode cell. Reprinted from Ref. 69 with permission from Marcel Dekker, Inc.

2. The total cell current I.

3. The ohmic potential drop $\Delta\Phi_{ohm}$ across the porous matrix.

Solution

The design equations for a flow-through porous electrode under mass transfer limitations have been considered by Bennion and Newman [86]. Their results can be expressed as

$$1 - X = \exp(-Y) \qquad (1.186)$$

$$\Delta\Phi^* = X + Y(1 - X) \qquad (1.187)$$

with

$$Y \begin{pmatrix} \text{dimensionless} \\ \text{electrode thickness} \end{pmatrix} = \frac{s k_{jm} l}{v_s} \tag{1.188}$$

$$\Delta\Phi^* \begin{pmatrix} \text{dimensionless ohmic} \\ \text{potential drop across} \\ \text{the porous bed} \end{pmatrix} = \frac{\Delta\Phi_{\text{ohm}} s k_{jm} \kappa_{\text{eff}}}{nF v_s^2 C_0} \tag{1.189}$$

Here C_0 refers to the concentration of DPA in the inlet electrolyte; k_{jm} refers to the mass transfer coefficient of the porous bed; and v_s is the superfacial velocity of the electrolyte in the porous electrode. The quantity κ_{eff} is the effective conductivity of electrolyte inside the porous bed; it can be estimated from the bulk solution conductivity, κ, using [87]

$$\kappa_{\text{eff}} = \kappa \epsilon^{1.5} \tag{1.190}$$

1. Using (1.187), the required dimensionless electrode thickness is found to be

$$Y = - \ln (1 - X) = - \ln (1 - 0.999) = 6.908$$

The superfacial velocity v_s is equal to volumetric flow rate divided by the cross-sectional area of the cell:

$$v_s = \frac{5.28 \times 10^{-6}}{1.76 \times 10^{-2}} = 3.0 \times 10^{-4} \, \text{m/sec}$$

To estimate the mass transfer coefficient, k_{jm}, we use a correlation suggested by Bird et al. [35] for flow in a porous bed

$$\frac{k_{jm}}{s D_j} = 0.91 \left(\frac{\rho v_s}{s \mu \Psi} \right)^{0.49} \Psi^2 \left(\frac{\mu}{\rho D_j} \right)^{1/3} \tag{1.191}$$

Here Ψ is a shape factor; for a bed composed of spherical particles, $\Psi = 1$. Thus

$$\frac{k_{jm}}{s D_j} = 0.91 \left[\frac{(0.79 \times 10^3)(3.0 \times 10^{-4})}{(1.53 \times 10^4)(3.67 \times 10^{-4})(1)} \right]^{0.49}$$

$$\times (1)^2 \left[\frac{3.67 \times 10^{-4}}{(0.79 \times 10^3)(1.64 \times 10^{-8})} \right]^{1/3} = 0.5882$$

Using (1.189), one obtains for the required thickness of the flow-through porous electrode:

$$l = \frac{Y v_s}{s k_{jm}} = \frac{(6.908)(3.0 \times 10^{-4})}{(1.53 \times 10^4)(1.476 \times 10^{-4})}$$

$$= 9.18 \times 10^{-4} \, \text{m (or 0.918 mm)}$$

2. The total cell current, I, can be calculated from the material balance, (1.28), by assuming 100% current efficiency.

$$I = nF(Av_sC_0X) \tag{1.193}$$

where A is the apparent electrode area (or the cross-sectional area of the cell). Thus

$$I = (2)(9.65 \times 10^7)(1.76 \times 10^{-2})(3 \times 10^{-4})(0.01)(0.999)$$
$$= 10.2 \text{ A}$$

3. Using (1.188), (1.190), and (1.191), one can calculate the ohmic potential drop across the porous electrode:

$$\Delta\Phi^* = 0.999 + (6.908)(1 - 0.999) = 1.006$$

$$\Delta\Phi_{\text{ohm}} = \frac{nF v_s^2 C_0 \Delta\Phi^*}{sk_{jm}k_{\text{eff}}}$$

$$= \frac{(2)(9.65 \times 10^7)(3 \times 10^{-4})(0.01)(1.006)}{(1.53 \times 10^4)(1.476 \times 10^{-4})(2.06 \times 0.73^{1.5})}$$

$$= 6.02 \times 10^{-2} \text{ V (or 60.2 mV)}$$

A more detailed analysis taking into account the kinetic information of the electrooxidation of DPA in a flow-through porous electrode has been made by Alkire and Gould [85]. This example demonstrates that proper mathematical modeling together with reasonable simplifications can lead to the design equations for the scale-up of electrochemical cells.

List of Symbols

a radius of RHE, m
a_j activity of j-species, dimensionless
A total electrode area of the cells, m^2
A_a area of anode, m^2
A_c area of cathode, m^2
A_{ex} area of heat exchanger, m^2
b distance between working electrode and countercurrent electrode, m
c_j concentration of j-species, $kg \cdot mole/m^3$
$C_{j,s}$ surface concentration of j-species, $kg \cdot mole/m^3$
$C_{j,\infty}$ bulk concentration of j-species, $kg \cdot mole/m^3$
C_K concentration of K-species, $kg \cdot mole/m^3$
$C_{K,i}$ initial concentration of K-species, $kg \cdot mole/m^3$
C_L concentration of L-species, $kg \cdot mole/m^3$
C_M concentration of M-species, $kg \cdot mole/m^3$
C_O concentration of O-species, $kg \cdot mole/m^3$
$C_{O,i}$ initial concentration of O-species, $kg \cdot mole/m^3$
$C_{O,s}$ surface concentration of O-species, $kg \cdot mole/m^3$
$C_{O,\infty}$ bulk concentration of O-species, $kg \cdot mole/m^3$

$C_{p,\text{HCl}}$ specific heat of HCl (aq), $J/kg°K$

C_{p,Cl_2} specific heat of Cl_2 (*l*), $J/kg°K$

$C_{R,s}$ surface concentration of R-species, $kg \cdot mole/m^3$

$C_{R,\infty}$ bulk concentration of R-species, $kg \cdot mole/m^3$

C_X concentration of X-species, $kg \cdot mole/m^3$

C_{v,H_2} specific heat of $H_{2(g)}$ at constant volume, $J/kg°K$

$C_{\pm}$ average concentration given by (1.98), $kg \cdot mole/m^3$

$C_{\$,\text{acid}}$ cost efficiency of HCl/Cl_2 storage, $\$/m^{1.4}$

$C_{\$,e}$ cost of electric energy, $\$/kWh$

$C_{\$,\text{ex}}$ cost efficiency of heat exchanger, $\$/m^{1.9}$

$C_{\$,f}$ capital cost of cells, $\$/m^2$

$C_{\$,\text{H}_2}$ cost efficiency of H_2 storage, $\$/m^{1.6}$

$C_{\$,j}$ cost efficiency of *j* equipment, $\$/unit$ capacity

$C_{\$,\text{pipe}}$ cost efficiency of pipe, $\$/m$

$C_{\$,\text{pump}}$ cost efficiency of pump, $\$/J^{0.98}S^{0.98}$

d_e equivalent diameter, m

D_j diffusion coefficient of j-species, m^2/sec

D_O diffusion coefficient of O-species, m^2/sec

D_R diffusion coefficient of R-species, m^2/sec

$D_{\pm}$ effective diffusion coefficient given in (1.99), m^2/sec

E close cell voltage, V

E^0 open cell voltage at a reference state, V

E_0 open cell voltage, V

f_y local friction factor given in (1.127), dimensionless

F Faraday's constant, $9.65 \times 10^7 \, J/(kg)(equiv)$

F_j design capacity of *j* equipment, unit capacity

ΔG Gibbs free energy, J

ΔG^0 Gibbs free energy at a reference state, J

ΔH enthalpy change, J

ΔH_f° standard heat of formation, $J/(kg)(mole)$

$\Delta H_{f,\text{HCl}}^\circ$ standard heat of formation of HCl (aq), $J/(kg)(mole)$

$\Delta H_{1,2,3}$ ethalpy change described in Fig. 1.5, J

i current density, A/m^2

i^0 standard exchange current density, A/m^2

i_0 exchange current density, A/m^2

i^* dimensionless current density given in (1.59), dimensionless

i_a current density at anode, A/m^2

i_a^0 standard current density at anode, A/m^2

i_c current density at cathode, A/m^2

i_c^0 standard current density at cathode, A/m^2

i_{lim} limiting current density, A/m^2

i_y local current density along *y* direction, A/m^2

I total current through the cell, A

J part in a unit operation

k_{jm} mass transfer coefficient, m/sec

$k_{jm,\text{av}}$ average mass transfer coefficient, m/sec

ΔK kinetic energy, J

l dimension of length, m

L length of planar electrode, m

L_f useful life of cells, year

M number of parallel pipes, dimensionless

n number of electrons transferred in the cell reaction, dimensionless

Δn_{HCl} change in the amount of HCl, kg·mole

N dimensionless limiting current density given in (1.149), dimensionless

N_j total flux of j-species, kg·mole/m^{2}sec

O   oxidized species

P pressure, Pa

Q heat absorbed from the surrounding, J

Q_{ex} heat input through the heat exchanger, J

Q_{H_2} heat absorbed by the hydrogen from the surrounding, J

r radial coordinate, m

R reduced species

Re   Reynolds number, dimensionless

Rey   local Reynolds number, dimensionless

R_g universal gas constant, 8314J/(kg)(mole)(°K)

R_{int} sum of all internal cell resistance, Ω

R_j rate of generation due to a homogenous bulk reaction in the control volume, kg·mole/(m^3)(sec)

s area per unit length along the flow direction in a plug flow cell, m

Sc   Schmidt number, dimensionless

Sh_{av} average Sherwood number, dimensionless

Sh_{loc} local Sherwood number, dimensionless

Sh_y local Sherwood number along y direction, dimensionless

ΔS entropy change, J/(kg)(mole)(°K)

t time, sec (hr)

T absolute temperature, °K

T_{max} maximum operating temperature, °K

T_{min} minimum operating temperature, °K

U_{av} average velocity, m/sec

U_j mobility of j-species, m^{2}kg·mole/(J)(sec)

U_{max} maximum velocity, m/sec

U_s constant velocity at a continuous moving electrode system, m/sec

U_+ mobility of cation, m^{2}kg·mole/(J)(sec)

U_- mobility of anion, m^{2}kg·mole/(J)(sec)

U_∞ characteristic velocity of a fully developed flow, m/sec

ΔU internal energy, J

v volumetric flow rate, m^3/sec

$\vec{v}$ flow velocity of electrolyte, m/sec

v_r radial velocity component, m/sec

v_y y-direction velocity component, m/sec

v_z z-direction velocity component, m/sec

v_θ latitudinal velocity component, m/sec

v_ψ tangential velocity component, m/sec

V volume of electrolyte in the cell, m^3

V_{acid} total volume of HCl (aq) and $Cl_{2(l)}$ required for storage, m^3

w width perpendicular to the flow of planar electrode systems, m

W work done by the system, J

W_{Cl_2} amount of liquid chlorine at time t, kg

$W_{Cl2.0}$ amount of liquid chlorine at time t_0, kg

W_{HCl} amount of electrolyte HCl solution at time t, kg

$W_{HCl,0}$ amount of electrolyte HCl solution at time t_0, kg
W_P brake power of pump, J/sec
X conversion, dimensionless
y y direction, m
Y dimensionless cell length given in (1.92), dimensionless
Y_P production cost, $/kg in (1.171), and $/kWh in (1.176)
z spatial coordinate normal to the electrode surface, m
Z_j charge number of j-species, dimensionless
Z_+ charge number of cation, dimensionless
Z_- charge number of anion, dimensionless

GREEK LETTERS

α transfer coefficient, dimensionless
α_{H_2} scaleup exponent of H_2 storage, dimensionless
α_j scaleup exponent of j-equipment, dimensionless
β_1, β_2 Tafel slopes, dimensionless
β_{1a}, β_{2a} Tafel slopes for anodic reaction, dimensionless
β_{1c}, β_{2c} Tafel slopes for cathodic reaction, dimensionless
δ distance between cathode and anode, m
δ_N Nernst boundary layer thickness, m
$\delta_{N,av}$ average Nernst boundary layer thickness, m
ϵ_f faradaic coefficient, dimensionless
ϵ_0 overall efficiency, dimensionless
ζ dimensionless radial coordinate given in (1.111)
η_{conc} concentration overpotential, V
$\eta_{conc,a}$ concentration overpotential at anode, V
$\eta_{conc,c}$ concentration overpotential at cathode, V
η_{ohm} ohmic potential drop between working electrode and reference electrode, V
η_s surface overpotential, V
$\eta_{s,a}$ surface overpotential at anode, V
$\eta_{s,c}$ surface overpotential at cathode, V
η_{sc}^* dimensionless surface overpotential given in (1.59)
$\eta_{total,a}$ total overpotential at anode, V
$\eta_{total,c}$ total overpotential at cathode, V
θ latitude coordinate, rad
κ the conductivity of the electrolyte, $(\Omega\text{-m})^{-1}$
κ_{av} average conductivity of the electrolyte, $(\Omega\text{-m})^{-1}$
μ_j chemical potential of j-species, J/(kg)(mole)
μ_j^0 chemical potential of j-species at a reference state, J/(kg)(mole)
ν kinematic viscosity, m^2/sec
ν_j stochiometric coefficient of j-species, dimensionless
ν_+ stochiometric coefficient of cation, dimensionless
ν_- stochiometric coefficient of anion, dimensionless
ξ dimensionless exchange current density given in (1.150), dimensionless
$\Delta\pi$ potential energy change, J
τ dimensionless time given in (1.58)
ϕ closed circuit electrode potential, V
ϕ_0 reversible electrode potential, V
ϕ^0 reversible electrode potential at a reference state, V
$\phi_{0,app}$ apparent "reversible" electrode potential, V

Φ_r electrode potential at the reference electrode, V
Φ_s electrode potential at the surface of working electrode, V
$\nabla\Phi$ potential gradient, V/m
χ_i weight fraction of HCl in the solution, dimensionless
χ_f final weight fraction, dimensionless
ψ azimuthal coordinate, rad
Ω angular velocity, liter/sec
ω_1, ω_2 reaction order, dimensionless
ω_{1a}, ω_{2a} anodic reaction order, dimensionless
ω_{1c}, ω_{2c} cathodic reaction order, dimensionless

References

1. N. L. Weinberg, *AIChE Symp. Ser., Electroorganic Synthesis Technology*, No. 185, **75**, 30 (1979).
2. R. L. Taylor, *Chem. Met. Eng.*, **44**, 588 (1973).
3. J. H. Prescott, *Chem. Eng.*, **72** (23), 238 (1965).
4. J. A. May and K. A. Kohe, *J. Electrochem. Soc.*, **97**, 183 (1950).
5. K. S. Udupa, G. S. Subramanian, and H. V. K. Udupa, *Indian Chem.*, **May**, 238 (1963).
6. P. C. Condit, *Ind. Eng. Chem.*, **48**, 1252 (1956).
7. H. Suter, H. Nohe, F. Beck, and A. Hsubesch, U.S. Patent 3,471,381 (Oct. 7, 1969).
8. J. W. Drew and G. J. Moll, *Ind. Eng. Chem.*, **53** (9), 48A (1961).
9. K. Natarajan, K. S. Udupa, G. S. Subramanian, and H. V. K. Udupa, *Electrochem. Technol.*, **2** (5-6), 151 (1964).
10. J. A. M. LeDuc, U.S. Patent 3,342,717 (Sept. 19, 1967).
11. W. Kronig and J. Grolig, British Patent 1,090,006, (Nov. 8, 1967).
12. R. W. Foreman and J. W. Sprague, *Ind. Eng. Chem. Prod. Res. Dev.*, **2**, 303 (1963).
13. H. F. Conway, and V. E. Sohns, *Ind. Eng. Chem.*, **51**, 637 (1959).
14. A. I. Kamneva, M. Ya. Fioshin, L. I. Kazakova, and S. M. Itenberg, *Neftekhim*, **2**, 550 (1962).
15. J. L. Fitzjohn, *AIChE Symp. Ser., Electroorganic Synthesis Technology*, No. 185, **75**, 64 (1979).
16. R. E. Balzhiser, M. R. Samuels, and J. D. Eliassen, *Chemical Engineering Thermodynamics*, Prentice Hall, Englewood Cliffs, NJ, 1972.
17. R. H. Perry and C. H. Chiton, *Chemical Engineers' Handbook*, McGraw-Hill, New York, 1973.
18. E. C. Potter, *Electrochemistry*, Cleaver-Hume Press, London, 1970.
19. R. C. Reid, J. M. Prausnitz, and T. K. Sherwood, *The Properties of Gases and Liquids*, McGraw-Hill, New York, 1976.
20. J. S. Newman, *Electrochemical Systems*, Prentice Hall, Englewood Cliffs, NJ, 1973.
21. K. J. Vetter, *Electrochemical Kinetics*, Academic Press, New York, 1967.
22. J. M. Smith and H. C. Van Ness, *Introduction to Chemical Engineering Thermodynamics*, McGraw-Hill, New York, 1975.

23. A. Beaufrere, R. S. Yeo, Srinivasan, J. McElroy, and G. Hardt, "A Hydrogen-Halogen Energy Storage System for Electric Utility Applications," Proceedings of the 12th Inter-Society Energy Conversion Engineering Conference, Washington, DC, (1977), paper No. 779148.

24. R. S. Yeo, J. McBreen, A. C. C. Tseung, S. Srinivasan, and J. McElroy, *J. Appl. Electrochem.*, **10**, 393 (1980).

25. K. L. Hsueh, M. S. thesis, Clarkson College, Potsdam, NY, 1979.

26. K. L. Hsueh, D-T. Chin, J. McBreen, and S. Srinivasan, *J. Appl. Electrochem.*, accepted for publication.

27. A. S. Gendron, Ph.D Dissertation, Massachusetts Institute of Technology, Cambridge, MA, 1971.

28. O. Levenspiel, *Chemical Reaction Engineering*, Wiley, New York, 1972.

29. R. Aris, *Introduction to Analysis of Chemical Reactors*, Prentice Hall, Englewood Cliffs, NJ, 1965.

30. J. Newman and W. Tiedemann, *AIChE J.* **21**, 25 (1975).

31. D. J. Pickett. *Electrochemical Reactor Design*, Elsevier Science, New York, London, Amsterdam, 1977.

32. V. Levich, *Physicochemical Hydrodynamics*, Prentice Hall, Englewood Cliffs, NJ, 1962.

33. H. Schlichting, *Boundary Layer Theory*, McGraw-Hill, New York, 1968.

34. J. C. Slattery, *Momentum, Energy and Mass Transfer in Contina*, McGraw-Hill, New York, 1972.

35. R. B. Bird, W. E. Stewart, and E. N. Lightfoot, Transport Phenomena, Wiley, New York, 1960.

36. W. Nernst, *Z. Phys. Chem.*, **47**, 52 (1904).

37. D-T. Chin, *J. Electrochem. Soc.*, **118**, 1434 (1971).

38. D-T. Chin, *J. Electrochem. Soc.*, **118**, 1764 (1971).

39. D-T. Chin, *J. Electrochem. Soc.*, **120**, 631 (1973).

40. K. Nisancioglu and J. Newman, *J. Electrochem. Soc.*, **121**, 241 (1974).

41. W. G. Cochran, *Proc. Cambridge Philos. Soc.*, **30**, 365 (1934).

42. W. H. H. Banks, *Z. Angew. Math. Phys.*, **16**, 780 (1965).

43. A. P. Colburn, *Trans. AIChE*, **29**, 174 (1933).

44. R. G. Deissler, *NACA Rep.*, (**1955**) 1210.

45. C. S. Lin, R. W. Moulton, and G. L. Putnam, *Ind. Eng. Chem.*, **45**, 636 (1953).

46. W. R. Vieth, J. H. Porter, and T. K. Sherwood, *Ind. Eng. Chem., Fundamen.*, **2**, 1 (1963).

47. D-T. Chin, *J. Electrochem. Soc.*, **119**, 1699 (1972).

48. R. Manohar, *Z. Angew. Math. Phys.*, **18**, 320 (1967).

49. J. Newman, *J. Electrochem. Soc.*, **119**, 69 (1972).

50. D-T. Chin, *AIChE J.*, **20**, 245 (1974).

51. H. S. Carslaw, J. C. Jaeger, Conduction of Heat in Solids, Oxford, Clarendon Press, 1959.

52. C. L. Mantell, *Electrochemical Engineering*, McGraw-Hill, New York, 1960.

53. W. E. Hoare and E. S. Hedges, *Tinplate*, Edward Arnold, London, 1945.

54. A. Tvarusko, *J. Electrochem. Soc.*, **119**, 43 (1972).

55. D-T. Chin, *J. Electrochem. Soc.*, **122**, 643 (1975).

56. K. Viswanathan and D-T. Chin, *J. Electrochem. Soc.*, **124**, 709 (1977).

57. G. Knudson and D. L. Katz, *Fluid Dynamics and Heat Transfer*, McGraw-Hill, New York, 1958.
58. P. W. Bridgman, *Dimensional Analysis*, Yale University Press, New Haven, 1931.
59. R. E. Johnstone and M. W. Thring, Pilot *Plants, Models and Scale up in Chemical Engineering*, McGraw-Hill, New York, 1957.
60. J. Newman, *Ind. Eng. Chem.*, **60**, 12 (1968).
61. C. W. Tobias and R. G. Hickman, *Z. Phys. Chem. (Leipz.)*, **229**, 145 (1965).
62. P. Van Shaw and T. J. Hanratty, *AIChE J.*, **10**, 475 (1974).
63. P. Van Shaw, L. P., Reiss and T. J. Hanratty, *AIChE J.*, **9**, 362 (1963).
64. N. Ibl. *Electrochem. Acta*, **22**, 465 (1977).
65. N. Ibl and E. Schalch, *Chem. Ing. Tech.*, **41**, 208 (1969).
66. T. R. Beck, *Techniques of Electrochemistry*, Vol. 3, Wiley, New York, 1978.
67. G. S. G. Beveridge and R. S. Schechter, *Optimization Theory and Practice*, McGraw-Hill, New York, 1970.
68. R. W. Foreman and E. Veach, U.S. Patent 3,119,760 (Jan. 28, 1964).
69. M. M. Baizer, *Organic Electrochemistry*, Dekker, New York, 1973.
70. C. L. Mantel, *Chem. Eng.*, **74**, (12), 128 (1967).
71. D. E. Danley and R. W. McWhorter, Belgium Patent 699,284 (May 31, 1967).
72. F. Beck and H. Guthke, *Chem. Ing. Tech.*, **41**, 943 (1969).
73. F. Wenisch, H. Nohe, H. Hannebaum, R. K. Horn, M. Stroezel and D. Degner, *AIChE Symp. Ser.*, *Electroorganic Synthesis Technology*, No. 185, **75**, 14 (1979).
74. K. I. Kryschenko, M. Ya, Fioshin, N. G. Bakhchisarayt'syan, and G. A. Kokarev, *Sov. Chem. Ind.*, **7** (July 1969).
75. H. V. K. Udupa, *AIChE Symp. Ser.*, *Electroorganic Synthesis Technology*, No. 185, **75**, 26 (1979).
76. P. M. Robertson, F. Schwager, and N. Ibl, *J. Electroanal.*, **65**, 883 (1975).
77. P. M. Robertson, P. Cettou, D. Matic., F. Schwager, A. Storck, and N. Ibl, *AIChE Symp. Ser.*, *Electroorganic Synthesis Technology*, No. 185, **75**, 115 (1979).
78. D. G. Braithuraite, U.S. Patent 3,391,067 (July 2, 1968).
79. B. Fleet and S. D. Gupta, *Nature*, **263**, Sept. 9 (1976).
80. M. Fleischmann, J. W. Oldfield, C. L. K. Tennakoon "The Bipolar Particulate Cell," paper presented in at the Electrochemical Engineering Symposium, Institution of Chemical Engineers, University of Newcastle upon Tyne, England, March 30–April 1, 1971.
81. R. Thangappan and H. V. K. Udupa, *Trans. of Saest*, **9**, 59 April–June (1974).
82. J. R. Backhurst, J. M. Coulson, F. Goodridge, and R. E. Plimley, *J. Electrochem. Soc.*, **116**, 1600 (1969).
83. J. R. Backhurst, M. Fleischmann, F. Goodridge, and R. E. Plimley, British Patent 1,194,181 (June 10, 1970).
84. R. E. Sioda, *J. Phys. Chem.*, **72**, 2322 (1968).
85. R. C. Alkire and R. M. Gould, *J. Electrochem. Soc.*, **127**, 605 (1980).
86. D. N. Bennion and J. Newman, *J. Appl. Electrochem.*, **2**, 113 (1972).
87. R. E. de la Rue and C. W. Tobias, *J. Electrochem. Soc.*, **106**, 827 (1959).

Chapter II

PERFORMANCE OF
TWO-PHASE-ELECTROLYTE ELECTROLYSIS

Hans Feess and Hartmut Wendt

Institut für Chemische Technologie der TH Darmstadt, D610-Darmstadt
Federal Republic of Germany

1 INTRODUCTION

Very often electrochemical conversions of organic substances have to be performed in aqueous electrolytes as dictated by reaction techniques, reaction selectivity, yield, and so on. Whenever this is necessary, the chemist must face the fact that the electrosynthesis is hampered by the low solubility of the organic depolarizer in the aqueous phase. Low solubility means two handicaps in the performance of electrolytic processes:

1. The depolarizer supply that is stored in a given volume of electrolyte is relatively low. To maintain a stationary current in the electrolyzer, either the electrolyte holdup must be very high or rapid recirculation with continuous depolarizer resaturation is required.

2. Because of the low depolarizer concentration the maximally obtainable diffusion limited current densities i_d are relatively low (2.1).

$$i_d = k_m c_{dep}^\infty z F \quad ; \quad k_m = \text{mass transfer coefficient} \tag{2.1}$$

The diffusion limited current density as defined by (2.1), the volume specific electrode area a (cm²/cm³) of the used electrolyzer, the current yield γ of the electrosynthesis, and the molecular weight MW of the electrolysis product define the space time yield [kg/(h)(m³)] of the electrolyzer (2.2).

$$\eta = \gamma \left(\frac{i_d}{zF} \right) \cdot a \cdot MW \tag{2.2}$$

This space time yield may become technically and economically unattractive at current densities below some 10^{-2} A/cm² and is reached according to (2.1) at depolarizer concentrations of about 10^{-1} to about 10^{-2} moles/liter if mass transfer coefficients of 10^{-3} to 10^{-2} cm/sec are assumed to be established in the electrolyzer. A mass transfer coefficient of 10^{-3} cm/sec is quite normal for technical electrolyzers, whereas 10^{-2} cm/sec is obtained under practical conditions where special measures (very high pumping rates, special cell constructions, etc.) are taken.

Many different designs have been proposed to enhance the rate of mass transfer and hence diffusion limited current densities. Most of them imply enhancement of the depolarizer solubility by changing the solvent composition. Some others aim at the same goal by changing the physical conditions in the electrolyte or electrolyzer.

Improvement of depolarizer solubility may be achieved by one of the following:

1. Addition of acids or bases by which means basic (e.g., amines) or acidic depolarizers, respectively (e.g., carbonic acids, and phenols) can be dissolved in much higher concentrations.

2. Addition of organic solvents to the aqueous electrolyte (mixed solvents).

3. Change from mainly aqueous to mainly organic solvents together with use of lipophilic electrolytes.

4. Addition of McKee salts to the aqueous electrolyte (what we now know to be micellar solubilization of the sparsely soluble organic depolarizer).

Improvement of mass transfer conditions for aqueous solutions of sparsely soluble organic depolarizers can be obtained without changing from purely aqueous solvents by one of the following:

1. Use of porous electrodes with controlled slow supply of the depolarizer to the outer electrolyte/electrode phase boundary through the electrode pores.

2. Preparation of "paste electrodes," which are a thorough mixture of powdered electrode material (mainly carbon), depolarizer, and electrolyte.

3. Processing highly turbulent dispersions of the electrically nonconducting depolarizer phase (which mostly is a liquid but might as well be a gaseous phase, e.g., ethylene) with the aqueous electrolyte that constitutes the coherent phase of the dispersion.

4. Processing of an electrically conducting depolarizer phase together with the aqueous electrolyte (the conductivity of the nonaqueous depolarizer phase may result from addition of lipophilic electrolytes to the organic phase or from ionic species being generated in the course of the electrochemical synthesis). In this case the electrically conducting aqueous and nonaqueous phases may be streaming in counter or parallel current across a particulate electrode structure such as perforated plates and mesh.

5. Addition of "mediating" redox systems to the aqueous electrolyte, that constitutes the coherent phase of a thoroughly stirred aqueous/organic dispersion. Such added mediating systems are also known as "oxygen carriers" (for mediated anodic oxidations) or as oxidation or reduction "promotors." By use of a mediating system mass transfer of the sparsely soluble depolarizer to the electrode is substituted by mass transfer of the mediating system to the electrode, which is not limited by the low solubility of the depolarizer.

6. The use of mediating systems phase transfer catalysts that allow the transfer of the mediating redox system into the dispersed organic depolarizer phase and hence shift the redox reaction between organic depolarizer and water-soluble mediating system out of the aqueous into the organic phase.

It is the aim of this chapter to describe the various types of electrolytic techniques using heterogeneous two-phase systems of aqueous electrolytes with organic depolarizers (gaseous, liquid, and solid) or with electro-chemically inert nonaqueous solvents containing the organic depolarizer. It should be pointed out that the use of heterogeneous electrolytes is not only of interest because sparsely soluble depolarizers are to be processed and mass transfer is intended to be enhanced, but, additionally, this may be the method of choice for chemical reasons too. Quite often the dispersed organic phase extracts an intermediately produced species out of the aqueous phase and thus helps to protect this intermediate against further electrochemical attack and conversion. This is demonstrated by a few examples.

1. The cathodic reduction of nitro compounds in basic aqueous elec-trolytes yields mainly azo or hydrazo compounds. By use of two-phase mix-tures, however, the reduction may be stopped at the azoxy stage.

2. With the anodic oxidation of benzene to benzoquinone the reaction product is protected against further chemical and electrochemical degrada-tion by extraction into the organic phase (benzene).

3. The cathodic separation of uranium and plutonium in the PUREX process by simultaneous reduction and phase transfer demands the presence of the organic phase as carrier for the two actinides that are to be separated.

Finally, it should be stressed that the alternative option, to process the organic depolarizer in a single-phase electrolyte based on an organic solvent instead of an aqueous/nonaqueous heterogeneous system, may not be desirable. This is because (1) water itself is needed as a reactant and (2) the use of an aqueous/nonaqueous two-phase electrolyte implies the use of a low-resistance electrolyte, which allows a remarkable amount of electrical energy to be saved that otherwise would be wasted by current passage through the highly resistive organic solvent.

2 SCOPE OF THE CHAPTER

Section 3 describes the different modes in which two-phase-electrolyte electrolysis-processes are performed. The underlying physical mass transport mechanisms are discussed, fundamental investigations of these mechanisms and their results are reported, and their role in the engineering of respective electrolyzers and the process engineering of respective electroorganosynthesis is described. Included in this section is the case of electrochemical processing of micelle-solubilized solutions of sparsely soluble organic substances in aqueous electrolytes, which represents the borderline case between true molecularly dispersed solutions and heterogeneous two-phase dispersions.

The following section (i.e., section 4) is devoted to a literature survey of the vast number of organoelectrosynthetic reactions that have been reported since the early days of organic electrochemistry and in which the oxidation of arenes and alkylarenes and the reduction of nitroarenes are by far the two most important reaction types. Section 3 also includes a description of data published on porous or on "wick" electrodes, work on processing of gaseous organic substances, and studies on electrochemical conversion of micelle-solubilized depolarizers. Mediated electrochemical conversions of dispersed organic depolarizers or depolarizer solutions in organic solvents and the use of phase transfer catalysts for such electrolysis processes are treated as well.

The material in section 4 is arranged according to reaction type (oxidations and reductions) and to the chemical structure of the respective depolarizers with details on synthetic processes in organic electrochemistry based on the use of two-phase electrolytes that gained some technical and economical success either in the past or that are still being operated successfully.

3 TECHNIQUES, PERFORMANCE, AND MASS TRANSFER MECHANISMS IN TWO-PHASE-ELECTROLYTE ELECTROLYSIS

Fundamentals: Stationary and Nonstationary Mass Transfer Toward Phase Boundaries (Boundary Layer Models)

Electrochemical conversion of any depolarizer, due to the depolarizer consumption at the electrode, is always accompanied by buildup of concentration profiles that extend from the electrode into the electrolyte solution. By application of appropriately high working potentials, the depolarizer concentration just at the electrode drops down to zero, and the rate of conversion becomes mass transfer controlled. Hence a diffusion limited (maximal) current is measured that is determined by the extent of mass transfer of the depolarizer toward the electrode and that is defined by the concentration gradient at zero distance [$y = 0$, see (2.1), (2.3)].

$$i_d = zF\dot{n}_{y=0} = zFD\left(\frac{\partial c}{\partial y}\right)_{y=0} \qquad (2.3)$$

In unstirred solutions concentration changes with time are determined by the second Fick equation and—under diffusion limited current conditions—by the boundary condition $c_{y=0} = 0$. Thus the changing concentration profile, as time passes, extends farther into the solution and the gradient $(\partial c / \partial y)_{y=0}$, which determines the diffusion limited current, decreases with $(t)^{-1/2}$ (Fig. 2.1a) as:

$$\dot{c} = D \cdot \frac{\partial^2 c}{\partial y^2}; \qquad \text{diffusion limited current: } c_{y=0} = 0 \qquad (2.4a)$$

$$\left(\frac{\partial c}{\partial y}\right)_{y=0} = \frac{2c^{\infty}}{\sqrt{\pi Dt}} \tag{2.4b}$$

The situation is quite different if steady convection is superimposed on molecular diffusion either by continuous stirring of the electrolyte or by continuous flow of the electrolyte through a channel whose walls constitute the working electrodes. When any transients at the beginning of the electrolysis experiment are neglected the situation becomes stationary ($dc/dt = 0$) and is described by the differential equation (2.5a), which includes convective mass transfer.

$$\text{div}\,(wc) + D\,\text{div grad}\,(c) = 0 \tag{2.5a}$$

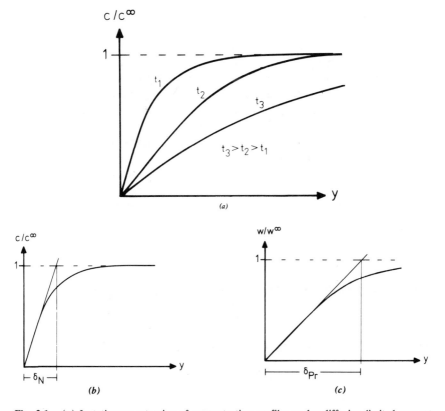

Fig. 2.1. (a) Instationary extension of concentration profiles under diffusion limited current conditions [$c(y = 0) = 0$] in unstirred solution. (b) Definition of Nernstian diffusion layer thickness δ_N by linear approximation of concentration profile. (c) Definition of laminar sublayer under turbulent flow conditions in the bulk of flowing fluid.

The solution of (2.5a) is a stationary concentration profile, the spatial extension of which strongly depends on the applied flow conditions. Very important to note is the fact that the ratio of the extension of the concentration profile as defined by the thickness δ_N of the diffusion layer (Figs. 2.1b and 2.1c) and the extension of the velocity boundary layer, which is characterized by a similiarly defined velocity boundary layer (or Laminar sublayer) thickness δ_{Pr} (Prandtl-layer thickness), is given for turbulent flow by the third root of the ratio of the diffusion coefficient D of the depolarizer to the kinematic viscosity, ν, of the solvent [1].

$$\frac{\delta_N}{\delta_{Pr}} = \left(\frac{D}{\nu} \right)^{1/3} \qquad (2.5b)$$

Thus, for instance, when water is used as the electrolyte where the ratio of (D/ν) amounts to approximately 10^{-3}, the Nernstian diffusion layer has only one-tenth of the extension of the laminar Prandtl sublayer. Under laminar flow conditions similarly simple relations do not exist and the relative extensions of the Nernstian diffusion and the velocity boundary layer depend on flow conditions and on the special geometry of the flow and electrolysis device. Nevertheless, it can be stated for laminar as well as for turbulent flow that the faster the flux, or the steeper the velocity gradients at the electrode/electrolyte phase boundary, the steeper are the concentration profiles. Hence mass transfer rates and diffusion limited currents are enhanced by increased convection or flow rates.

In most cases concentration gradients cannot be measured easily. Diffusion limited currents and mass transfer rates, however, are directly measurable quantities. Thus it is advisable to introduce the mass transfer coefficient k_m instead of $D \cdot (\partial c / \partial y)_{y=0}$ into the fundamental equation (2.3), which yields

$$i = zFk_m (c^\infty - c_{y=0}) \qquad (2.6a)$$

and which under diffusion limited conditions ($c_{y=0} = 0$) becomes

$$i_d = zFk_m c^\infty \qquad (2.6b)$$

By introducing the linear approximation

$$D \left(\frac{\partial c}{\partial y} \right)_{y=0}^{limit} \doteq D \left(\frac{c^\infty}{\delta_N} \right) \qquad (2.6c)$$

one obtains an expression for k_m relating the measured mass transfer coefficient k_m to the diffusion coefficient D and the thickness δ_N of the Nernstian diffusion layer:

$$k_m = \frac{D}{\delta_N} \tag{2.6d}$$

The value of δ_N is mainly determined by the flow conditions, whereas D is determined by the species that is transported by superimposed convection and diffusion toward the electrode (often called convective diffusion) and that is also determined by the nature of the electrolyte and solvent respectively.

Techniques for Electrochemical Conversion of Sparsely Soluble Substances in Aqueous Solutions

Electrolysis of Micelle-Solubilized Solutions of Sparsely Soluble Substances

It is a well-established observation that many organic hydrophobic substances of medium molecular weight (MW < 300) that normally possess a low water solubility of say 10^{-5} moles/liter or less gain a much higher solubility in the presence of micelle-forming substances. These may be long-C-chain, arenecarboxylic-, or sulfonic acids and their alkali salts (anionic surfactants), long-C-chain-alkylammonium salts (cationic surfactants), or polyoxyolefins and polyoxyolefin esters (nonionic surfactants). This remarkably enhanced solubility, which very often amounts to more than a 1000 times the value of the initial very low solubility, is due to the incorporation of depolarizer molecules into the micelles of the surfactant. However, in most cases the mean number of incorporated (solubilized) depolarizer molecules per micelle is relatively small (between two and five approximately), so that to obtain a reasonable gain in solubility very high amounts of the micelle-forming surfactant must be added. With respect to mass transfer it is very important to note that compared to normal molecules, micelles are very large entities and thus diffuse only very slowly. Their diffusion coefficients very often are approximately 100 times smaller than diffusion coefficients of normal molecules.

Thus the mass transfer coefficients (being approximately defined by $k_m \approx D/\delta_N$) under identical hydrodynamic conditions become about 2 orders of magnitude smaller for micelles and micelle-solubilized molecules than for molecular species. Consequently, according to (2.6b) and (2.6d), mass transfer rates are much less enhanced by micelle solubilization than are the effective solubilities.

In a study published by Yeh and Kuwana [2] the electrochemistry of micelle-solubilized ferrocene was investigated. The solubilizing agent was a nonionic surfactant (Tween 20) with an average molecular weight of 1650. This surfactant forms micelles of a mean molecular weight of 143,000 that possess an average radius of 40 Å and are able to solubilize up to three ferrocene molecules per micelle. Table 2.1 contains the most important results of this study and compares the relevant quantity $D_{eff} \cdot c^{\infty}$ for molecularly dissolved and solubilized ferrocene solutions in water.

The data of Table 2.1 show an increase in solubility for ferrocene by a factor of 10^3, whereas the quantity $D_{eff} \cdot c^{\infty}$, which is relevant for mass transfer rates, is only increased by a factor of 40 by solubilization. An important implication of this interdependence of increased solubility and decreased effective diffusive mobility is that saturation concentrations of micelle-solubilized depolarizers in technical processes have to be approximately 25 to 30 times as high as the practical lower concentration limit of molecularly dissolved depolarizers processed in usual electrolysis. Under normal conditions diffusion limited current densities of at least some 10^{-2} A/cm^2 are required to obtain an acceptable space time yield. For molecular species a value of the mass transfer coefficient (k_m) of 10^{-2} cm/sec would account for this postulated current density if a depolarizer concentration of at least 10^{-2} moles/liter could be established. With micellar-solubilized substances, however, this minimal value of the effective solubility would have to be higher at least by a factor of 25 to 30. This means that if micelle solubilization does not increase the effective solubilities up to values of 0.2 to 0.5 moles/liter this method will not enable the chemical engineer to improve space time yield up to the necessary minimal value.

Electrolysis of Emulsions of Organic Liquids in Aqueous Electrolytes: Techniques and Mass Transfer Mechanisms

Since the early beginnings of organic electrochemistry the problem of limited mass transfer and of too low a depolarizer holdup for sparsely soluble substances was intuitively tackled by using emulsions of the organic depolarizer in aqueous solutions. To keep these emulsions stabilized and to prevent them from demixing, they generally have been processed under highly turbulent flow conditions with vigorous stirring. In 1900 Kempf [3] claimed in a patent the use of emulsions of benzene in aqueous H_2SO_4 to produce quinone and hydroquinone (2.7) by anodic benzene oxidation.

$$C_6H_6 + 2H_2O \rightarrow C_6H_4O_2 + 6H^+ + 6e \qquad (2.7)$$

This method has since been repeated by numerous electrochemists with eventual industrial application in the work of Udupa [4] and in two newly developed benzoquinone processes [5, 6]. In this industrial version, Udupa made use of the simple principle of a vigorously stirred beaker or stirred tank reactor in so far as rotating electrodes served as stirrers. Fremery and coworkers [5] and Millington [6], however, used turbulent channel-flow to keep the benzene emulsions in the aqueous phase in a well-defined and highly

Table 2.1. Solubility Data and Mass Transfer Characteristics for Molecularly Dissolved and for Micelle-Solubilized Ferrocene in Water [2]

Nonsolubilized Aqueous Ferrocene Solutions			Tween-20 Solubilized Aqueous Ferrocene Solutions			
c^∞ (moles/cm^3)	D (cm^2/sec)	$D \cdot c^\infty$ [moles/(cm)(sec)]	c^∞ (moles/cm^3)	$\bar{r}$ (micelles) (Å)	D_{eff} [cm^2/sec]	$D_{eff} \cdot c^\infty$ [moles/(cm)(sec)]
10^{-8}	10^{-5}	10^{-13}	10^{-5}	42	4×10^{-7}	4×10^{-12}

dispersed state. Although the use of two-phase emulsion electrolysis is now a routine procedure, hardly any work has been performed to elucidate the mass transfer mechanisms in such emulsions until recently.

Dworak and Wendt [7] investigated the question of whether mass transfer in such emulsions is improved at all, or whether the use of highly turbulent emulsions only serves to (1) increase the depolarizer holdup in the electrolysis cell and (2) provide continuous resaturation of the electrolyte with the sparsely soluble depolarizer. This maintains the saturation concentration of the depolarizer throughout the total electrolysis cell apart from the small volume fraction belonging to the Nernstian diffusion layer where the depolarizer is consumed at the electrode.

The authors assumed that mass transfer in emulsions could be enhanced by two different mechanisms.

1.   The presence of droplets of the finely dispersed organic phase near and within the hydrodynamic boundary layer results in viscous and energetic heterogeneities, as well as in local differences of momentum, since the aqueous electrolyte and the dispersed droplets possess different densities and velocities. This may result in a different response of the dispersed non-aqueous phase and the aqueous electrolyte on viscous drag and deceleration within the laminar sublayer of the two-phase flow along the electrode. It is quite obvious that this different dynamic response of the dispersed non-aqueous phase and of the coherent aqueous supporting electrolyte on turbulent acceleration and viscous drag may lead to enhanced eddy formation at the outer parts of the laminar viscous sublayer. This may result further in a compression of the sublayer (see Fig. 2.2a) and consequently in a compression of the underlying Nernstian diffusion layer, that is, improved mass transfer will be observed (Fig. 2.2b). This mechanism, yielding improved mass transfer of the depolarizer from the continuously resaturated aqueous solution to the electrode surface, is called a "boundary-layer compression mechanism."

2.   The second—quite different—alternative mechanism for improved mass transfer with the use of a highly dispersed organic depolarizer phase in an aqueous electrolyte might be due to a direct contact between droplets of the dispersed phase and the electrode. Such direct contact may be assumed to occur if the droplets of the organic phase are able to penetrate through the hydrodynamic boundary layer to the electrode surface where they may adhere for some time before being swept away (Fig. 2.3). Then mass transfer could be largely enhanced because the depolarizer molecules would be supplied to the free, uncovered parts of the electrode surface by diffusion across much shorter distances than the extension of the Nernstian diffusion layer or even by surface diffusion. The mechanism for improved mass transfer due to

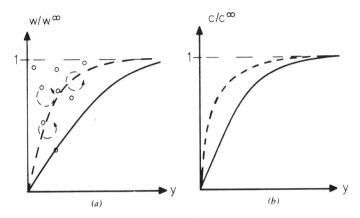

Fig. 2.2. (*a*) Compression of velocity boundary layer and (*b*) subsequent compression of concentration profile induced by dispersed droplets or particles in a fluid streaming along a wall.

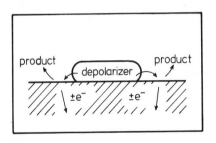

Fig. 2.3. Demonstration of enhanced mass transfer by "wetting mechanism" due to contact between droplets of sparsely soluble liquid depolarizer and electrode surface.

physical contact between droplets of the dispersed organic phase and the electrode was termed "wetting mechanism" by Wendt and Dworak.

To investigate the possibility of a compression of the boundary layer in the presence of dispersed liquid phases they measured the mass transfer of ferrocyanide anions (which are soluble in the aqueous phase only) for turbulent channel flow by comparing (a) single-phase aqueous electrolytes and (b) 30-vol % emulsions of *n*-hexane (which possesses a 40% lower density than the aqueous phase) and of carbon tetrachloride (which possesses a 60% higher density than the aqueous electrolyte).

Figure 2.4 shows the results of these measurements obtained by plotting the logarithm of the mass transfer coefficients for one- and two-phase channel flow versus the Reynolds number (as based on flow velocity and kinematic viscosity of the aqueous phase). Two-phase emulsion flow was investigated

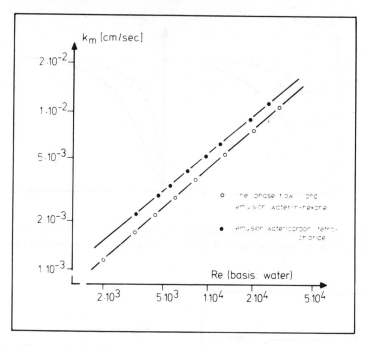

Fig. 2.4. Mass transfer coefficients versus Reynolds number for single-phase and two-phase channel flow (dispersed organic phase 30 vol %) [7].

where mean drop diameters were well defined by preparing the emulsion in a premixer. In this mixer 0.1- and 0.15-mm mean droplet diameters of the dispersed phase could be prepared. The results in Fig. 2.4 demonstrate very clearly that the "boundary layer compression mechanism" enhances homogeneous mass transfer only to a relatively minute amount (no more than 20–30%), and this only if the dispersed phase possesses a higher density than the aqueous electrolyte. Thus boundary layer compression is not an important factor for enhancing mass transfer in emulsion electrolysis.

The "wetting mechanism" is also not operative under highly turbulent conditions. By use of microelectrodes 0.05 mm in diameter that were inserted into the walls of the quadratic duct in which mass transfer for channel flow was investigated, Wendt and Dworak showed that under the chosen conditions no droplets were able to touch the electrode surface because of their strong deceleration, which causes demixing of the dispersion in the hydrodynamic boundary layer. The authors calculated that under not excessively high turbulence ($5000 < \text{Re} < 8000$) in 1 cm^2 cross-section channels mean drop diameters have to be around 1 mm or more to ensure penetration of the droplets through the boundary layer. This drop diameter, however, is beyond the stable value under the flow and turbulence conditions used [8-11].

It is thus quite clear that the main effect of the use of liquid/liquid emulsions for processing sparsely soluble organic depolarizers arises from the following:

1. Increase of depolarizer holdup in the electrolyte.

2. Maintainance of saturation concentration.

3. Possibility of extracting the product of the electrochemical synthesis into the dispersed phase and by this action, protecting it from further electrochemical attack or chemical degradation. Recently such extractive protection has been patented [12].

WATER SOLUBILITY OF DIFFERENT SPARSELY SOLUBLE DEPOLARIZERS AS
A LIMITING FACTOR FOR THE APPLICATION OF EMULSION ELECTROLYSIS

If mass transfer, as pointed out, is not remarkably enhanced by the use of emulsions of the depolarizer in the aqueous electrolyte, then the saturation concentration of the depolarizer is a very important and decisive factor that determines the feasibility (with respect to space time yields) of an electrochemical conversion based on emulsion electrolysis. For such emulsions space time yield is defined by a mass transfer coefficient that is nearly the same as for one-phase flow under otherwise identical hydrodynamic conditions (2.8).

$$\eta_{max} = a \cdot k_m^{(disp)} \cdot c^{\infty} \cdot \gamma \cdot MW$$

with

$$k_m^{(disp)} \approx k_m^{(homog.)} \tag{2.8}$$

As is noted above, for technical electrolysis cells ($k_m \approx 10^{-3}$ to 10^{-2} cm/sec) attractive space time yields are obtained only with depolarizer concentrations of 10^{-2} mole/liter (10^{-5} mole/cm^3) or more. Thus it is quite clear that a lower solubility limit of 10^{-2} moles/liter defines the solubility borderline below which the use of emulsion electrolysis will be of little or no help for the electrochemical conversion of sparsely soluble organic depolarizers in aqueous electrolytes. Table 2.2 shows the water solubilities of a multitude of arenes, alkylarenes, nitroarenes, aminoarenes, aromatic aldehydes, aromatic ketones, and phenols at 25°C. Solubility values above 10^{-2} moles/liter are rather scarce except for phenols and the oxocompounds.

From the data of Table 2.2 it is quite obvious that apart from the oxo and hydroxy compounds, only benzene, fluorobenzene, nitrobenzene, and aniline may be processed as emulsions if a direct conversion of the respective

Table 2.2. Solubilities[a] of Sparsely Soluble Organic Substances in Water at 25°C

Substance	Solubility (moles/liter)	Ref.
Benzene	2.0×10^{-2}	13–23
Toluene	6.0×10^{-3}	15, 17–20, 24
Ethylbenzene	1.6×10^{-3}	15, 17, 18, 20, 25
i-Propylbenzene	4.9×10^{-4}	17, 20, 25
n-Propylbenzene	4.7×10^{-4}	25, 26
t-Butylbenzene	2.5×10^{-4}	25
s-Butylbenzene	2.3×10^{-4}	25
n-Butylbenzene	9.2×10^{-5}	25
n-Pentylbenzene	7.1×10^{-5}	25
o-Xylene	1.7×10^{-3}	15, 17, 19, 20
m-Xylene	1.6×10^{-3}	15, 18, 19
p-Xylene	1.8×10^{-3}	15, 18, 19
Mesitylene	8.1×10^{-4}	25
1,2,4-Trimethylbenzene	4.8×10^{-4}	17
Biphenyl	4.3×10^{-5}	18, 19
Diphenylmethane	8.9×10^{-6}	19
Naphthalene	2.5×10^{-4}	18, 19, 23
Anthracene	4.2×10^{-7}	27
Phenanthrene	8.0×10^{-6}	19, 27
Phenol	9.0×10^{-1}	13, 14
o-Cresol	2.0×10^{-1}	13, 14, 28
m-Cresol	2.1×10^{-1}	13, 14, 28
p-Cresol	1.9×10^{-1}	13, 14, 28
α-Naphthol	3.0×10^{-4}	14
Benzyl alcohol	3.7×10^{-1}	14, 29
Benzaldehyde	3.8×10^{-1}	14, 30
Acetophenone	4.5×10^{-2}	14, 30
Benzophenone	4.4×10^{-4}	14
Nitrobenzene	1.6×10^{-2}	13, 14, 31, 32
o-Dinitrobenzene	7.8×10^{-4}	14, 33
m-Dinitrobenzene	4.8×10^{-3}	14, 33
p-Dinitrobenzene	3.2×10^{-3}	14, 33
o-Nitrotoluene	4.8×10^{-3}	14, 34
m-Nitrotoluene	3.7×10^{-3}	14, 34
p-Nitrotoluene	3.2×10^{-3}	14, 34

Table 2.2 (continued).

Substance	Solubility (moles/liter)	Ref.
Aniline	4.0×10^{-1}	13, 14
β-Naphthylamine	1.3×10^{-6}	14
Fluorobenzene	1.6×10^{-2}	14, 35, 36
Chlorobenzene	4.4×10^{-3}	14, 35, 36
Bromobenzene	2.6×10^{-3}	14, 35, 36
Iodobenzene	8.8×10^{-4}	14, 35, 36

[a]Where different solubility data have been published by different authors, the "most probable" value has been chosen according to data accumulation. This choice, however, is not very critical for the discussion of dispersion electrolysis since different solubility data in general do not disagree by more than 20%.

depolarizer at the working electrode is required in aqueous solutions. Toluene as the first member of the alkylarenes is just a borderline case for which direct electrochemical conversion can perhaps still be performed in aqueous solutions by use of emulsion electrolysis.

Kuhn and co-workers [39] studied the anodic oxidation of benzene, toluene, and anisole at lead dioxide anodes. The arenes were dissolved in the aqueous electrolyte in various concentrations up to their respective saturation concentrations. The authors showed that at low anode potentials, anodic currents are determined by heterogeneous chemical kinetics (chemical oxidation of benzene by PbO_2). At sufficiently high anodic potentials and with concentrations near the depolarizer's saturation value, they measured limiting currents that very nearly reflected the order of solubilities of these three aromatic compounds [i_{limit} (benzene) $\approx$ 100 mA/cm^2, i_{limit} (anisole) $\approx$ 30 mA/cm^2, and i_{limit} (toluene) $\approx$ 15 mA/cm^2] and additionally gave the correct order of magnitude for diffusion limited currents for the mass transfer conditions prevailing in the cell.

For the estimation of mass transfer for organic compounds one should take into account that in most cases water solubility can be improved by increasing the working temperature of the electrolyte from ambient temperature up to 70 or 80°C. This is demonstrated in Fig. 2.5a, where the saturation concentration of benzene in water is plotted versus temperature [21, 22]. Figure 2.5b shows the influence of electrolyte concentration on benzene solubility [37, 38] and demonstrates the detrimental salting-out effect of nearly any electrolyte. The effect is especially pronounced for those electrolytes whose ions are strongly solvated either because of higher charge (see

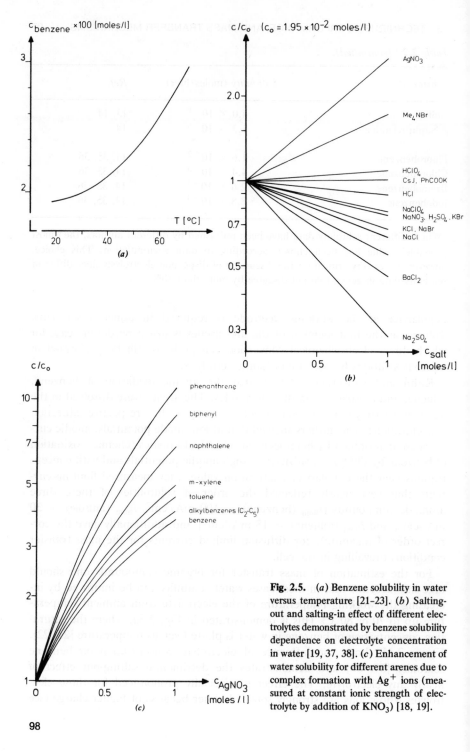

Fig. 2.5. (*a*) Benzene solubility in water versus temperature [21-23]. (*b*) Salting-out and salting-in effect of different electrolytes demonstrated by benzene solubility dependence on electrolyte concentration in water [19, 37, 38]. (*c*) Enhancement of water solubility for different arenes due to complex formation with Ag^+ ions (measured at constant ionic strength of electrolyte by addition of KNO_3) [18, 19].

e.g. two-valent ions Ba^{2+}, SO_4^{2-}, etc.) or because of small ionic radii (OH^-, Na^+). The use of salts composed of ions that exhibit some organic character improves the solubility of benzene (for tetramethylammonium bromide this is not due to micelle solubilization since tetramethylammonium cations do not form micelles). As a special peculiarity the expressed solubility enhancement in solutions that contain silver ions should be mentioned. This effect is caused by complex formation between conjugated π-electron systems and silver cations. Figure 2.5c demonstrates this effect for a number of different arenes [18, 19, 25]. Although this effect is worthwhile mentioning it is likely of no practical value because of the costs associated with high silver ion requirements.

Two-Phase-Electrolyte Electrolysis with Porous Electrodes

About 1950, Ghosh and co-workers [40, 41] published some papers in which they described the use of hollow cylindrical carbon electrodes with porous walls through which benzene or other sparsely water-soluble liquid depolarizers could be fed to the outer mantle of the cylinder, which was immersed in the aqueous electrolyte (Fig. 2.6a). The outer cylinder wall served as the working electrode. The depolarizer was forced through the porous

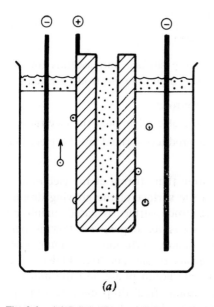

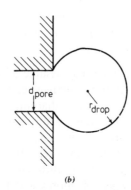

(a)   (b)

Fig. 2.6. (a) Schematic presentation of porous electrode. The liquid depolarizer is supplied by gravity through the pores of the hollow cylinder to the outside of cylinder wall acting as anode [40, 41]. (b) Schematic presentation of a droplet growing at the mouth of a pore until buoyancy forces ($^3/_4 \pi r^3 / \Delta \rho \cdot g$) outrun surface force ($2 \pi r \sigma$).

electrode by gravity. Its flux rate in general exceeded the electrochemical consumption at the outer side of the porous cylinder, so that part of the depolarizer ascended as droplets along the cylinder wall and gathered on top of the electrolyte (provided its specific weight was lower than that of the aqueous electrolyte) or descended and was collected at the bottom of the vessel if it possessed a higher density than the electrolyte. The unconverted depolarizer extracted the product of the electrochemical conversion and preserved it from any further electrochemical or chemical degradation.

Unfortunately Ghosh, although mentioning the porosity (50.7%) of the porous carbon used, did not give any additional information to characterize the microstructure of the material. All information concerning pore diameters are missing in his paper. The physical model of such a porous electrode seems nonetheless quite clear: it is in principle a device that makes use of the aforementioned "wetting" mechanism. As is depicted schematically in Fig. 2.6b, a drop forms at the mouth of a pore, which opens into the electrolyte because of the surface tension at the organic liquid/aqueous electrolyte interface, just as a mercury drop forms at the end of the capillary of the dropping mercury electrode. After having reached a critical radius (r_{crit}), which is determined by (1) the surface tension σ, (2) the density difference $\Delta\rho$ between the organic liquid and the aqueous phase and (3) the pore diameter d_p:

$$r_{crit} = \left(\frac{3d_p \cdot \sigma}{4\Delta\rho \cdot g} \right)^{1/3} \tag{2.9a}$$

the organic-phase droplet detaches from the surface and ascends (or descends depending on its density). On its way along the electrode surface it

1. Stirs the stagnant solution at the electrode.

2. Dissolves partially and resaturates the electrolyte near the electrode.

3. Extracts at least a fraction of the products formed by the electrode reaction.

One may assume as a first approximation that stirring at the electrode is relatively ineffective. Mass transfer from the surface of the growing droplet toward the adjacent parts of the free electrode surface is based mainly on nonstationary diffusion, or, since the process of depolarizer flow is performed with constant rate, on quasistationary diffusion. Assuming pore diameters between 10^{-3} and 10^{-2} cm, surface tensions of 20 to 40 dynes/cm and density differences between the organic liquid phase and the aqueous phase of approximately 0.5 g/cm^3 one obtains final drop diameters of approximately 10^{-1} to 10^{-2} cm. From this critical drop diameter, mean diffusion lengths $\bar{x}$ (as averaged over the total lifetime of the drop) of some 10^{-2} cm can be estimated. Thus effective mass transfer rates of

$$\overline{k}_m \text{ (porous electrode)} \approx \frac{D}{x} \approx 10^{-3} \text{ cm/sec} \qquad (2.9b)$$

may be expected for the above situation.

Ghosh reported current yields of 50% for the anodic formation of benzoquinone for low current densities up to a limiting value of 40 mA/cm^2, and linearly decreasing current yields upon further increase of the current density for the anodic benzene oxidation. This observation indicates that even higher mass transfer rates than estimated in (2.6d) are observed.

Based on (2.9a) and (2.9b) and on the benzene solubility data for 25°C given in Table 2.2 one would expect a diffusion limited current density for benzene oxidation at porous carbon anodes with pore diameters of 0.01 to 0.1 mm, which amounts to only:

$$i_d = \overline{k}_m \cdot c^{sat}_{benzene} \cdot 6 \cdot F \approx 10^{-3} \text{ A/cm}^2 \qquad (2.9c)$$

This value is one order of magnitude smaller than the maximal current density obtained by Ghosh. The current yields are as high as 50% at 40 mA/cm^2, because stirring of the boundary layer by the moving organic droplets may become so effective that it enhances mass transfer remarkably.

Furthermore, the very simple mass transfer estimation of (2.9b) and (2.9c) based on the assumption of unperturbed droplet growth, did not take into account the fact that droplets forming at the mouths of neighboring pores will merge into one another and may detach from the surface as soon as the sum of their radii approaches the mean pore distance. This effect obviously will improve mass transfer quite significantly. The effective mass transfer coefficient at the porous electrodes used by Ghosh were of the order of 10^{-2} cm/sec without external stirring. Thus porous electrodes seem to be very promising for processing liquid organic substances whose solubility is at least 10^{-2} moles/liter. It seems worthwhile to investigate this device of Ghosh's thoroughly to learn something more about its potentialities and to exploit the porous electrode in appropriate cell constructions.

Winslow and Heise [42, 43] reported the use of porous carbon electrodes immersed in nitrobenzene on top of which, aqueous electrolyte is spread. The porous hydrophobic carbon is soaked with nitrobenzene which at the openings of the pores may diffuse across short diffusion paths to the free carbon surface, where it may be converted. Mass transfer conditions may be not too poor, but they are uncontrolled since no additional means of stirring or a controlled exchange of the depolarizer within the pores is employed. Although Winslow reported relatively good current yields for p-aminophenol, it seems very doubtful whether such an electrode with a stagnant or very slowly moving organic phase has any advantage compared to other process techniques.

Paste Electrodes for Processing Liquid or Solid Organic Depolarizers of Very Low Water Solubility

As is discussed above, processing of any depolarizer in the form of vigorously agitated emulsions or slurries or by passing a liquid depolarizer or the solution of a solid depolarizer in a suitable organic solvent through a porous electrode becomes obsolete whenever the saturation concentration of the respective depolarizer in the aqueous electrolyte lies below the critical limit of 10^{-2} mole/liter. In such cases the solubility of the depolarizer might be enhanced by using mixed solvents. This, however, always implies difficulties in the workup procedure that might be prohibitive with respect to process engineering and costs for an envisaged technical process. In the use of paste electrodes there exists an alternative option that allows one to avoid any organic solvent/water mixture as solvent for the electrolyte. Further, paste electrodes allow one to process substances with solubilities as low as 10^{-6} to 10^{-5} mole/liter. Such paste electrodes were employed in 1929 by Lowy and co-workers [44–46] and later by Bionda and Civera [47–50]. They consisted of a mixture of finely ground electrode material (generally carbon) and an appropriate electrolyte and the depolarizer (either in the form of finely ground crystalline material or—if it is a liquid—just added as such). These mixtures were pressed into a metallic particulate structure (grid or net) to form solid coherent electrodes. On contacting with aqueous electrolyte solutions the electrolysis device was completed (see Fig. 2.7).

The internal structure of these paste electrodes very closely resembles the structure of a halfway-discharged lead battery plate. Inserted into an elec-

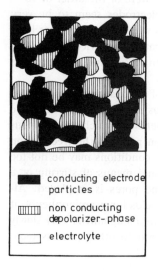

conducting electrode particles

non conducting depolarizer-phase

electrolyte

Fig. 2.7. Micro structure of electrode/depolarizer/electrolyte mixture being used in paste electrodes for conversion of depolarizers with extremely low water solubility.

trically conducting matrix of the electrode material are very small lumps of low-solubility material (which in the case of the lead battery is $PbSO_4$, which on charging is converted to metallic lead and lead dioxide at the anode and cathode, respectively). Mass transfer in this compound structure is largely enhanced by the very narrow distances that the dissolving organic molecules have to cross to reach the surface of the electrode particles.

Vetter [51] discussed the mass transfer problem of low-solubility materials in highly divided electrode structures. He showed that because of the very low diffusion distances (200 to 300 Å) and because of the very high ratio of electrode area to volume in such structures (10^3 m^2/cm^3) for externally applied formal current densities of as high as 1 A/cm^2 a saturation concentration of 4×10^{-6} mole/liter is sufficient to provide the necessary mass transfer rate for depolarizer supply (porosity 50%, electrode thickness 0.1 cm, diffusion coefficient $D \approx 10^{-5}$ cm^2/sec). If only 0.1 A/cm^2 of external current density were applied, a solubility of the depolarizer of 4×10^{-7} moles/liter would be sufficient, so that even the least soluble substances listed in Table 2.2 could very well be processed in paste electrodes.

However, in paste electrodes, the problem of potential and current distribution arises. The theoretical description of current and potential distributions in porous electrodes has been well developed; however, no approach has been made to extend this treatment toward paste electrodes. Interest in paste electrodes has been revived [52–54], and it might well be that paste electrodes will find a wider use than before.

Electrochemical Conversion of Gaseous Organic Compounds with Low Solubilities in the Solvent/Electrolyte System

Electrochemical conversion of gaseous substances very often involves mass transfer problems because of their relatively low solubilities in polar solvents. The light hydrocarbons such as methane, ethane, propane, ethylene, and propylene offer problems in electrochemical processing because of their low saturation concentrations under ambient pressures. Therefore, a decrease of working temperature and continuous resaturation by feeding the gaseous substrate through porous frittes or other suitable gas distributors, situated at the lower parts of the electrolyzer, are the usual means employed for electrochemical processing of gases. Figure 2.8 shows a cell construction used by Schmeisser and Sartori [55] for the electrofluorination of the lighter hydrocarbons and of hydrogen sulfide. It is noteworthy that the use of gas injection at electrodes does improve mass transfer to a very remarkable extent. If electrochemical conversion of the gas itself is desired, injection serves two purposes: steady resaturation and perturbation of the boundary layers at the electrodes. Tobias and co-workers [56] showed that stirring with gas bubbles at electrode surfaces promotes mass transfer much more effectively than me-

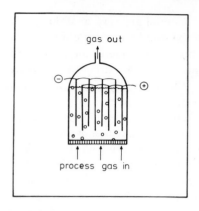

Fig. 2.8. Schematic presentation of electrolysis setup for conversion of gaseous substrates in aqueous electrolytes (fluorination of hydrocarbons) [55].

chanical stirring, because gas bubbles perturb the electrolyte just at the point where it should be stirred, that is, very close to the phase boundary. Figure 2.9 depicts the improvement of mass transfer by gas bubble stirring—gases (e.g., H_2) being generated electrochemically.

Use of Mediating Redox Systems for the Electrochemical Conversion of Sparsely Soluble Organic Substances Being Dispersed in an Aqueous Electrolyte

The problem of very low mass transport rates caused by too low a solubility of the depolarizer can often be solved very effectively by using mediating redox systems. It should, however, be stressed that apart from reasons concerning mass transfer there exist chemical reasons for the use of mediated electrolysis, too. Very often direct and mediated electrochemical conversions follow different reaction routes and hence yield different products. Such mediating systems have been used nearly as far back as dispersion electrolysis has been applied, and they are not used exclusively to process low-solubility substrates in the form of two-phase dispersions, but also in electrochemical conversion of homogeneously dissolved substrates.

In mediated electrochemical conversions electron transfer from the electrode to the substrate (reduction) or in the opposite direction (oxidation) proceeds in two steps. First, charge is exchanged between the electrode and the mediating system. In the second step the mediator reacts with the substrate. In general, mass transfer hindrance can be ruled out for the first charge transfer between the electrode and the homogeneously dissolved mediating system because mediating systems generally are present in high concentrations (e.g., 10^{-1} up to 1 mole/liter). As is discussed below, mass transfer may become a problem for the second step whenever a reactant with very low solubility in a homogeneous redox reaction is to be converted in the aqueous

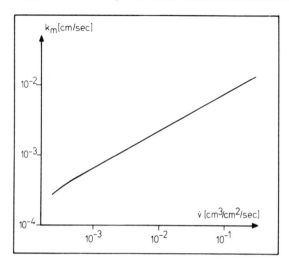

Fig. 2.9. Mass transfer enhancement by gas bubble evolution (mass transfer coefficient versus surface specific volume flow rate of evolved gas) [56].

phase. Because of this very different situation for the first and second step with respect to mass transfer it is not unusual to solve the respective reaction engineering problems by using two spatially separated reactors. Indeed, this is very often done, and for many technical applications it is advisable to circulate the mediating system without addition of dispersed substrate through the electrolyzer to separate spatially the first electrochemical step from the second chemical redox reaction. The final (chemical) redox reaction between mediator and substrate may then be performed in a separate reactor (mostly a stirred tank reactor) after admixture of the well-dispersed substrate phase (2.10a and 2.10b).

$$(\mathrm{Med}) \pm ne^- \xrightarrow{\text{electrode}} (\mathrm{Med})^{\mathrm{Red/Ox}} \qquad (2.10a)$$

$$m \cdot (\mathrm{Med})^{\mathrm{Red/Ox}} + \text{substrate} \xrightarrow{\text{electrolyte}} \text{product} + m \cdot (\mathrm{Med}) \qquad (2.10b)$$

After completion of the second step and after separation from still unconverted substrate (which very often then contains the reaction products), the reconverted mediating system may then be recycled through the electrolyzer (see Fig. 2.10).

It is the second "chemical" redox process that very often is relatively slow and may be either limited in its rate by mass transfer or (sometimes being strongly kinetically hindered) by the rate of the chemical redox reaction in the aqueous phase.

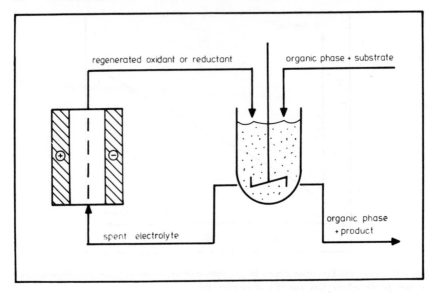

Fig. 2.10.   Schematic presentation of technical performance of mediated electrolysis with spacial separation of electrolyzer and two-phase redox reactor (stirred tank reactor).

One important step in the chemical conversion of a dispersed phase is phase transfer, that is, passage of the substrate molecules across the phase boundary from the surface of a dispersed droplet into the aqueous phase. Although in some cases, especially in the presence of adsorbed surface active substances at the phase boundary, phase transfer might be slow [57], very often this step is not especially hindered. Therefore, the phase transfer step is left out of any further consideration concerning the rates of the chemical redox reaction. With respect to mass transfer control of the redox reaction, quite different cases may exist as described below and as schematically depicted in Figs. 2.11a to 2.11c.

1.   The homogeneous reaction rate [moles/(cm^3)(sec)] may be much lower than the maximal rate of mass transfer for the transport of the substrate molecules from the surface of the dispersed droplets into the bulk of the aqueous solution. This maximal volume-specific rate of mass transfer is given by the maximal driving concentration difference (equal to the saturation concentration $c_{sat}^{substrate}$), by the mass transfer coefficient k_m for mass transfer between the dispersed droplets and the bulk of the electrolyte, and by the volume-specific phase boundary a [cm^2/cm^3]. The value of k_m mainly depends on the hydrodynamics of the dispersion, that is, it is a function of

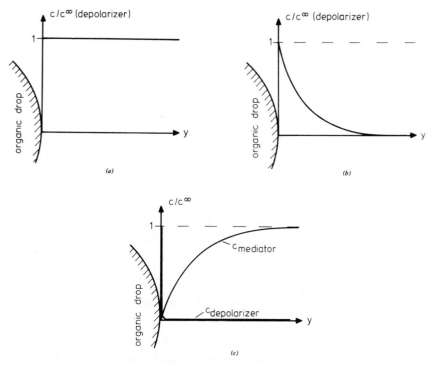

Fig. 2.11. Schematic presentation of concentration profile in front of dispersed droplets being converted by a homogeneously dissolved mediating redox system. (a) Very slow reaction: bulk phase is fully saturated with depolarizer; (b) Moderately slow reaction: depolarizer is consumed in reaction layer ($\delta_R < \delta_N$) in front of dispersed droplet; (c) Very fast reaction (nearly or truly surface reaction): depolarizer concentration drops to zero in front of drop surface.

Reynolds and Weber number mainly, whereas a depends on the hydrodynamics, as well as on the volume fraction of the dispersed phase.

$$\text{volume specific chemical rate} \ll k_m \cdot c_{sat}^{substrate} \cdot a \qquad (2.11a)$$

If condition (2.11a) holds, then saturation concentration of the substrate is established throughout the reactor (Fig. 2.11a). In this case the total rate of conversion does not change over a wide range of mass transfer coefficients, that is, stirring intensity and volume ratio of dispersed versus aqueous phase.

2.   If, however, the rate of chemical reaction increases, mass transfer begins to compete with the chemical rate (Fig. 2.11b). Under steady-state conditions the following relation holds:

$$\text{volume specific chemical rate} \approx k_m \cdot (c_{sat} - c^\infty)^{substrate} \cdot a \qquad (2.11b)$$

Mass transfer with superimposed chemical reaction is a well-known and often treated case in the framework of mass transfer calculations in bubble columns, spray towers, extractors, and similar phase transfer devices [58]. If the reaction of the substrate in the diffusion boundary layer can be treated as a pseudomonomolecular reaction (in large excess of the mediating system), then according to reaction-enhanced mass transfer the volume-specific rate of substrate conversion can be expressed as

$$\text{volume specific chemical rate} = k_m \cdot \left(\frac{D \cdot k_{chem}}{k_m^2} \right)^{1/2} \cdot c_{sat}^{substrate} \qquad (2.11c)$$

where k_{chem} is a pseudo-monomolecular rate constant.

3. If the chemical redox reaction is very fast, the extension of the diffusion/reaction layer for the substrate shrinks to zero and mass transfer of the mediating system to the surface of the droplets becomes rate limiting (Fig. 2.11c):

$$\text{volume specific chemical rate} \leq k_m \cdot c^\infty \text{ (mediator)} \cdot m \qquad (2.11d)$$

where m is the number of mediating molecules consumed per molecule of substrate.

It is up to the chemical engineer to match electrochemical and chemical conversion rates in the electrolyzer and stirred tank reactor—a task that cannot be treated here in a generalized form but that demands detailed knowledge of the chemistry and chemical kinetics of the respective process. Additionally, special technical and economic boundary conditions (e.g., purity of the product, necessary degree of conversion, investment costs, etc.) have to be fulfilled.

It should be pointed out that mass transfer conditions in dispersions have not been investigated thoroughly enough to make generalized statements. Nevertheless, some valuable reviews exist on this topic (see for instance Ref. 59).

Table 2.3 provides values of standard potentials of some important redox systems that are frequently used as mediators for anodic oxidations as well as for cathodic reduction.

Use of Mediating Redox Systems Together with Phase Transfer Catalysts for Mediated Electrochemical Conversions in Two-Phase Emulsion Electrolytes

Phase transfer catalysts have been used as a valuable tool in preparative organic chemistry [60]. They typically are used for synthetic reactions in two-phase liquid/liquid dispersed systems when two reactants with preferential solubility in opposite phases participate in the reaction. Such phase transfer

Table 2.3. Standard Potentials of Some Redox Couples Being Frequently Used as Mediating Systems

Redox Couple	Standard Potential (V)	
$[Co(CN)_6]^{4-}/[Co(CN)_6]^{3-}$	-0.83	Reduction
Cr^{2+}/Cr^{3+}	-0.41	
Ti^{2+}/Ti^{3+}	-0.37	
Sn^{2+}/Sn^{4+}	$+0.15$	
$[Fe(CN)_6]^{4-}/[Fe(CN)_6]^{3-}$	$+0.36$	
U^{4+}/U^{6+}	$+0.40$	
MnO_4^{2-}/MnO_4^{-}	$+0.54$	
$VO^{2+}/V(OH)_4^{+}$	$+1.00$	
$3Br^{-}/Br_3^{-}$	$+1.05$	
$Cr^{3+}/Cr_2O_7^{2-}$	$+1.36$	
Ce^{3+}/Ce^{4+}	$+1.44$	
Mn^{2+}/Mn^{3+}	$+1.51$	
Mn^{2+}/MnO_4^{-}	$+1.52$	
Pb^{2+}/Pb^{4+}	$+1.69$	
Co^{2+}/Co^{3+}	$+1.84$	Oxidation

catalysts or reagents may lead to conversion of an organic substrate, dispersed in aqueous solutions either as a liquid or dissolved in an organic solvent of low water solubility with water-soluble oxidants or reductants. Thus in the presence of phase transfer catalysts, MnO_4^{-} and $Cr_2O_7^{2-}$ as oxidizing, or Cr^{2+} and Ti^{3+} as reducing agents may lend themselves to very efficient and fast redox reactions in the organic phase.

Phase transfer catalysts are ionic in nature and, apart from their electrical charge, which provides for dipolar solvation in water, are pronouncedly lipophilic. This is because long-C-chain aliphatic or aromatic substituents are incorporated in their molecular structure. Typically, tetraalkyl or alkylaryl ammonium salts are used when lipophilic cations are needed, and long-C-chain carbonic acids, alkyl sulfates and sulfonates, and arene sulfonates are used when an anionic phase transfer catalyst is needed.

Such large lipophilic organic ions may form ion pairs with counterions of inorganic character. It is this ion pair formation, that is, neutralization of charge together with formation of a complex that exhibits expressedly lipophilic character, that renders finite solubility of the bound inorganic ions in the organic phase (Fig. 2.12a). In some cases where a high solubility of the complex ion pair in the organic phase may not be accomplished, at least preferential adsorption of the ion pair at the phase boundary may result (Fig. 2.12b).

Fig. 2.12.   Action of phase transfer catalysts in mediated redox reactions. (*a*) Formation of organic-phase-soluble ion pair from mediator and charge transfer catalyst ions. (*b*) Adsorption of ion pair formed from mediator and charge transfer catalyst at the aqueous/organic interface.

In 1974 Eberson published a paper on the application of phase transfer catalysts in an electrochemical synthesis [61, 62]. Recently Lee and Freedman [63] studied phase transfer catalyzed oxidations of alcohols and amines by hypochlorite (which may be produced by electrolysis of dilute neutral aqueous alkali chloride solutions).

Shortly afterward Pletcher and co-workers published results on phase transfer catalyzed oxidation of benzyl alcohol with electrogenerated hypobromite [64]. They also described phase transfer catalyzed oxidation of alcohols and aromatic hydrocarbons with dichromate ions, which after their reduction to water-soluble Cr^{III} ions, might be electrochemically regenerated and recycled [65, 66]. Since the oxidizing species were anionic, a cationic species, namely, tetrabutylammonium ion was used by Pletcher as charge transfer catalyst. Carbon tetrachloride or methylene chloride was used as the organic phase.

Pletcher's results can be summarized as follows:

1.   The anionic oxidant is really transferred into the organic phase on addition of the phase transfer.

2.   The redox reaction in the organic phase is orders of magnitude faster than the same reaction proceeding as a homogeneous reaction in the aqueous phase.

3.   High sulfuric acid concentrations in the aqueous phase (which are necessary if the redox reaction with dichromate is performed without phase transfer catalysts) may be avoided (1–3 moles/liter H_2SO_4 instead of 10 moles/liter!).

4.   If performed with phase transfer catalysts, the oxidation of alcohols to aldehydes becomes more selective and better yields may be obtained.

Application of phase transfer catalysts for mediated electrochemical conversions of sparsely water-soluble organic substances is still at the very beginning of a promising scientific and technical development. It is not improbable that if technical organic electrosynthesis has a real future, it will be due partially to the possibilities that are now opened by application of phase transfer catalysts. For continuously operating production cycles the stability of the phase transfer catalyst against chemical and electrochemical attack and degradation is an important question. Accordingly, further advances in production and commercial availability of very stable fluorinated ionic surfactants will be of some importance for significant developments in this field.

Performance of Electrolysis with Two-Phase Electrolytes with Finite Electrical Conductance of the Nonaqueous Phase

Depolarizer solubility in the aqueous electrolyte and mass transfer through the aqueous phase to the electrodes by convective diffusion is not a limiting factor with respect to current densities and space time yields if (as schematically depicted in Fig. 2.13).

1.   Droplets of the organic phase are brought in physical contact with the electrode.

2.   Charge transfer is no longer restricted to the electrode/aqueous electrolyte interface, but may also proceed at the organic phase/electrode interface.

Performing charge transfer and electrochemical conversion at the organic phase/electrode interface demands a finite and not too low electrical conductivity of the organic phase. This would allow one to perform the electrochemical conversion in one phase, choosing the organic phase alone as the electrolyte.

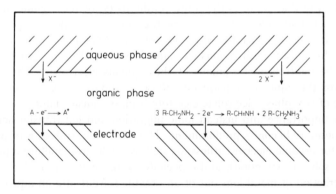

Fig. 2.13.   Charge transfer at organic phase/electrode interface and ionic charge and mass transport as precondition for electrochemical conversion of depolarizers within the organic phase in two-phase electrolysis. (*a*) One-electron oxidation of a neutral species (A) produces a cation (A^+), the charge of which is neutralized by anions (X^-) crossing the aqueous/organic phase boundary. (*b*) Two-electron oxidation of amines produces imines and two ammonium ions being neutralized by anions (X^-).

The simultaneous use of the two phases, however, in an appropriately constructed electrolyzer is not only advantageous but may be indispensable:

1.   Water is very often needed as a reactant. For cathodic reductions water serves as proton donor and for anodic oxidations water supplies the oxygen for the functional groups ($-OH$, $C=O$) of the oxygenated compound.

2.   The aqueous phase in general possesses much higher conductivities than the organic phase. Therefore, it is desirable to fill the interelectrode gap with conducting aqueous electrolyte and to have only a small fraction of this gap filled with the poorly conducting organic phase.

3.   It might be necessary to perform the electrochemical conversion and the consecutive transference of the reaction product into the aqueous phase simultaneously whenever the product is preferentially soluble in water and not, however, in the nonaqueous phase.

In general an appropriate lipophilic electrolyte is to be used in the organic phase in order to supply the necessary electrical conductivity for the organic phase. This is the case whenever the distribution equilibria do not favor a high enough electrolyte content in the nonaqueous phase and too much lipophilic electrolyte would thus be wasted and lost with the aqueous phase. In other cases, the lipophilic nonaqueous phase—as its constituents are organic ions—may not be absolutely inert against electrochemical attack.

In all these cases electrical conductivity of the organic phase must be supplied by the adducts or products of the electrochemical conversion. As the reaction products are at least partially ionic, they may serve as constituents of the electrolyte that supplies the finite conductivity for the organic phase. As is shown schematically in Fig. 2.13 for the anodic oxidation of species A to A^+ and for the two-electron oxidation of an alkylamine to the respective imine (Schiff base), positive ions (A^+ and $R-CH_2-NH_3^+$) are generated that have to be neutralized by anions of the aqueous electrolyte. These anions must cross the aqueous/nonaqueous phase boundary. It is evident that for this type of two-phase electrolyte electrolysis mass transfer limitations may be of minor importance and instead ionic migration and ion transfer across the phase boundary becomes predominant. This sort of electrolysis is of potential interest for organic electrosynthesis as well as for hydrometallurgic processes (i.e., for solvent extraction), and especially in the final step of electrochemical reextraction of metal ions out of an organic solvent into an aqueous electrolyte.

Feess and Wendt [67] investigated as a model reaction electrochemical extraction "using" the anodic oxidation of ferrocene. Ferrocene is dissolved in an organic solvent and, in a process parallel to the anodic oxidation, is transferred simultaneously into the aqueous phase in the form of ferricinium salts.

The authors showed that the most important parameters governing the rate of anodic conversion of the neutral ferrocene molecules at the organic phase/electrode interface are as follows:

1.   Electrical conductivity supplied to the organic phase by the anodically produced ferricinium salt.

2.   Low difference of surface tensions at the electrode/organic phase interface and the electrode/aqueous phase interface that can be brought about by using appropriate surfactants, especially anionic surfactants.

3.   Small values of the free enthalpy of ion-transfer of the anions of the supporting electrolyte for transfer from the aqueous into the organic phase.

The theory of ion exchange across liquid/liquid phase boundaries was developed on a semiquantitative basis [68]. This theory by Koryta and co-workers is consistent with the results obtained by Feess and Wendt. Table 2.4 [69, 70] contains values of the free enthalpy of phase transfer for different cations and anions from nitrobenzene into water. The data clearly demonstrate that among the cations it is especially trimethyl-ammonium and among the anions it is perchlorate which are easily able to cross the organic/aqueous phase-boundary in either direction. The latter authors used their results as a basis to construct a so-called trickle cell with stacks of particulate horizontal working electrodes (nets, perforated plates, etc.) (Fig. 2.14a). The organic phase is continuously supplied to these electrodes through frittes. The electrodes work as electrochemically active plates of a perforated plate column for electrochemical extraction processes (Fig. 2.14b). Figure 2.15a shows a current/voltage curve as obtained for ferrocene oxidation on a single perforated plate that is continuously supplied with fresh ferrocene solution in methylene chloride. The current–voltage curve is rather distorted (because of

Table 2.4.   Free *standard-transfer-enthalpies* of some selected anions and cations for the transfer from the aqueous into the nitrobenzene phase [69, 70]

Cations	$\Delta G°_{transf}/kJ/mole$	Anions	$\Delta G°_{transf}/kJ/mole$
Li^+	38.2	dipycrylaminate	-39.4
Na^+	34.2	I_5^-	-38.8
H^+	32.5	$PhyB^-$	-35.9
Cs^+	15.4	I_3^-	-23.4
Me_4N^+	3.4	$C_2O_4^-$	8
Et_4N^+	-5.7	I^-	18.8
But_4N^+	-20.0	Br^-	28.4

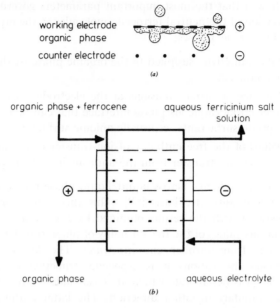

Fig. 2.14. Continuous operation of two-phase electrolyte electrolysis based on charge transfer at organic phase/electrode interface [67]. (*a*) Perforated plate anode being wetted by the organic phase, which contains the depolarizer. The organic phase is supplied from above and leaves the electrode as large drops. A very coarse grid that does not interfere with the flow of the organic phase is used as cathode. (*b*) Schematic description of a column for continuous electrochemical extraction of oxidizable (ferrocene) or reducible depolarizers from the organic into the aqueous phase.

slow and hindered ionic charge transfer across the phase boundary and through the organic phase), but still shows a diffusion limited current at high potentials. Figure 2.15*b* shows the current density versus volumetric flow rate of the organic phase for one, two, and a stack of three perforated plates together with the conversion/flow rate dependence (Fig. 2.15*c*). This sort of electrochemical reactor was described by both a physical and a mathematical model. It is not impossible that this device for two-phase electrolysis, because of its technical simplicity, may gain greater importance in hydrometallurgical solvent-extraction processes.

Two-phase dispersion electrolyte electrolysis or any other suitable process technique that is devised to improve mass transfer in solutions of sparsely soluble substances has been widely used on a laboratory scale. In the past an intuitive approach was used to solve mass transfer problems, however, the fundamental phenomena involved in two-phase-electrolyte electrolysis are better understood today. However, detailed knowledge, numerical data, and

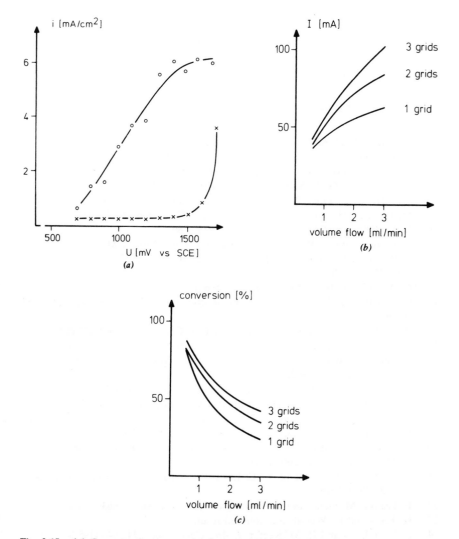

Fig. 2.15. (*a*) Current voltage curve for ferrocene oxidation at an electrochemical extractor anode (ferrocene dissolved in methylene chloride), (○) ferrocene oxidation, (×) oxygen evolution. (*b*) Total current versus volumetric flow rate of organic phase for anodic extraction of ferrocene (0.05 mole/liter) from methylene chloride into the aqueous phase for single electrode plate and stacks of two and three plates, respectively. (*c*) Conversion versus flow rate for anodic ferrocene extraction (other conditions same as in *b*) [67].

dimensionless relations are nearly totally missing. Nonetheless, there exist already cases of process development that may find technical application in the future. Section 4 of this chapter deals with a literature survey on the work published hitherto on two-phase electrolyte electrolysis.

References to Section 1, 2 and 3

1. W. Vielstich, *Z. Elektrochem.*, **57**, 646 (1953).
2. P. Yeh and T. Kuwana, *J. Electrochem. Soc.*, **123**, 1334 (1976).
3. T. Kempf, German Patent 117,251 (1899); *Chem. Zentr.* **1901/I**, 348.
4. K. S. Udupa, G. S. Subramanian, and H. V. K. Udupa, *Bull. Acad. Polon. Sci., Sér. Sci. Chim.*, **9**, 45 (1961) [*Chem. Abstr.* **58**, 13443 (1963)].
5. M. Fremery, H. Höver, and G. Schwarzlose, *Chem. Ing. Tech.*, **46**, 635 (1974).
6. J. P. Millington, *Chem. Ind. (Lond.)*, **1975**, 780.
7. R. Dworak and H. Wendt, *Ber. Bunsenges. Phys. Chem.*, **81**, 729 (1977).
8. S. B. Collins and J. G. Knudsen, *AIChE J.*, **16**, 1072 (1970).
9. J. O. Hinze, *AIChE J.*, **1**, 289 (1955).
10. C. A. Sleicher, Jr., *AIChE J.*, **8**, 471 (1962).
11. G. Narsimhan, J. P. Gupta, and D. Ramkrishna, *Chem. Eng. Sci.*, **34**, 257 (1979).
12. P. A. Wehrli (F. Hoffmann-La Roche & Co), German Patent 2,032,081 (1971).
13. *International Critical Tables*, Vol 3, McGraw-Hill, New York 1928, p. 389.
14. Landolt-Börnstein, Vol II, Part IIb, 6th ed., Springer-Verlag, Berlin, Göttingen, Heidelberg, 1962.
15. J. Polak and B. C.-Y. Lu, *Can. J. Chem. Eng.*, **51**, 4018 (1973).
16. P. J. Leinonen and D. Mackay, *Can. J. Chem. Eng.*, **51**, 230 (1973).
17. C. McAuliffe, *J. Phys. Chem.*, **70**, 1267 (1966).
18. R. L. Bohon and F. W. Claussen, *J. Am. Chem. Soc.*, **73**, 1571 (1951).
19. L. J. Andrews and R. M. Keefer, *J. Am. Chem. Soc.* **71**, 3644 (1949).
20. C. McAuliffe, *Nature*, **200**, 1092 (1963).
21. D. S. Arnold, C. A. Plank, E. E. Erickson, and F. P. Pike, *Chem. Eng. Data Ser.*, **3**, 253 (1958).
22. F. Franks, M. Gent, and H. H. Johnson, *J. Chem. Soc.*, **1963**, 2716.
23. H. Feess and H. Wendt, unpublished results.
24. L. J. Andrews and R. M. Keefer, *J. Am. Chem. Soc.*, **74**, 640 (1952).
25. L. J. Andrews and R. M. Keefer, *J. Am. Chem. Soc.*, **72**, 5034 (1950).
26. H. Fühner, *Chem. Ber.*, **57**, 510 (1924).
27. W. W. Davis, M. E. Krahl, and G. H. A. Clowes, *J. Am. Chem. Soc.*, **64**, 108 (1942).
28. N. V. Sidgwick, W. J. Spurrell, and T. E. Davies, *J. Chem. Soc.*, **107**, 1202 (1915).
29. W. Hückel, M. J. Niesel, and L. Büchs, *Chem. Ber.*, **77**, 334 (1944).
30. H. S. Booth and H. E. Everson, *Ind. Eng. Chem.*, **41**, 2627 (1949).
31. H. E. Vermillion, B. Werbel, J. H. Saylor, and P. M. Gross, *J. Am. Chem. Soc.*, **63**, 1346 (1941).

32. P. M. Gross, *J. Am. Chem. Soc.*, **53**, 1744 (1931).
33. M. Shikata and N. Hozaki, *Chem. Zentr.*, **1937/II**, 2817.
34. P. M. Gross, J. H. Saylor, and M. A. Gorman, *J. Am. Chem. Soc.*, **55**, 650 (1933).
35. A. Klemenc and M. Löw, *Rec. Trav. Chim. Pays-Bas*, **49**, 629 (1930).
36. L. J. Andrews and R. M. Keefer, *J. Am. Chem. Soc.*, **72**, 3113 (1950).
37. W. F. McDevit and F. A. Long, *J. Am. Chem. Soc.*, **74**, 1773 (1952).
38. J. H. Saylor, A. I. Whitten, I. Claiborne, and P. M. Gross, *J. Am. Chem. Soc.*, **74**, 1778 (1952).
39. J. S. Clarke, R. E. Ehigamusoe, and A. T. Kuhn, *J. Electroanal. Chem.*, **70**, 33 (1976).
40. J. C. Ghosh, S. K. Bhattacharyya, M. S. Muthanna, and C. R. Mitra, *J. Sci. Ind. Res. (India)*, **11B**, 356 (1952) [*Chem. Abstr.*, **47**, 2064e (1953); *Curr. Sci.*, **16**, 87 (1947) [*Chem. Abstr.*, **41**, 4725c (1947)].
41. J. C. Ghosh, S. K. Bhattacharyya, M. R. A. Rao, M. S. Muthanna, and R. B. Patnaik, *J. Sci. Ind. Res. (India)*, **11B**, 361 (1952) [*Chem. Abstr.* **47**, 2064e (1953)].
42. N. M. Winslow, *Trans. Electrochem. Soc.*, **88**, 81 (1945).
43. N. M. Winslow and G. W. Heise, U.S. Patent 2,427,433 (1947) [*Chem. Abstr.*, **42**, 46f (1948)].
44. C. H. Rasch and A. Lowy, *Trans. Electrochem. Soc.*, **56**, 477 (1929).
45. E. G. White and A. Lowy, *Trans. Electrochem. Soc.*, **62**, 223 (1932).
46. E. G. White and A. Lowy, *Trans. Electrochem. Soc.*, **62**, 305 (1932).
47. G. Bionda, *Atti Reale Accad. Sci. Torino, Classe Sci. Fis. Mat. Nat.*, **80**, 213 (1944/45) [*Chem. Abstr.*, **41**, 7276f (1947)].
48. G. Bionda, *Ann. Chim. applicata*, **36**, 204 (1946) [*Chem. Abstr.*, **41**, 3701d (1947)].
49. G. Bionda and M. Civera, *Ann. Chim.*, **43**, 11 (1953) [*Chem. Abstr.*, **47**, 5821b (1953)].
50. G. Bionda and M. Civera, *Ann. Chim.*, **41**, 814 (1951) [*Chem. Abstr.*, **46**, 7908f (1952).
51. K. J. Vetter, *Chem. Ing. Tech.*, **45**, 213 (1973).
52. D. Bauer and P. Gaillochet, *Electrochim. Acta*, **19**, 597 (1974).
53. M. Lamache, D. Kende, and D. Bauer, *Nouv. J. Chim.*, **1**, 377 (1977).
54. M. Lamache, *Electrochim. Acta*, **24**, 79 (1979).
55. R. Schmeisser and P. Sartori, *Chem. Ing. Tech.*, **36**, 9 (1964).
56. C. W. Tobias, M. Eisenberg, and C. R. Wilke, *J. Electrochem. Soc.*, **99**, 395 (1952).
57. W. Nitsch, R. Wiedholz, and M. Raab, *Chem. Ing. Tech.*, **43**, 465 (1971).
58. W. Otto and K. Schügerl, *Chem. Ing. Tech.*, **45**, 563 (1973).
59. K. Bauckhage, H. D. Bauermann, E. Blass, H. Sauer, M. Stölting, J. Tenhumberg, and H. Wagner, *Chem. Ing. Tech.*, **47**, 169 (1975).
60. E. V. Dehmlow, "Phase Transfer Catalyzed Two-Phase Reactions in Preparative Organic Chemistry." In *New Synthetic Methods*, Vol. 1, Verlag Chemie, Weinheim, 1975, pp. 1–28.
61. L. Eberson and B. Helgée, *Chem. Scr.*, **5**, 47 (1974).

62. L. Eberson and B. Helgée, *Acta Chem. Scand.*, **29B**, 451 (1975).
63. G. A. Lee and H. H. Freedman, *Tetrahedron Lett.*, **20**, 1641 (1976).
64. D. Pletcher and N. Tomov, *J. Appl. Electrochem.*, **7**, 501 (1977).
65. D. Pletcher and S. J. D. Tait, *Tetrahedron Lett.*, **18**, 1601 (1978).
66. D. Pletcher and S. J. D. Tait, *J. Chem. Soc. Perkin Trans. II*, **1979**, 788.
67. H. Feess and H. Wendt, "Performance of Electrolysis with Two-Phase Electrolytes II," *Ber. Bunsenges. physik. Chem.*, **85** (1981) in press.
68. J. Koryta, *Electrochim. Acta*, **24**, 293 (1979).
69. J. Rais, *Collect. Czech. Commun.* **36**, 3253 (1971).
70. J. Rais, P. Sehicki, M. Kyrs, *J. Inorg. Nucl. Chem.*, **38**, 7376 (1976).

4 PERFORMANCE OF ORGANOELECTROSYNTHETIC RE-ACTIONS BASED ON ELECTROCHEMICAL CONVERSION OF SPARSELY WATER SOLUBLE DEPOLARIZERS

Oxidation of Arenes

Anodic Oxidation of Benzene Dispersed in Aqueous Electrolytes

The anodic oxidation of benzene in emulsions, patented in 1900 by Kempf [1], has been extensively studied. The electrolysis products, hydroquinone and benzoquinone, are of commercial interest. In divided cells the reaction product, which is formed in separate steps (Fichter and Stocker [2]: benzene → phenol → hydroquinone → benzoquinone) is benzoquinone:

$$C_6H_6 + 2H_2O \rightarrow C_6H_4O_2 + 6H^+ + 6e \tag{2.12}$$

Literature surveys covering all published work and patents up to 1961 were reported in Fichter's papers in 1914 [2], 1925 [3], and 1929 [4] and in a paper by Udupa [5]. The authors of this chapter do not intend to repeat here a review on the historical development, but report the published data in an accompanying table and only discuss the more fundamental results of different authors.

In their paper, "Anodic Oxidation of Aromatic Hydrocarbons and Phenols," Fichter and Stocker [2] investigated the influence of the process parameters: anode material, electrolyte composition at constant working temperature (ambient temperature), and current density (10 mA/cm²). The results are summarized in Table 2.5 and show very clearly that lead dioxide anodes and aqueous sulfuric acid (1 M) are the electrode/electrolyte combination of choice. This combination is adopted, too, in the technical processes on which we report in the last part of this section.

Fichter showed that the first oxidation product of intermediate stability is hydroquinone, which, however, is easily oxidized to benzoquinone so that only very low stationary hydroquinone concentrations are observed in the aqueous anolyte phase. Very minute amounts of catechol are found too, but

Table 2.5. Current Yields for Anodic Benzene Oxidation at Different Anode Materials and in Different Electrolytes at 25°C; current density, 10 mA/cm^2 [2]

Anode	Electrolyte	Current Yield (%)
PbO$_2$	2 N H$_2$SO$_4$	21.3
Graphite	2 N H$_2$SO$_4$	11.0
Pt	2 N H$_2$SO$_4$	7.4
PbO$_2$	4 N H$_3$PO$_4$	5.0
Pt	10% HClO$_4$	7.8
Pt	2 N HNO$_3$	5.0

catechol is so sensitive toward further anodic oxidation that it is ultimately oxidized to maleic and oxalic acid and to CO$_2$ and H$_2$O.

Kinetic studies of this important electrosynthesis were not performed until recently. In 1976 Kuhn and co-workers [6] published a careful kinetic study on the anodic oxidation of benzene, toluene, and anisole at PbO$_2$ anodes. Rather than using emulsions of these compounds in aqueous solutions, they used homogeneous solutions of the arenes in aqueous 1 M sulfuric acid to obtain well-defined mass transfer conditions. They showed that at very low current densities and hence low anode potentials, 100% current efficiencies could be obtained with respect to benzoquinone production (2.12). On increasing the anode potential, the amount of maleic acid increases steadily, because benzoquinone is not stable to further anodic oxidation and is degraded to maleic acid and oxalic acid:

$$C_6H_4O_2 + 6H_2O \rightarrow \underset{\text{Maleic acid}}{C_4H_4O_4} + \underset{\text{Oxalic acid}}{C_2H_2O_4} + 10\,H^+ + 10e \qquad (2.13)$$

This successive anodic oxidation of benzoquinone lowers the current yield remarkably (see Fig. 2.16). Current yields for benzoquinone production decrease from nearly 100% at 1.1 V versus Hg/Hg$_2$SO$_4$ to only 60% at 1.5 V. Above an anode potential of 1.55 V mainly oxygen is evolved.

Kuhn and co-workers observed that at 25°C limiting currents were somewhat lower than would correspond to diffusion limited currents and concluded that the reaction is rate limited by the heterogeneous oxidation of benzene by lead dioxide. Below the limiting current value they obtained a formal reaction order of 2 for benzene. Remarkable as these results are, they nevertheless do not predict optimal working conditions for conducting the reaction (2.12) in aqueous emulsions for the following reasons:

1. Extraction of the benzoquinone into the dispersed benzene phase and protection of the product against further oxidation plays a dominant role if two-phase emulsion electrolytes are used.

2. Short term measurements (as electrode kinetic measurements usually are) reflect only to a very limited extent the conditions that are valid for a long term technical electrolysis with respect to electrode-surface conditioning and influence of impurities on heterogeneous kinetics.

3. A wider range of working temperatures had not been covered because of experimental restrictions. However, it is very likely that enhanced working temperatures above ambient temperature are optimal.

The statements of different authors concerning working temperature are confusing and contradictory. Values of 15°C [5, 7, 8] and 25°C [9, 10] are reported. Reports that include experience on technical processes indicate that higher working temperatures, namely, 40 [11] or 50°C [5], are preferable because of the enhanced benzene solubility in water.

A divided cell must be chosen since with the omission of a diaphragm or membrane severe losses of current yield are to be expected because of electrochemical short-circuiting of the cell by the benzoquinone/benzohydroquinone redox couple. Vigorously stirred two-phase emulsions are used in general. [There is only one exception: see a patent by Horrobin and New [9], who (to circumvent Kempf's patent?) proposed cycling saturated aqueous benzene solutions through the cell].

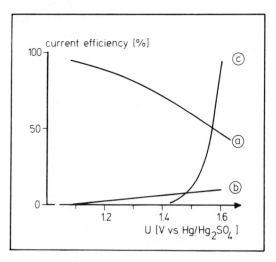

Fig. 2.16. Current efficiencies versus anode potential in anodic benzene oxidation at lead dioxide at 25°C [6] for (*a*) benzoquinone, (*b*) maleic acid, and (*c*) oxygen.

Udupa used a rotating electrode as stirrer [5]. The cell construction of Ghosh et al. [10] made use of a porous electrode (see Fig. 2.6) through which benzene flowed to the front of a PbO_2-covered carbon electrode.

In almost all cases applied current densities are reported to be around 20 to 30 mA/cm^2. It is very likely that the authors approached by trial and error the current density value that is determined by the solubility-limited mass transfer of benzene to the anode.

Very important to note, but not at all understood, is the fact that the addition of a relatively small amount of potassium ferricyanide to the electrolyte results in remarkable improvements of current yields [5, 10]. It is not unlikely that ferricyanide plays a role in mediated oxidation of some radical intermediate that otherwise would react to form an oligomeric or polymeric material.

Generally, the applied process technique involves accumulation of the benzoquinone product in the dispersed benzene phase. Either batchwise workup to obtain the benzoquinone, or transference of the benzene solution of benzoquinone into the cathode compartment where it is cathodically reduced in a second two-phase emulsion electrolysis to hydroquinone is applied to recover the respective product (2.14).

$$C_6H_4O_2 + 2H^+ + 2e \rightarrow C_6H_6O_2 \qquad (2.14)$$

Pilot-scale processes perform this combined oxidation and reduction continuously. To improve current and mass yields the benzoquinone content of the benzene phase in the anode compartment has to be kept rather low (i.e., below 3 wt %).

Current efficiencies vary from 100% [6] to very low values. For pilot-scale electrolysis an optimal performance of the process with respect to total costs may be obtained with 40% current yield. Mass yields were not reported by the different authors in most cases, but under practical conditions a mass yield of 80% with respect to converted benzene seems to be feasible [11].

For a more detailed survey on the particular electrolysis conditions of the papers cited above and other work dealing with the oxidation of benzene see Table 2.6. Further anodic reactions of benzene performed in two-phase electrolytes, such as chlorination and nitration, are added at the end of Table 2.6.

Anodic Oxidation of Naphthalene in Two-Phase Electrolytes or Appropriately Dispersed State with Aqueous Electrolytes

Relatively little work has been published on anodic oxidation of naphthalene to naphthoquinone and even less on anodic oxidations using two-phase mixtures as electrolytes.

$$C_{10}H_8 + 2H_2O \rightarrow C_{10}H_6O_2 + 6H^+ + 6e \qquad (2.15)$$

Table 2.6.   Survey on Electrolysis Conditions and Mass Yields for Anodic Oxidation of Benzene to Hydroquinone or Benzoquinone

Electrolysis Conditions	Anode	Yield (%)	Type	Ref.[a]
10% H_2SO_4, 4 V, 10 mA/cm^2	PbO_2	45-65	Emulsion	1
2 N H_2SO_4, 2 N H_3PO_4, 2 N $HClO_4$, 2 N HNO_3, 10-30 mA/cm^2	Pt, PbO_2, graphite	5-21	Emulsion	2, 4
4% Na_2SO_4 + 1% NaOAc in H_2O, $T < 20°C$, 10-20 mA/cm^2	PbO_2	77	Emulsion	8, 12
2% Na_2SO_4 + 1% NaOAc in H_2O, 5.5 V, 15°C, 30 mA/cm^2	PbO_2	75	Emulsion	7
10% H_2SO_4, 25-30°C 23.5 mA/cm^2	PbO_2	?	Saturated solution	9
2% H_2SO_4 + 6% $K_3[Fe(CN)_6]$, 14 mA/cm^2, 27°C	PbO_2 on graphite	51	"wick" Electrode	10
5% H_2SO_4 + $K_3[Fe(CN)]_6$, 15-50°C	PbO_2 on graphite	45-60	Emulsion	5
15% H_2SO_4, 4 V, 160 mA/cm^2, 40°C	PbO_2	80	Emulsion	11
H_2SO_4, 20 mA/cm^2	PbO_2	70	Emulsion	13

[a] For further references to oxidation of benzene see Refs. 14 to 21; for chlorination of benzene see Refs. 88 to 90; for bromination of benzene see Ref. 89; for nitration of benzene see Refs. 91 and 92; and for amination of benzene see Ref. 93.

De Bottens published some results on one-phase electrolysis in 1902 [22], and in the same year, in a German patent the anodic production of naphtho-quinone from naphthalene was claimed [23] based on Ce^{IV}-mediated oxidation of dilute H_2SO_4/naphthalene dispersions. Current and mass yields, however, were not published in this patent, but the selectivity was mentioned with respect to naphthoquinone versus further oxidation to o-phthalic acid. Schetty in work toward his doctoral thesis [24] tried to repeat this electrosynthesis but was unable to obtain more than traces of naphthoquinone by Ce^{IV}-mediated oxidation. It should be stressed that because of the relatively restricted commercial interest in the product, naphthoquinone, the published results on mediated anodic oxidation are too few, are not conclusive, and do not really allow a realistic estimation with respect to the potential value of this procedure or an optimal process technique for such process.

Of special interest as an unusual process technique is the work of White and Lowy [25], who used a paste electrode composed of powdered carbon

(40%) and naphthalene (60%). This mixture was pressed as a thoroughly mixed cake on a platinum net that served as current collector and mechanical support. This porous mixed electrode was immersed in an aqueous electrolyte, and anodic current densities of 10 mA/cm^2 were applied. After workup by ether extraction and steam distillation of residual naphthalene and naphthoquinone, the latter was determined titrimetrically. The authors reported a current efficiency of some 35% as the best yield with a mass yield of no more than 40%. Low current and mass yields were due to further anodic oxidation of the product, by which a remarkable amount of polymeric material was formed. The decreasing amount of missing weight of residual naphthalene, naphthoquinone, and polymers with an increasing amount of electricity consumed is clear evidence for an increasing percentage of total anodic oxidation to H_2O and CO_2. Although, as pointed out in section 3, the use of paste electrodes meets the demands very well of improved mass transfer of low solubility substances, it has the great disadvantage of keeping the product in contact with the anode for a longer time, during which it is oxidized further.

Interest in mediated naphthalene oxidation has been revived in a number of patents by Rennie [26]. By using Ce^{IV} as mediating agent, he obtained almost quantitative yields of naphthoquinone, whereas Joo and Bryan found only 57% naphthoquinone by Cr^{VI}-mediated oxidation [27]. This result was confirmed by the work of Pletcher and Tait [94], who showed that the yields of naphthoquinone from oxidation with dichromate are relatively poor even in the presence of phase transfer catalysts. Millington [28] described the oxidation of naphthalene at PbO_2-coated titanium wire gauzes by addition of cosolvents to improve the solubility of the depolarizer.

Anodic Conversion of Anthracene to Anthraquinone by Mediated and Direct Anodic Oxidations in Two-Phase Electrolytes

Because of the great importance of anthraquinone in the dye industries, the first patent on the production of this compound by anodic oxidation of anthracene was issued in 1902 [23]. Since then interest in this type of electroorganic synthesis has been kept alive, although a purely chemical route for the production of this compound by ring closure of o-benzylbenzoic acid is the basis of a long established chemical process [29]. Rasch and Lowy, in their paper published in 1929 [30] cited five patents and six further papers. Publishing has continued since then. From the very beginning, research workers preferred mediated anodic oxidation of finely dispersed anthracene in acidified aqueous solutions to direct anodic oxidation of anthracene homogeneously dissolved in organic or mixed organic/aqueous solvents [31].

In 1957 Shirai and Sugino [32] stated that they could convert anthracene to anthraquinone at PbO_2 anodes by using dispersions in 40% H_2SO_4 solutions with addition of 2% V_2O_5 or $Ce(SO_4)_2$ with approximately 60% current yield and up to 90% mass yield.

A paper by Udupa and co-workers published in 1965 [33] summarizes the situation at that time by stating that both electrochemical and chemical oxidation processes are described in the literature, and that according to different workers the electrochemical oxidation of anthracene could be performed using electrolytically regenerated:

1. Chromic acid (dichromate in aqueous sulfuric acid) in two-stage and single-stage processes [34–37].
2. Ce^{IV} in a single-stage process [23, 32, 33, 37, 38].
3. Mn^{III} in a single-stage process [33].
4. V^V in a single-stage process [32, 37].

The earlier patents, however, did not mention the relevant details about current efficiencies, yield and purity of the product, and optimal process parameters such as current densities, working temperature, and optimal fluid dynamic conditions of the dispersions.

Udupa and co-workers (see below: Industrial Application for details) investigated different "oxygen carriers," that is, mediating redox systems, with respect to the optimal conditions for anthraquinone production. They used a dispersion of 200 g of powdered anthracene in 2000 ml of aqueous sulfuric acid and 0.25 M solutions of the mediating system that were stirred either by the rotating rod anode or by a stirrer in the undivided cell. Because of the relatively rapid mediated oxidation the stationary concentration of the mediating oxidant was obviously kept low enough to avoid excessive reduction of the mediator at the cathodes in the undivided cell since current efficiencies varied from 30 to 50%. The current efficiency is obviously limited by the anodic regeneration efficiency, which is limited by mass transfer of the reduced form of the mediator to the anode on one hand and by anodic coevolution of oxygen on the other hand. Both effects are evident, since on increasing the anode potential beyond a critical value of $+1.8$ V versus Hg/Hg_2SO_4 results in a sharp decrease of current efficiency, and the current voltage curve measured by the authors at the PbO_2 anode clearly shows that just at this voltage O_2 evolution starts. Unfortunately, the hydrodynamic conditions in the undivided electrolyzer are so poorly defined that a generalized evaluation of the results of Udupa and co-workers cannot be given.

However, noteworthy and of general validity is their result with respect to the comparison of different mediators. They compared Ce^{IV}, Mn^{III}, and V^V and concluded that vanadium (V) is very ineffective and should not be used. Mass yields obtained with Ce^{IV} and Mn^{III} are good, approaching 90 to 98% in small-scale electrolysis (20 g of anthracene converted) and 95% in medium-scale preparations (200 g of anthracene converted). For an efficient use of Mn^{3+}, an H_2SO_4 concentration of 60% should be chosen. Nevertheless, Mn^{III} oxidation of anthracene seems to be superior to Ce^{IV}-mediated oxidation in 30% H_2SO_4, since in contact with the latter electrolyte the lead dioxide anodes deteriorate and the anthraquinone produced is of poor quality.

Udupa and co-workers did not investigate Cr^{VI} as a mediating system. Cr^{VI}, however, was included in a study performed by Japanese workers [37] published in 1969, according to which dichromate oxidation causes lower yields than pervanadyl or ceric ions.

Summarizing Udupa's results, Mn^{III} in 60% H_2SO_4 seems to be the mediating system of choice and Ce^{IV} in 30% H_2SO_4 very likely would work as well provided an electrode (lead dioxide deposited on titanium) could be found that is corrosion resistant in this medium. The rates of the oxidation of the dispersed phase and the mediator have not been measured quantitatively. The observation that the color of the oxidant changes only at the end of the reaction indicates that the reaction is very rapid. This example is very likely a borderline case where the conversion of the dispersed phase is a true or very nearly a true heterogeneous surface reaction at the dispersed phase/aqueous electrolyte interface.

The results concerning the use of V^V as oxidizing mediating system are controversial. Shirai and Sugino in 1957 [32] and Niki, Sekine, and Sugino in 1969 [37] reported current yields for the V^V-mediated anthracene oxidation that were as high as those for Ce^{IV}-mediated oxidation. But Udupa's pervanadyl system gave only poor current yields.

Discrepancies also exist with respect to dichromate as the mediator. Whereas the Japanese workers reported that dichromate was inferior to Ce^{IV} and V^V, Florenski and Metelkin [34] reported satisfactory results in the use of dichromate. However, they did not obtain higher mass yields than 80% for anthraquinone.

Thus summarizing all published work one could state that very likely Mn^{III} in 60% H_2SO_4 is the most effective mediator with respect to current efficiency and mass yield.

Finally, it should be mentioned that Rasch and Lowy [30] 1929 published a report on anthracene oxidation in a paste electrode made of a mixture of 75 wt % PbO_2 and 25% anthracene that was inserted into a lead-battery grid. The added electrolyte was 18% H_2SO_4 and in some cases $Ce(SO_4)_2$ was added as mediator too. The results obtained with the paste electrode cannot compete with the Mn^{III}-mediated oxidation of dispersed anthracene, as no more than 60% mass yield and 26 to 40% current efficiencies were obtained.

Table 2.7 summarizes the data published by different authors on the anodic oxidation of anthracene to anthraquinone.

Table 2.7. Survey on Electrolysis Conditions and Mass Yields for Anodic Oxidation of Anthracene[a] to Anthraquinone

Electrolysis Conditions	Anode	Yield (%)	Type	Ref.
40% H_2SO_4 + 2% V_2O_5 or $Ce(SO_4)_2$, 16 mA/cm^2, 3-4 V, 90°C	PbO_2	90	Suspension	32
60% H_2SO_4 + 2% $Mn_2(SO_4)_3$, 30% H_2SO_4 + 2% $Ce(SO_4)_2$, 20 mA/cm^2, 2.5-4 V, 90°C	PbO_2	90	Suspension	33
20% H_2SO_4 or NaOH + salts of Cr, Mn or Ce, 10-50 mA/cm^2, 70-90°C, 2.5-3.5 V	PbO_2	80-100	Suspension	23,31
75 g PbO_2 + 25 g anthracene + 40 g H_2SO_4 (20%), 10 mA/cm^2, 80°C,	PbO_2	43	Paste	30
by addition of $Ce(SO_4)_2$		60		
35% H_2SO_4 + $K_2Cr_2O_7$ + little acetone or HOAc, 90-100°C, 100 mA/cm^2, 3-5 V		> 80	Suspension	34

[a]For further information on oxidation of anthracene see Refs. 26, 28, 35-38, and 94.

Anodic Oxidation of Condensed Arenes

There has been little interest in the anodic oxidation of condensed arenes such as phenanthrene. Only the very early patent of Farbwerke Hoechst [23] in 1902 reports on Ce[IV]-mediated oxidation of phenanthrene yielding phenanthrenequinone and diphenic acid and benzoic acid as products of further mediated anodic degradation of phenanthrenequinone.

Oxidation of Substituted Arenes

Anodic Oxidation of Toluene in Emulsion Electrolysis

INTRODUCTION AND SURVEY

Before the beginning of this century, Renard [39] subjected toluene to anodic oxidation. Later Elbs oxidized nitrotoluene in homogeneous one-phase electrolysis and showed that the main product (*p*-nitrobenzylalcohol or its acetate ester) was due to anodic side-chain oxidation of nitrotoluene [40]. Puls [41] electrolyzed alcoholic solutions of sulfuric acid containing toluene as depolarizer. By simultaneous sulfonation and oxidation of toluene, *p*-sulfobenzoethyl ester was formed. This result again showed that the anodic attack preferentially occurred at the side chain. Thus the mechanism for anodic toluene oxidation seemed quite clear, at least for the use of single-phase electrolytes. In the early 1900s Lang [42] produced in two patents benzaldehyde with 80% mass yield from toluene by using as oxidants Mn^{III} ions produced by anodic oxidation of aqueous manganese–ammonium sulfate solutions.

Law and Perkin tried direct anodic oxidation of toluene in two-phase emulsion electrolysis using acetone as an additive to improve the solubility of toluene in the aqueous electrolyte [43]. Current efficiencies (15%) and mass yields (14%) were rather poor, but the main products again were benzaldehyde and benzoic acid. Two papers by Fichter in 1913 [44] and 1914 [2] confused the situation. The authors claimed that hydroxylation of the nucleus was the main reaction. They obtained toluquinone as the main product after electrolyzing toluene emulsions in aqueous sulfuric acid at PbO_2 anodes under conditions of very poor mass transfer and completely uncontrolled anode potential. Later in 1920, Fichter and Uhl [45] found that toluquinone was obtained at less than 1% current yield and was only regarded to be the main product because it was formed as the most stable side product. Benzaldehyde and benzoic acid, which normally are the main reaction products, were completely oxidized to CO_2 and H_2O under the conditions used by Fichter.

Today the issue of the overall reaction appears to be settled: toluene whether oxidized directly or by mediators is oxidized at the side chain and forms benzylalcohol, benzaldehyde, and benzoic acid in distinct steps. Benzylalcohol, however, is so sensitive toward further oxidation that it cannot be isolated. Benzaldehyde, too, is easily oxidized to benzoic acid, but its oxidation is retarded since the anodic oxidation of oxo groups demands their dehydration, which is a slow reaction for benzaldehyde. Therefore, the production of benzaldehyde is possible in high yields by extraction of the intermediate compound into the organic toluene phase, where it is protected against further anodic attack.

$$Ph\text{-}CH_3 \xrightarrow[-2H^+]{-2e + H_2O} Ph\text{-}CH_2OH \xrightarrow[-2H^+]{-2e} Ph\text{-}CHO \xrightarrow[-2H^+]{-2e + H_2O} Ph\text{-}COOH \quad (2.16)$$

KINETIC STUDIES ON DIRECT AND MEDIATED ANODIC CONVERSIONS OF TOLUENE

Several studies have been published on the kinetics of toluene oxidation by different mediating systems, including MnO_4^- [46], Ce^{IV} [47], Mn^{III} [48, 49], and Co^{III} [50, 51]. All these measurements were performed in single-phase systems (mostly acetic acid/water mixtures and acetonitrile in one case [47]). Therefore, they cannot be used to calculate rates and degrees of conversion in two-phase mixtures, that is, for mediated anodic toluene oxidation in emulsions of toluene in aqueous electrolytes. Kuhn and co-workers [6] investigated the electrode kinetics of the anodic oxidation of toluene homogeneously dissolved in aqueous sulfuric acid. For current densities substantially lower than the observed limiting current densities, they obtained a rate law that was second order with respect to toluene at PbO_2 anodes. Their results indicate that limiting current densities are very likely diffusion controlled.

There exists, however, a report of an interesting kinetic study by Udupa [52] on the anodic oxidation of toluene, dispersed in aqueous H_2SO_4 solutions containing Mn^{III} ions. There are two important results:

1. The rate law is first order with respect to the oxidant (Mn^{III}).

2. The apparent activation energy is relatively high (133 kJ/mole).

Because of the latter result, one may conclude that mass transfer does not interfere with the chemical reaction in this case, that is, mass transfer is much faster than the redox reaction. Since, additionally, a first-order reaction with respect to the oxidant is observed, there could exist three different situations, which would be consistent with the observed first-order rate law.

1. The reaction is so fast that it is a surface reaction taking place at the interface of the dispersed toluene droplets and the electrolyte (see Fig. 2.11c).

2. The bulk concentration of the oxidant is established up to the surface of the dispersed droplets, but the toluene concentration depletes to zero within the reaction diffusion layer.

3. The reaction is so slow that the aqueous phase is always saturated (see Fig. 2.11a) with toluene so that the reaction rate becomes

$$\text{rate} = k_{\text{homog.}} \cdot c^{\text{sat}}_{\text{toluene}} \cdot c_{Mn^{III}} \qquad (2.17)$$

There is a good reason to assume that the third case holds, since the reaction half times range between 2 and 140 min. This time scale seems to be well above the half time for saturation of the aqueous phase in the well-stirred emulsion, which is approximately 1 to 10 secs.* It is clear then that the apparent activation energy includes the ethalpy of solution of toluene in the electrolyte solution, which contained 48% H_2SO_4 and 42 g $MnSO_4 \cdot 5H_2O$/liter.

---

*Saturation proceeds according to an exponential rate law: $c(t) = c^{\text{sat}} \cdot [1 - \exp(-t/\tau)]$ with $1/\tau = a \cdot k_m$; k_m is of the order of 10^{-3} to 10^{-2} cm/sec in well-stirred dispersions, where the mean drop radius amounts to 100 μm. If the volume ratio of aqueous to toluene phase is approximately 1, then the density of exchange surface amounts to approximately 100 cm²/cm³, so that τ is 1 to 10 sec.

INVESTIGATIONS CONCERNING OPTIMAL PROCESS TECHNIQUES FOR THE ANODIC
CONVERSION OF TOLUENE TO BENZALDEHYDE IN TWO-PHASE ELECTROLYTES

The anodic conversion of toluene in one-phase mixed solvents as reported by different authors does not give acceptable current or mass yields nor a pure product. The presence of an organic cosolvent gives rise to polymerization reactions that cause great difficulties in the final workup and purification procedure [40, 41, 43]. In contrast, toluene emulsions in aqueous electrolytes (although introducing serious limitations because of the relatively low solubility of the organic substrate) provide improved selectivity of the oxidation reaction and hence purity of the produced benzaldehyde product. Because of mass transfer limitations, the direct anodic conversion of toluene in two-phase emulsion electrolytes and in the absence of any mediating redox system is not a very successful process technique. Japanese workers [53] oxidized toluene emulsions in 40% HNO_3 at PbO_2 anodes and obtained a product mixture containing benzaldehyde, benzoic acid, nitrotoluene (from chemical nitration), dinitrocresol, and other compounds.

Kato and Sakuma claimed in a patent [54] to have obtained a remarkable improvement of current and mass yields (60%) in direct anodic conversion of dispersed toluene by use of an AC-modulated DC electrolysis where the AC to DC currents had a ratio of 1:8.

Mann and Paulson again tried to solve the problem of mass transfer limitations caused by low solubility of toluene in direct anodic conversion of dispersed toluene by addition of an organic cosolvent. The cosolvent possessed some solubility in water and hence was expected to enhance the water solubility of toluene [55]. Acetic acid seemed to be very suitable for this purpose, but the current yields did not exceed 20%.

In 1925 Kawada [56] reinvestigated the procedure of Lang [42] by performing Mn^{II} to Mn^{III} oxidation and mediated toluene conversion in one single reactor. He obtained promising current and mass yields.

Mitchell [57] was the first to consider systematically the question of mass transfer limitation in anodic conversion of aqueous toluene emulsions. He conducted some disappointing experiments to improve mass transfer in direct anodic conversion by improved agitation and by use of micelle solubilization. Mitchell then rediscovered the concept of Lang [42] involving use of mediators and substitution of the severely hampered mass transfer to the electrodes by mass transfer between the dispersed phase and the bulk of the electrolyte. Mitchell investigated the ability of different mediating systems to perform the toluene oxidation to benzaldehyde smoothly, rapidly, and selectively. $Ni^{II,III}$ oxide and Ag^{II} in alkaline solutions and dichromate and Ce^{IV} and Mn^{III} in strongly acidic solutions were investigated. All solutions except Mn^{III} in 55% H_2SO_4 failed to give satisfactory current yields or a sufficiently pure product (With dichromate benzoic acid was obtained rather than benzaldehyde). Mitchell's work left no doubt that Mn^{3+} in 50 to 60% sulfuric acid really is the mediator of choice for the anodic conversion of toluene to benzaldehyde.

Dey and Maller [58] further showed that by appropriately conducting the Mn^{III}-mediated toluene oxidation in undivided cells at temperatures of 60 to 70°C and with addition of 2% ceric sulfate as a "promoter," a further increase of mass yield (up to 90%) and of current efficiency (up to 45%) for benzaldehyde production could be demonstrated.

With respect to industrial application of the Mn^{III}-mediated toluene oxidation (2.18a and 2.18b), Udupa and co-workers [59] made the salient point that the continuous and steady performance of mediated toluene oxidation in a one-cell/one-reactor system does not work at all, because

$$Ph\text{-}CH_3 + 4Mn^{3+} + H_2O \longrightarrow Ph\text{-}CHO + 4H^+ + 4Mn^{2+} \quad (2.18a)$$

$$Mn^{2+} \longrightarrow Mn^3 + e \quad (2.18b)$$

on prolonged performance of the electrolysis the accumulation of different organic degradation products results in a strong decrease of the anodic oxygen overpotential. As a result current efficiencies for Mn^{2+} reoxidation decline after a short time and become quite low. Therefore, the authors returned to the idea of Lang [42] of separating Mn^{2+} reoxidation (2.18b) and toluene oxidation (2.18a) spatially and recycling the electrolyte into the electrolysis cell only after a purification step that removes all traces of deteriorated organic matter. Furthermore, they proposed the use of Mn^{II} salt concentrations that are so high that at the beginning of the anodic oxidation that Mn^{III} sulfate precipitates out on complete conversion and a slurry of this salt may be used for the chemical redox reaction (2.18a). By this measure the volume of cycled sulfuric acid solutions (55%) is significantly decreased.

Table 2.8 summarizes the most important data published on direct and mediated anodic oxidation of toluene.

Anodic Conversion of Alkyl-, Hydroxy-, Alkoxy-, Halide-, and Pseudohalide-Substituted Benzenes Performed in Two-Phase Electrolytes

Between 1913 and 1937 Fichter and his co-workers published much work on the anodic oxidation of substituted benzenes. This work was mainly devoted to the task of clarifying whether anodic oxidation of these compounds occurs by oxidation of the aromatic nucleus or of the alkyl side chain. The anodic oxidation of the dispersed organic compound in thoroughly agitated aqueous sulfuric acid solutions was used most often. Fichter himself observed that this procedure generally gave poor current efficiencies and mass yields. This occurred because the primary products were exposed to further anodic at-

Table 2.8.   Survey on Electrolysis Conditions and Mass Yields for Anodic Oxidation of Toluene[a] to Benzaldehyde (all work performed in emulsions of toluene in the electrolyte)

Electrolysis Conditions	Anode	Yield (%)	Ref.
50-60% H_2SO_4 + Mn(NH$_4$)(SO$_4$)$_2$ two-step process		80	42
50-60% H_2SO_4 + 10% Mn$_2$(SO$_4$)$_3$, 50-60°C, 5-100 mA/cm^2, 2.8 V, one- or two-step process	PbO_2	60-80	54, 56, 57, 59
60% H_2SO_4 + 1% Ce(SO$_4$)$_2$ + 10% Mn$_2$(SO$_4$)$_3$, 6-7 V, 60-80 mA/cm^2, 60-70°C	PbO_2	90	58
20% HNO_3 + 10% HOAc, 20 mA/cm^2, 4.5 V	Pt	20	55
0.5-2 N H_2SO_4, 20°C, 10 mA/cm^2	PbO_2	Traces of phenol, toluquinone, benzoquinone, and benzaldehyde	2, 4, 44
0.5 N H_2SO_4, 16 mA/cm^2, 20°C, anionic surfactant	PbO_2	13	60
H_2SO_4, H_2O, acetone	Pt	14	43, 61

[a]For further information on toluene oxidation see Refs. 26, 28, 39, 41, 53, 62, and 63.

tack, and the greater part of the already converted arenes was in many cases degraded down to CO_2 and H_2O. These results are then only of very limited value. Fichter therefore frequently supplemented his work on emulsion electrolysis with anodic oxidation of the respective compound in single-phase electrolytes (which generally was an acetic acid/sulfuric acid mixture). Usually the anode material was PbO_2, and in some cases Pt. Anodic degradation was more severe with Pt anodes than with PbO_2.

Table 2.9 and Fig. 2.17 summarize the results of Fichter. As is pointed out above, these results do not really show the relative yield for primary anodic attack at the aromatic nucleus or at the alkyl substituent, but rather reflect the relative stabilities of quinones and other products against consecutive oxidation. This especially holds where total mass yields fall below 20%. Table 2.9 suggests that an increase in the number of attached alkyl groups enhances the ease of anodic attack of the aromatic nucleus [2, 64, 66, 67]. Furthermore the presence of alkoxy and hydroxy groups facilitates nucleus oxidation

Table 2.9. Survey on Products and Product Yields Obtained for the Anodic Oxidation of Different Arenes.

substrate	product, material yield	ref.
(benzene)	(dihydroxybenzene) HO– / –OH 1%	2 44
(toluene)	CHO 23% + COOH 10% + (cresol) OH 6% + (biphenyl diol) HO– 45% + (methyl-benzoquinone) O...O 13% + (methyl-quinone) O...O 1%	64
(m-xylene)	CHO PbO₂: 0.9% Pt: 4% + COOH / COOH 2.8% 2.3% + (quinone) O...O few	4 65
(p-xylene)	COOH 5% + CHO 26% + COOH / COOH 1% + HO– (biphenyl) –OH 45% + –OH 0.5% + (quinone) O...O 10%	2 66 67

Electrolysis conditions: emulsion of depolarizer in dilute sulfuric acid, divided or undivided cell, lead dioxide anodes.

132

Table 2.9. (Continued)

substrate	product, material yield	ref.

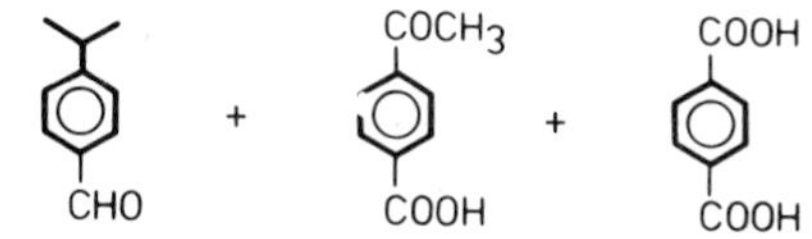

Row 1: substrate (p-isopropyl toluene structure) →
- (toluene with CHO) 8% + (COCH₃ / COOH substituted) 12% + (terephthalic acid COOH/COOH) 12%
ref. 68

Row 2: substrate (pseudocumene structure) →
traces of aldehydes, carbonic acids, quinones, phenols
most pseudocumene converted to CO_2 and H_2O
ref. 69

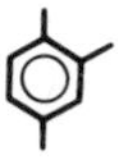

Row 3: substrate (mesitylene structure), results obtained by addition of micelle solubi-lizing agent →
CHO 13% + COOH 3% + COOH/COOH 1% + COOH OH 1% + OH (traces) + OH/OH
ref. 60

Row 4: substrate (o-tolunitrile, CH₃ with CN) →
COOH CN 6%
ref. 66

Row 5: substrate (p-methyl benzonitrile, CH₃ / CN) →
COOH CN 25% + COOH COOH 19%
ref. 66

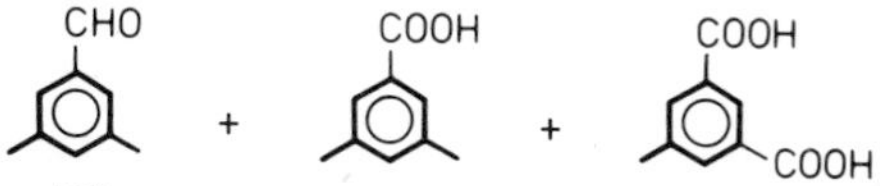

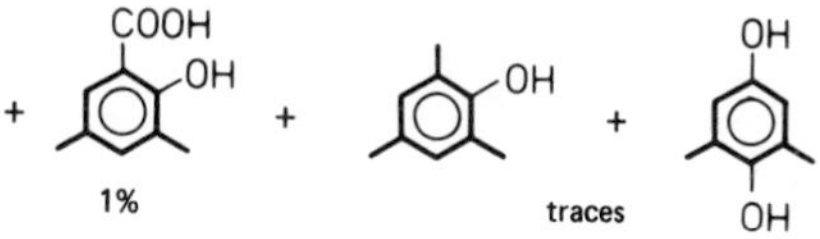

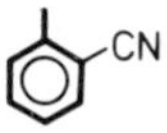

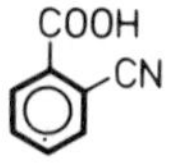

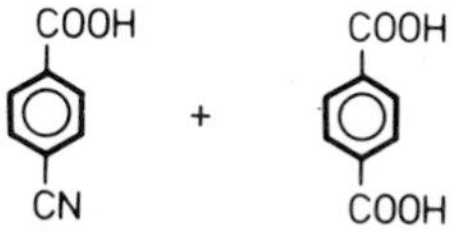

Row 6: substrate (2,4-dimethyl benzonitrile, CN) →
CN COOH 12%, by addition of micelle solubilizing agent
ref. 66

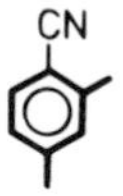

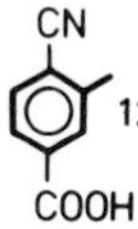

133

Table 2.9. (*Continued*)

substrate	product, material yield	ref.
Cl — benzene (chlorobenzene, para)	Cl-C₆H₄-COOH (para) 34% + Cl-C₆H₄-CHO (para) 10%	71
	up to 100%	72
Cl, Cl — benzene (dichloro, para + ortho)	Cl-C₆H₃(Cl)-COOH 6.5%	71
J — benzene (para)	J-C₆H₄-COOH + J-C₆H₄-CHO few	73
J — toluene (ortho)	J-C₆H₄-COOH (ortho) + J-C₆H₄-CHO (ortho) few	73
NO₂ — toluene (ortho)	degradation	74
NO₂ — toluene (meta)	degradation	74
NO₂ — toluene (para)	C₆H₄(COOH)(NO₂) para, COOH 2%	74
	34%	72

Table 2.9. (*Continued*)

substrate	product, material yield	ref.
CHO ⬡ NO$_2$	COOH ⬡ 60% NO$_2$	74
CH$_2$OH ⬡ NO$_2$	COOH ⬡ 10% NO$_2$	74
⬡–NO$_2$ NO$_2$	COOH ⬡–NO$_2$ 30% NO$_2$ 75%	75 76
O$_2$N–⬡–NO$_2$ NO$_2$	COOH O$_2$N–⬡–NO$_2$ 30% NO$_2$	75
⬡–OMe	O=⬡=O + HO–⬡–⬡–OMe 20% 30%	77
⬡–OMe	O=⬡=O + HO–⬡–⬡–OMe 30% 20%	77
⬡ OMe	CHO ⬡ + COOH ⬡ + ⬡–⬡–OMe OMe OMe HO 35% 8% 35%	77

Table 2.9. *(Continued)*

substrate	product, material yield	ref.
	6% few	2 78
	few	2 78
	9% 88%	67
	33%	65
in 0–xylene	20% 6% 76%	64
	6%	2

Table 2.9. (*Continued*)

substrate	product, material yield	ref.
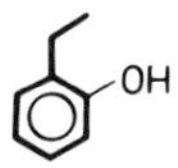	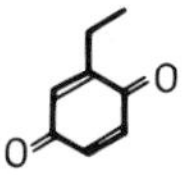 10%	79
	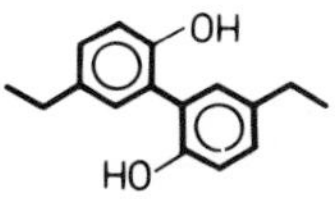 12%	79
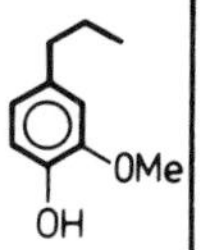	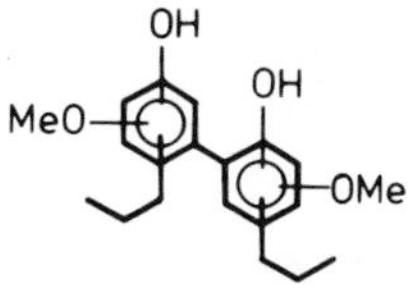 very few	80
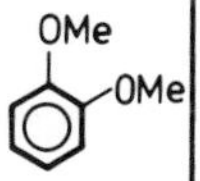	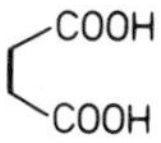 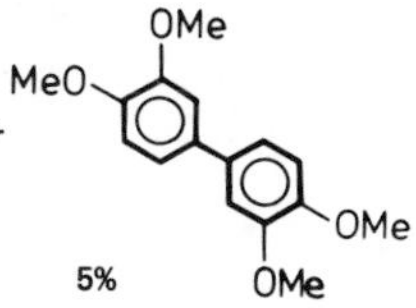 20% 5%	81
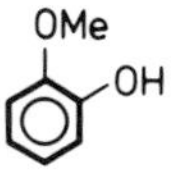	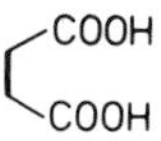 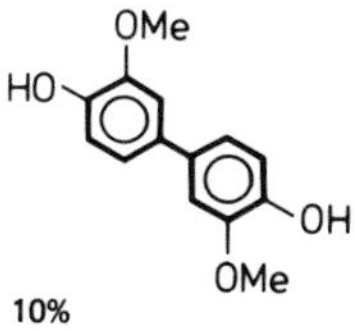 25% 10%	81
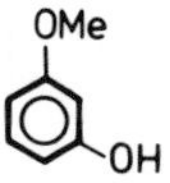	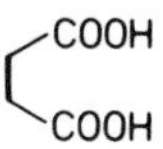 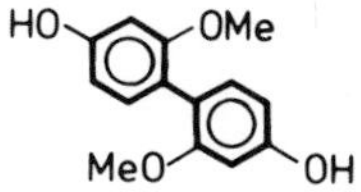 15% 5%	81

137

Table 2.9. (*Continued*)

substrate	product, material yield	ref.
(ethylbenzene structure)	CHO (structure) + COCH$_3$ (structure) few	79
CN (structure)	COOH (structure) 24%	66
(propenyl-OMe benzene structure) OMe	CHO (structure) OMe 52% + COOH (structure) OMe 25%	80
(propenyl-OMe benzene structure) OMe	CHO (structure) OMe OMe 78% + COOH (structure) OMe OMe 12%	80
(structure) COOH	(structure) COOH HO	82
CN (structure)	CN (structure) OH HO traces	66
CHO (structure)	COOH (structure)	Pt: 11% PbO$_2$: 60% 97%	45 83
COCH$_3$ (structure)	COOH (structure) COOH + COOH (structure) HOOC few	84

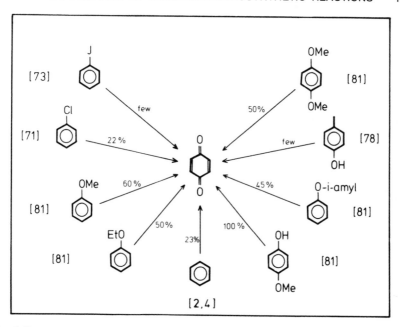

Fig. 2.17. Mass yields for anodic benzoquinone production as reported by Fichter and co-workers.

even more. Biphenyl compounds are formed frequently in addition to quinones.

It is now well known that hydroxy- and alkoxy-substituted benzenes are much more accessible to anodic oxidation than benzene, so that the anodic working potential for their oxidation may be kept much lower than that for benzene or alkylbenzenes. Consequently, yields for the anodic oxidation of hydroxy- and alkoxybenzenes generally are better. Because of the relative stability of radical intermediates, anodic dimerization products are often obtained in relatively high yields [2, 64–67, 77, 81].

In general, the anodic oxidation of methylbenzenes (even under conditions favoring the production of quinones as most stable products) yields products that are formed by methyl group oxidation [64–71]. If compounds in which the aromatic nucleus is preoxidized (i.e., hydroxy- and alkoxybenzenes) are used as starting materials then further oxidation of the nucleus is favored and quinones or phenols are obtained in relatively good yield. Nucleus oxidation with biphenyl formation seems to be favored for *o*- and *p*-xylene.

Alkylbenzenes with alkyl groups other than methyl are oxidized at the benzylic carbon atom. This is expected, because the product of the primary charge transfer in oxidation of any alkylbenzene is a benzyl radical cation or—after proton loss—a benzyl radical (2.19). Further oxidation of this radical yields a ketone, or, by subsequent C-C fission, eventually leads to the formation of benzaldehyde and benzoic acid [68, 79]. Unsaturated side chains with π-electron systems that are in conjugation to the arene π-system are even more vulnerable to C-C fission because of the ease of oxidation of these C-C double bonds. Their oxidation yields γ, β-diols, which are easily oxidized by C-C bond fission [80]. Only benzaldehyde or benzoic acid is found as the oxidation product. Benzyl cyanide is also oxidized to benzoic acid, but hydrocinnamic acid is quite unexpectedly oxidized at the nucleus [66, 82].

Because of the enhanced water solubilities of benzyl alcohol and benzaldehyde (compared with arenes and alkylarenes) and because of the relative stability of benzoic acid to anodic attack and degradation, it is not surprising that using Fichter's emulsion electrolysis gives good mass yields for the formation of benzoic acid by anodic oxidation of benzyl alcohol and benzaldehyde [45, 83].

$$(2.19)$$

The anodic attack of halobenzenes is very inefficient and results in anodic fission of the carbon/halogen bond and the formation of quinones (or hydroquinones if the anodic oxidation is performed in undivided cells). Benzonitrile is relatively inert toward oxidation and is oxidized to the cyano benzohydroquinone (although only in minute amounts) [66].

A comparison of the different results for mediated and direct anodic oxidation of the alkyl chains of alkylarenes based on processing emulsions of the arene substrates in aqueous electrolytes suggests clearly that mediated oxidation is the method of choice from any point of view. Higher mass and current yields and better selectivities may be obtained. This may be demonstrated by one final example (one of many published). Sachs and Kempf prepared 2,4-dinitrobenzoic acid from 2,4-dinitrotoluene with 30% mass yield and approximately 8% current efficiency according to the method later used by Fichter [75]. For the same synthesis Brown and Brown [76] compared dichromate-mediated and direct anodic conversions of 2,4-dinitrotoluene and found a mass yield of more than 70% and a current yield of approximately 50% for the mediated oxidation.

Finally, as a case of misleading and "misled process development" the "electrochemical" nitration of naphthalene dispersed in nitric acid of medium strength [85, 86] should be mentioned. Nitration of naphthalene is not based on anodic oxidation, but as Fichter and Plüss showed [87], is simply a chemical nitration induced by an increase of HNO_3 concentration in the anolyte due to ion migration. Furthermore, the rate of chemical nitration is increased because of dissipation of Joule heating in the electrolysis cell [87].

Examples of anodic conversions of aromatic compounds that are conducted by micelle solubilization of the depolarizer are dealt with in a later section.

Cathodic Reduction of Aromatic Nitro Compounds in Thoroughly Agitated Aqueous Electrolyte/Nitro Compound Mixtures

General Considerations

The cathodic reduction of aromatic nitro compounds has been under investigation since the end of the last century. Because of the potential use of this reaction to produce a wide variety of products that are of great interest as raw materials for the dye or pharmaceutical industries, it remains the most extensively studied electroorganic synthetic reaction. Most of the theoretical studies (electrode kinetics as well as preparative work) have been performed in single-phase electrolytes. As the results are only of limited value for the performance of electrolysis in two-phase electrolytes [88-94], we do not attempt here to summarize them all, and the reader is referred to some recent reviews published elsewhere [95-98].

In this section we describe mechanistic details only as far as it is necessary to clarify process techniques and process engineering considerations.

Nitrobenzene, the most important aromatic nitro compound, has only a relatively low solubility in water. Nevertheless, at elevated temperatures (Fig. 2.18) the solubility is sufficient to process this compound at moderate current densities and space time yields as a dispersed phase in thoroughly agitated aqueous electrolytes. As early as 1904, F. Haber [101] and C. Schmidt proposed a consistent model for the different reaction paths for nitrobenzene reduction (Fig. 2.19). Although established on the basis of synthetic experience gained in water/alcohol single-phase electrolytes the proposal holds for the process in two-phase electrolysis. According to this scheme [102-106] nitrobenzene is reduced first to nitrosobenzene and subsequently to phenylhydroxylamine. The hydroxylamine is a relatively stable intermediate, being reduced further only upon double protonation:

$$Ph-NO_2 \xrightarrow[-H_2O]{+2e, +2H^+} Ph-NO \xrightarrow[+2H^+]{+2e} Ph-NH\,OH \qquad (2.20)$$

Although phenylhydroxylamine is relatively stable to further cathodic attack it is not stable chemically. In acidic solutions it undergoes a rearrangement to p-aminophenol (2.19), which then is stable against both chemical

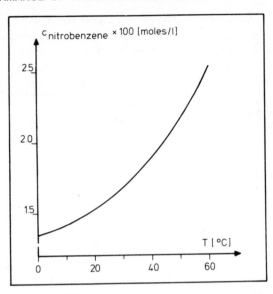

Fig. 2.18.   Dependence of nitrobenzene solubility in water on temperature [99, 100].

conversion and cathodic attack. Thus solutions of low pH yield *p*-amino-phenol or aniline. Phenylhydroxylamine may be obtained if the pH of the electrolyte is kept buffered between 5 and 9, and a relatively high current density and a low ratio of electrolyte volume to electrode surface are maintained. In this manner the saturation concentration of phenylhydroxylamine is quickly reached and the product is protected from further chemical and electrochemical conversion by precipitating out [107–110].

$$\text{Ph-NH-OH} \xrightarrow{\text{H}^+} p\text{-HO-Ph-NH}_2 \qquad (2.21)$$

In basic solutions the reaction scheme of Haber shows the typical hyroxyl-ion-catalyzed condensation of phenylhydroxylamine and nitrosobenzene (2.22a) to azoxybenzene, which subsequently may be further reduced to azo-benzene (2.22b) and hydrazobenzene (2.22c). The acid-catalyzed rearrangement of hydrazobenzene to benzidine made nitrobenzene reduction to hydrazobenzene economically very attractive.

$$\text{Ph-NH-OH} + \text{Ph-NO} \longrightarrow \text{Ph-N=N(O)-Ph} + \text{H}_2\text{O} \qquad (2.22a)$$

$$\text{Ph-N=N(O)-Ph} + 2\text{e} + 2\text{H}^+ \longrightarrow \text{Ph-N=N-Ph} + \text{H}_2\text{O} \qquad (2.22b)$$

$$\text{Ph-N=N-Ph} + 2\text{e} + 2\text{H}^+ \longrightarrow \text{Ph-NH-NH-Ph} \qquad (2.22c)$$

$$\text{Ph-NH-NH-Ph} \xrightarrow{\text{H}^+} \text{H}_2\text{N-Ph-Ph-NH}_2 \qquad (2.22d)$$

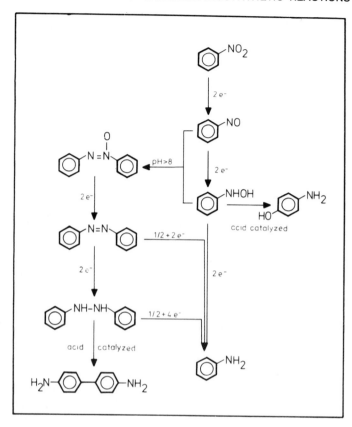

Fig. 2.19. Haber scheme for cathodic reduction of nitrobenzene [101, 107].

There are some doubts about the Haber mechanism for N-N coupling in strongly alkaline electrolytes supplied with nitrobenzene by intensely agitated nitrobenzene emulsions. From investigations in aqueous/alcoholic solutions it is well known that the coupling process (2.22a) is a relatively slow reaction [109]. Furthermore, azoxybenzene and azobenzene are less soluble in water than in nitrobenzene and thus would be expected to precipitate out and clog the electrode. Such precipitation would interrupt the cathodic process. This behavior, however, is not observed at all. Instead, only clogging by the hydrazobenzene, the final product, becomes a problem.

An interesting investigation by Russel and Geel (although performed in alcoholic solutions and hence with only limited reference to reactions in aqueous emulsions) points toward a quite different reaction path leading to N–N coupling [110]. They showed that base-catalyzed condensation of arene hydroxylamines and arene nitroso compounds proceeds by way of the formation of arene nitroso anions, which readily dimerize and dehydrate to azoxybenzene. This compound is reduced immediately to azobenzene and hydrazobenzene. Since nitrosobenzene possesses a cathodic half wave potential that is more anodic than that of nitrobenzene [111], one would expect that in alkaline solutions the reduction of nitrobenzene would proceed to the nitrosobenzene anion, from which dimerization and further cathodic reduction would yield azobenzene and hydrazobenzene (2.23c). This hypothesis is supported by the investigations of Heyrovsky and Vavricka, who observed direct formation of N–N coupled products at mercury cathodes.

$$2\text{Ph–NO}^- \longrightarrow \text{Ph–N(O}^-)\text{–N(O}^-)\text{–Ph} \xrightarrow[-\text{H}_2\text{O}]{+2\text{H}^+} \text{Ph–N(O)=N–Ph} \quad (2.23a)$$

$$\text{Ph–N(O)=N–Ph} + 4\text{H}^+ + 4e \longrightarrow \text{Ph–NH–NH–Ph} \quad (2.23b)$$

Starting reaction: $\text{Ph–NO}_2 + 2\text{H}^+ + 3e \longrightarrow \text{Ph–NO}^- \cdot + \text{H}_2\text{O} \quad (2.23c)$

It should be pointed out that all these reduction products of nitrobenzene and nitroarenes respectively could be (and have been) obtained by performing the electrolysis in single-phase electrolytes.

At first glance, it does not seem necessary to use two-phase electrolytes. This requires processing nitrobenzene emulsions in aqueous electrolytes and dealing with the difficulties involved in mass transfer hindrance and stabilization of two-phase mixtures by intense stirring. The reasons for adopting this seemingly inconvenient choice, however, are compelling:

1. Higher electrical conductivity and hence reduced cell voltages for aqueous electrolytes.

2. Improvement of selectivity of the electrosynthesis, because a higher range of pH and temperature is accessible in aqueous as compared to nonaqueous solutions.

3. Ease of workup and purification of products; by simple decantation or filtering for poorly soluble products, or by extraction or steam distillation to remove coproduced aniline.

For aniline synthesis, there is little chance to substitute the well-established nitrobenzene reduction to aniline by iron (which of course is also based on an electrochemical reaction, see (2.24) [112], by an electrochemical process that needs externally applied electrical current.

$$\text{Ph–NO}_2 + 2\text{Fe} + \text{H}_2\text{O} \longrightarrow \text{Ph–NH}_2 + \text{Fe}_2\text{O}_3 \quad (2.24)$$

However, cathodic aminophenol and benzidine synthesis might have a chance for industrial application. Indeed, as an emergency measure during World War I, p-aminophenol is reported to have been produced electrochemically at Kodak in the United States [113]. Dey and co-workers worked out the fundamental data and the flow sheets for cathodic p-aminophenol and benzidine production [115, 116]. Whether Dey's work on hydrazobenzene (or benzidine) production is economically promising is somewhat doubtful, because a well-established chemical hydrazobenzene process for the production of benzidine exists [114]. It might, however, be attractive to install and run electrochemical processes to produce important materials for the dyestuff and pharmaceutical industries whenever only small-scale production is necessary because of limited demand. This very certainly is the basis for the remarkable efforts that have been made by Dey and co-workers (Indian Central Electrochemical Research Institute) [115–131] and by the Japanese group of Sekine [132–140] on cathodic nitrobenzene reduction processes.

Cathodic Production of Phenylhydroxylamine

As is pointed out above, in acidic aqueous electrolytes the first product that is obtained and that is relatively stable against further cathodic conversion is phenylhydroxylamine. However, it is not stable chemically and is rearranged by an acid-catalyzed reaction to p-aminophenol. Ingold and co-workers [141], who investigated this isomerization reaction more closely, stated that this reaction involves a quinoid intermediate that allows one to introduce different nucleophiles into the ortho and para positions of the nucleus. Such a nucleophile might be the hydroxy group, but other nucleophiles are known to participate in this reaction also (2.26). Thus the phenylhydroxylamine/p-aminophenol rearrangement is only one of several different possible reactions that follow the dehydration of the protonated phenylhydroxylamine.

The kinetic data for the phenylhydroxylamine/aminophenol rearrangement have been discussed by Ingold and co-workers, who state that the rate of this rearrangement is first order with respect to the substrate and to protons. Unfortunately, no data have been published on rate laws and rate constants. It may, however, be stated that phenylhydroxylamine can only be prepared by cathodic nitrobenzene reduction if:

1. The reduction is performed in buffered aqueous solutions the pH of which is not lower than 5 and not higher than 9 (because N–N coupling must be prevented too).

2. The processing-time is not too long.

3. The product is continuously separated from the electrolyte solutions (for instance by continuous precipitation of the slightly soluble phenylhydroxylamine).

Haber [101] reported the production of phenylhydroxylamine using an aqueous/alcoholic electrolyte with ammonia/ammonium chloride buffer. Five years later, Brand [108] reported the preparation of phenylhydroxylamine by use of cooled emulsions of 50 g of nitrobenzene in 300 to 400 ml of water, with 20 g of sodium acetate and 15 g of glacial acetic acid as buffer. The hydroxylamine crystallized in very pure form while the electrolysis still was being run.

Aminophenols by Cathodic Reductions of Nitroarenes

In 1893 Gattermann [142] showed that in a single-phase electrolyte (concentrated sulfuric acid) nitrobenzene can be converted to p-aminophenol. A plant that used this procedure was designed and operated by the Eastman Kodak Company during World War I [143]. The process is accompanied by product sulfonation and may not be used for products that are more reactive with respect to sulfonation than p-aminophenol. Therefore, Darmstädter proposed the use of more dilute 50% H_2SO_4 as electrolyte and the processing of thoroughly stirred nitrobenzene emulsions [144].

$$NH_2OH^+ \quad\quad NH \quad\quad NH_2$$

(2.26)

In 1932 Brigham and Lukens [145] published an investigation concerning optimal process conditions and reported on mass and current yields that may be obtained under different electrolysis conditions. They used 50% sulfuric acid, a volume ratio of dispersed nitrobenzene to aqueous phase of 0.31, and current densities that did not exceed 34 mA/cm^2. Current and mass yields were around 60% at working temperatures between 25 and 60°C. Side product was predominantly aniline. All other workers who also used comparably high or even higher sulfuric acid concentration were not able to solve the problem of reducing the amount of coproduced aniline (Imray [146–148], Shoji [149, 150], Thatcher [151], McKee and Gerapostolou [152], Mann et al. [153]). Dey and co-workers [115] returned to the idea of Darmstädter and significantly decreased the sulfuric acid concentration (20 to 40%), thereby reducing the degree of conversion of the phenylhydroxylamine to aniline. To compensate for the slower rearrangement rate of the hydroxylamine due to decreased acidity, they raised the working temperature up to 80 to 90°C and were thus able to obtain yields for different aminophenols ranging between 30 to 60%. Industrial-scale aminophenol production mainly carried out with this process technique is reported to produce the aminophenol in 73% mass yield [154].

Wilson and Udupa extended these investigations to 2-chloro-6-nitrotoluene (yielding 3-methyl-2-chloro-*p*-aminophenol and 2-methyl-3-chloroaniline as by-products). Enhancing mass transfer by using a rotating electrode enabled them not only to increase the total current efficiency to 100%, but additionally allowed them to improve the selectivity up to 75%. This compares to 50% selectivity obtained in an electrolysis cell equipped with a stationary electrode and being stirred externally [155]. At the rotating cathode a relatively broad range of current densities (10 to 40 mA/cm²) may be used without any great influence on mass yields. Current yields, however, are only as high as 90 to 95% if amalgamated rotating copper cathodes are used. Thus a relatively high overpotential for hydrogen evolution must be established to obtain high current efficiencies for cathodic aminophenol formation. Although the work of Wilson and Udupa covers the important process parameters needed to optimize reaction conditions, their paper does not contain fundamental data such as rearrangement kinetics, half-wave potentials of nitrobenzene and hydroxylamine in the electrolyte, mass transfer coefficient for the rotating cathode, and so on. Only through knowledge of these data would it be possible to make a sound electrochemical process-engineering approach and thus find optimum conditions for this electrosynthesis. The most important problem to be solved is matching of the electrochemical conversion rate, mass transfer rate, and rate of rearrangement of phenylhydroxylamine. Thus, phenylhydroxylamine must be removed as fast as possible from the cathode, and this is done by improving mass transfer versus electrochemical conversion rate. An excess of dispersed nitrobenzene should be avoided. Otherwise phenylhydroxylamine will be extracted into the organic phase. Being dissolved in the nitrobenzene phase it would be protected from the acid-catalyzed rearrangement but not from coupling to form azoxy, azo, and hydrazo compounds [155]. Table 2.10 summarizes the mass yields obtained by different authors for cathodic production of different *p*-aminophenols under optimum conditions by using dispersion electrolysis.

An interesting contribution of Winslow [156] concerning aminophenol synthesis makes use of a porous carbon electrode working as a wicking electrode for the cathodic reduction of nitrobenzene in acidic aqueous solutions. Without any additional means to improve mass transfer, nitrobenzene may be converted at the three-phase boundary (carbon cathode/nitrobenzene/electrolyte) at relatively high current densities. Winslow used a 2 *N* H_2SO_4 solution as catholyte, applied a constant formal current density of 46 mA/cm², and claimed 90% mass yield with an overall current efficiency of 50% at working temperatures of 29 to 42°C. His results, however, cannot be assumed to be very reliable because he did not isolate the products, but determined them by a relatively uncertain and questionable titration procedure.

Table 2.10.   Reduction of Aromatic Nitro Compounds to *p*-Aminophenols Electrolysis conditions: amalgamated monel or copper cathode; catholyte, 20–40% H_2SO_4; current density, 25–50 mA/cm^2; 70–90°C; substrate finely dispersed by stirring.

Substrate	Product	Material Yield (%)	Ref.[a]
Nitrobenzene	*p*-aminophenol	60	115
		43	149, 150
		72	155
		65	145
o-Nitrotoluene	4-Amino-*m*-cresol	46	125
		74	155
m-Nitrotoluene	4-Amino-*o*-cresol	55	126
o-Chloronitrobenzene	3-Chloro-*p*-aminophenol	29	128
		76	155
2,5-Dichloronitrobenzene	2,5-Dichloro-*p*-aminophenol	72	155
m-Dinitrobenzene	2,4-Diaminophenol	45	129, 130
4-Chloro-2-nitrotoluene	2-Chloro-5-methyl-*p*-aminophenol	82	155
2-Chloro-6-nitrotoluene	2-Chloro-3-methyl-*p*-aminophenol	72	155

[a]For further information on aminophenol formation see Refs. 144, 146 to 154, and 156.

The work of McKee and co-workers on the performance of cathodic reduction of micelle-solubilized nitroarenes in acidic, neutral, and basic electrolytes is discussed later.

Cathodic Reduction of Nitroarenes to Azoxy-, Azo-, and Hydrazo Compounds Using Two-Phase Electrolytes

Aromatic nitro compounds, if cathodically reduced in neutral or basic electrolytes, yield *N*-coupled species as the main products, namely, azoxy, azo, and hydrazo compounds. This holds for emulsions of the depolarizer in the electrolyte, as well as for single-phase electrolytes. Azoxy compounds, however, being readily reduced to azo and hydrazo compounds, can only be obtained if certain precautions and measures are taken:

1.   The starting water-insoluble nitro compound should neither be processed in large excess nor should it be dissolved in an inert water-insoluble organic solvent. In both cases the azoxy compound would be extracted into the dispersed organic phase where it would be temporarily stored and resupplied for further cathodic reduction to the electrolyte and cathode, respectively.

2. The azoxy compound should possess—because of the absence of hydrophilic substituents—low water solubility.

3. A low hydrogen overvoltage cathode should be used. Thus under normal cathodic working conditions evolution of hydrogen would occur rather than the reduction of azoxy compounds to azo and hydrazo compounds. Under these conditions the sparsely water-soluble azoxy compound precipitates out from the electrolyte, protecting it from further reduction. According to this process technique, *p*-nitroanisole is converted into azoxy-*p*, *p*′-anisidine in almost quantitative mass yield [157], whereas the reduction of *m*- and *p*-nitrotoluene yields only 40% of the respective azoxy compounds [158].

Löb suggested that the cathodic production of azoxybenzene is one step in a two-step process to produce benzidine [159]. The further reduction to hydrazobenzene is carried out after acidification of the electrolyte [160]. A detailed investigation by Kerns (which was performed in 25% caustic soda as electrolyte at 25 to 30°C and with a constant current density of 25 mA/cm^2 in well-stirred emulsions) showed that 80% mass yield and 50% current efficiency could be obtained for the production of azoxybenzenes at lead and nickel cathodes [161].

The papers of McKee and Brockmann on the synthesis of azoxy and azo arenes by cathodic reductions of micelle-solubilized nitrobenzene or nitroarenes are reported later.

Most preparative work aimed at the production of hydrazobenzene (or hydrazoarenes) and benzidines, respectively, avoided separate formation of azoxy- and azoarenes. This can be achieved by using cathodes with high hydrogen overvoltage, such as a lead [162], Monel, or zinc cathode [160, 163], or by using electrodes plated with tin or lead [164, 165] and adding an organic solvent with low water solubility to keep the azoxy or azo intermediate compounds in solution [166].

Detailed studies on the influence of numerous process parameters on mass and current yields for hydrazoarene production were carried out by Sekine and Sugino [132–140]. These authors report material yields as high as 95% and current efficiencies around 90% for the reduction of nitrobenzene to hydrazobenzene in 10% aqueous NaOH. According to their experience iron, zinc, or lead cathodes coated with spongy layers of tin or lead were most effective. As organic solvent benzene or toluene was recommended and vigorous stirring was said to be essential. The authors were convinced that the reaction mechanism of electrochemical reduction proceeds up to the azoxy stage. They assumed that the further reduction steps were based on chemical redox reactions of azoxy and azo compounds with the spongy metal layer, which is electrochemically deposited and chemically dissolved repeatedly [137].

Ter-Minasyan [167] investigated the same synthesis and found mercury, copper, and lead to be the most effective cathode metals.

Dey and co-workers reported excellent results whenever a spongy lead deposit on the cathode was formed. This spongy surface layer was deposited cathodically *in situ* or by addition of lead oxide to the alkaline catholyte [116–124]. The authors reduced numerous nitro compounds at iron cathodes activated by lead deposition in 10% NaOH. After initial activation the cathodes could be reused repeatedly without any further lead deposition. Furthermore, they reported that the nature of the cathode support metal has a strong influence on structure and activity of the spongy lead layer. Iron as support metal gave optimal performance. To avoid clogging of the cathode by the precipitating hydrazoarene, it was important to add benzene or toluene after 90% conversion. These solvents dissolved hydrazobenzene and prevented its precipitation at the cathode.

Table 2.11 summarizes the mass yields for cathodic hydrazobenzene synthesis as reported by Dey el al. and Sekine and co-workers. In a thorough study [116] of cathodic hydrazobenzene production from nitrobenzene, Dey and co-workers compared different process techniques. Nitrobenzene reduction in homogeneous alcoholic solutions and micelle-solubilized nitrobenzene conversion in aqueous electrolytes were disadvantageous. The organic solvent demands process operation with uneconomically high cell voltages, and according to Dey, the unavoidable losses of micelle-solubilizing salts of organic anions experienced on prolonged process operation make this process technique economically rather unattractive. Like *p*-aminophenol production, hydrazobenzene (and benzidine) production has been tested on an industrial scale. This industrial process is discussed in the last section of this chapter.

Cathodic Reduction of Nitroarenes to Anilines Using Two-Phase Electrolyte Emulsions

The electrochemical reduction of nitroarenes to aminoarenes is dealt with at the end of this section, since aminoarenes are the fully reduced *N*-substituted arene species and because the cathodic reduction of nitroarenes to aminoarenes is the least promising synthesis to produce anilines [169]. This statement is valid not only regarding the established synthetic techniques, but even more from an economical point of view. Today aniline is produced industrially from nitrobenzene either by heterogeneously catalyzed gas phase hydrogenation with elemental hydrogen as reductant, or by an older process: reduction of nitrobenzene in acidic solutions with elemental iron. The latter is performed by dispersing nitrobenzene together with iron filings in aqueous acidic $FeCl_2$ solutions. Batchwise processing of the reduction is usual. Iron (II) ions act as mediating reductants and thus favor complete conversion of the nitro group to the amino group (see below).

Table 2.11. Reduction of Aromatic Nitro Compounds to Hydrazobenzenes[a] Electrolysis conditions: spongy lead-plated zinc, iron, or lead cathode; catholyte, 10% NaOH; current density, 20–30 mA/cm^2; 70°C; substrate finely dispersed by stirring; organic solvent, benzene, toluene, or xylene.

Substrate	Product	Material Yield (%)	Ref.
Nitrobenzene	Hydrazobenzene	90[b]	116, 168
		95	132–137
o-Nitrotoluene	o,o'-Hydrazotoluene	65[b]	122
p-Nitrotoluene	p,p'-Hydrazotoluene	80	138
o-Nitroanisole	o,o'-Hydrazoanisole	78[b]	123
		60	138
o-Chloronitrobenzene	o,o'-Dichlorohydrazobenzene	80[b]	117, 124
		?	138
2,5-Dichloronitro-benzene	2,2', 5,5'-Tetrachloro-hydrazobenzene	85[b]	118, 120
2-Nitro-4-chloro-anisole	2,2'-Dimethoxy-5,5'-dichloro-hydrazobenzene	?	121
2-Nitro-4-chloro-phenetole	2,2'-Diethoxy-5,5'-dichloro-hydrazobenzene	72[b]	119

[a] For further information on hydrazobenzene formation see Refs. 162 to 168 and 189.
[b] These yields given by Dey et al. already include the conversion of the hydrazobenzenes to the respective benzidine derivatives.

For successful cathodic reduction of nitrobenzene to aniline (or of other nitroarenes to aminoarenes, respectively), a relatively low pH of the solution must be maintained to achieve reduction beyond the hydroxylamine stage. Because of the low pH values, rearrangement to aminophenol is unavoidable. The problem is optimal matching of chemical conversion rates with mass transfer and electrochemical conversion rates to get a high amine versus aminophenol ratio in the product mixture. Almost quantitative yields may be realized for para-substituted nitroarenes where the isomerization is prohibited sterically.

Irrespective of the particular choice of different process parameters, Brigham and Lukens [145] obtained an almost constant ratio of aniline to p-aminophenol of 1:2. They used 50% sulfuric acid and nickel cathodes. Imray [146–148] obtained a ratio of aniline to aminophenol of 3:2 at lead cathodes in strongly acidic sulfuric acid solutions. For reduction of para-substituted nitroarenes, however, nearly 100% mass yields for the amino product may be expected and are in fact obtained. Dey et al. [127] reported a mass yield of 82% for 2,4-diaminophenol by cathodic reduction of 2,4-dinitrophenol. Bradt and co-workers [170, 171] reported nearly quantitative mass yields for the amines in the cathodic reductions of nitrophenetole ($EtO-Ph-NO_2$) and 3,5-dinitro-o-cresol, respectively.

The hydroxylamine/p-aminophenol rearrangement can be avoided to a much larger extent if mediated reduction is used. In a combined direct and mediated reduction it is possible to reduce the phenylhydroxylamine intermediate on its diffusive path away from the electrode in a homogeneous redox reaction before rearrangement may proceed.

$$PhNHOH + 2Red^- + 2H^+ \longrightarrow PhNH_2 + H_2O + 2Red. \quad (2.27)$$

The most successful attempts to improve amine yields were reported by Chilesotti [172] and Otin [173]. Chilesotti more or less tackled the problem intuitively be testing "hydrogen carriers" such as $CuCl_2$ and $SnCl_2$. Some of these hydrogen carriers, however, do not work at all as mediators for cathodic reductions. Because of the positive standard potential of the Cu^I/Cu^{II} couple the Cu^I ion does not reduce phenylhydroxylamine to aniline. Copper deposited on the cathode may change the electrode kinetic conditions for cathodic phenylhydroxylamine reduction in a favorable way. Otin demonstrated that the conditioning of the cathode surface was of some importance. He reduced nitrobenzene in 95% mass yield at lead "wool" cathodes or with cathodes made of very fine nickel wires.

Dey and co-workers [125, 126, 128] reduced o- and m-nitrotoluene and o-nitrochlorobenzene to the respective amino compounds. Yields of more than 90% were obtained only for o-toluidine.

Anantharaman and Udupa were aware of the possibilities of using appropriate cathodic mediators. They were able to perform the conversion of aromatic amines from the respective nitrocompounds by Ti^{III}-mediated reduction in nitro-compound emulsions on a semiindustrial scale (1000 A) [174]. Material yields and current efficiencies reported by the authors are excellent (see Table 2.12). The results of Dey, Chilesotti, and Otin are included in Table 2.12.

In addition to metal ions, ketones (or ketyl-radical anions) may be used as mediators for cathodic reduction. Mann et al. [153] reduced 2-nitro-p-cymene in aqueous dispersions at carbon and graphite electrodes at current efficiencies of more than 90%. At metal electrodes, such as monel, copper, lead, and cadmium electrodes, these high yields could be verified only after addition of small amounts of aromatic ketones. In this case, however, the arylhydroxylamine reduction was not very successful, since a nearly 50 : 50 mixture of aminocymene and aminothymol was obtained.

The cathodic reduction of nitrobenzene dispersed in 37% hydrochloric acid at platinum cathodes was reported by Löb [175] to yield o- and p-chloro-

Table 2.12. Reduction of aromatic nitro compounds to anilines[a] Electrolysis conditions: copper or nickel cathode; catholyte, dilute sulfuric acid containing salts of Sn, Cu, Pb, Hg, Fe, or Cr; current density, 40 to 100 mA/cm^2; 25 to 75°C; substrate finely dispersed by stirring. Best results by mediated reduction using electrochemically generated Ti^{3+} [174]

Substrate	Product	Material yield (%)	Ref.[a]
Nitrobenzene	Aniline	85–95	172,173
		95	174
o-Nitrotoluene	o-Toluidine	97	172,173
		93	125
p-Nitrotoluene	p-Toluidine	96	172,173
m-Nitrotoluene	m-Toluidine	?	126
o-Chloronitrobenzene	o-Chloroaniline	24	128
		80	172,173
m-Nitroaniline	m-Phenylenediamine	99	172,173
m-Dinitrobenzene	m-Phenylenediamine	99	174
2,4-Dinitrotoluene	2,4-Diaminotoluene	97	174
o-Nitrophenol	o-Aminophenol	97	174
p-Nitrophenol	p-Aminophenol	99	174
p-Nitrophenetole	p-Aminophenetole	100	170
3,5-Dinitro-o-cresol	3,5-Diamino-o-cresol	100	171

[a]For further information on aniline formation see Refs 145 to 148, 153, and 169.

aniline. This synthesis is related to the acid-catalyzed phenylhydroxylamine/ p-aminophenol rearrangement. However, in concentrated hydrochloric acid, chloride anions act as nucleophiles instead of H_2O:

Although somewhat out of context with this section, the cathodically initiated aniline synthesis by mediated electrochemical generation of aminyl radicals and their addition to benzene [93] in aqueous emulsions should be mentioned here.

$$Cu^+ + 2VO^{2+} + 4H^+ \longrightarrow Cu^{2+} + 2V^{3+} + 2H_2O \qquad (2.30a)$$

$$2V^{3+} + 2\,NH_2\,OH \longrightarrow 2VO^{2+} + 2NH_2^\bullet + 2H^+ \qquad (2.30b)$$

$$2NH_2^\bullet + 2\,Ph + Cu^{2+} \longrightarrow 2\,Ph{-}NH_2 + Cu^+ + H^+ \qquad (2.30c)$$

Current yields for benzene and toluene amination of 60 and 30% respectively are reported.

Electrochemical Conversion of Micelle Solubilized Organic Substances with Low Solubility in Aqueous Solutions and Phase Transfer Assisted Electrochemical Conversions of Sparsely Soluble Substrates

Around 1930 McKee published his first results on the improvement of water solubilities of poorly soluble organic substances in concentrated aqueous solutions of alkali salts of some organic anions such as toluenecymene- and m-xylene sulfonate. He made very efficient use of this effect for the electrochemical conversion of sparsely soluble depolarizers [83, 152, 176, 177]. This solubilizing effect of lipophilic ions is now understood as being due to micelle formation of the respective lipophilic anions and the incorporation of molecules of sparsely water-soluble organic compounds in these micelles (see for instance Ref. 178). Table 2.13 summarizes some of McKee's results with respect to saturation concentrations of the micelle-solubilizing salts. The enhanced solubilities of the poorly soluble depolarizers in these solutions shows that the solubilizing effect is quite remarkable and changes a substance that is only slightly water solubile into a substance that is very solubile.

McKee and co-workers showed that solutions of micelle-solubilized substances can be subjected to both cathodic reductions and anodic oxidations. They also showed that workup is facilitated by the resulting from dilution of the solutions precipitation of the solubilized compounds [152, 176]. It should, however, be pointed out that most micelle-solubilizing agents withstand cathodic reduction much better than anodic attack and that permanent losses of McKee salts due to anodic degradation may occur. This holds for anodic oxidations as well as for cathodic reductions in undivided cells where the catholyte has direct access to the anode.

Table 2.14 summarizes the work of McKee and co-workers for nitroarene reduction and for anodic oxidation of benzyl alcohol and benzaldehyde. Also included are results obtained by other authors for electrochemical conversions of micelle-solubilized substances.

Table 2.13. Water solubilities of some McKee salts and saturation concentrations of some slightly soluble substances in aqueous saturated solutions of McKee salts [83, 152, 176, 177]

McKee Salt	Temp. (°C)	g salt/ 100 g H_2O	Solubilities (g in 1000 g solvent)				
			Nitrobenzene	Nitrotoluene	o-Nitro anisole	Benzylalcohol	Benzaldehyde
Sodium-salicylate	>50	135	—	—	200	—	—
Sodium-cymenesulfonate	25	100	200	—	200	—	—
Potassium-xylenesulfonate	25	53	18	17	25	—	—
Potassium-xylenesulfonate	50	100	110	95	150	—	—
Sodium-xylenesulfonate	25	132	280	175	270	—	—
Sodium-xylenesulfonate	50	148	365	270	385	—	—
Sodium-benzenesulfonate	85	—	—	—	—	2000	200

Table 2.14. Electrochemical Conversions Carried Out in the Presence of Micelle-Solubilizing Agents

Reduction of aromatic nitro compounds (McKee et al. [152, 176]). Electrolysis conditions: phosphor bronze or monel cathode; catholyte; concentrated solution of xylene-, toluene-, or cymenesulfonates; current density, 5 to 20 mA/cm^2; 70 to 90°C.

a. For reduction to p-aminophenols: catholyte acidified with H_2SO_4, pH < 1.
b. For reduction to anilines: catholyte "moderately acidic or slightly alkaline" [152].
c. For reduction to hydrazo and azo arenes: strongly alkaline catholyte (NaOH), pH > 10.
d. For reduction to azo and azoxy arenes: neutral catholyte.

	Conversion		Yield (%)
Nitrobenzene	→d	azobenzene	70
	→a	p-aminophenol	47
	→c	azobenzene	72
	→c	hydrazobenzene	88
o-Nitrotoluene	→d	o,o'-azotoluene	86
	→c	o,o'-hydrazotoluene	80
p-Nitrotoluene	→d	p,p'-azotoluene	90
o-Nitroanisole	→c	o,o'-hydrazoanisole	80
p-Nitroanisole	→d	p,p'-azoanisole	?
p-Nitrophenol	→d	p,p'-azoxyphenol	?
	→a,b	p-aminophenol	97
2,4-Dinitrophenol	→a,b	2,4-diaminophenol	82
m-Dinitrobenzene	→d	m,m'-azoxynitrobenzene	?
p-Chloronitrobenzene	→d	p,p'-dichloroazobenzene	75
o-Nitroaniline	→b	o-phenylenediamine	95
m-Nitroaniline	→c	m,m'-hydrazoaniline	79
p-Nitroaniline	→d	p-phenylenediamine	90
α-Nitronaphthol	→d	α,α'-azonaphthol	?
	→c	α,α'-hydrazonaphthol	?

Oxidation of benzyl alcohol and benzaldehyde (McKee and Heard [83]). Electrolysis conditions: nickel anode; anolyte; concentrated solution of benzene-, xylene-, or cymenesulfonates; current density, 20 mA/cm^2; 25 to 75°C; increased current and mass yields by addition of salts of Cu, Co, or Ni to the strongly alkaline catholyte.

Benzyl alcohol	$\longrightarrow$ benzoic acid	89
Benzaldehyde	$\longrightarrow$ benzoic acid	97

Oxidation of aromatic alcohols to aldehydes and ketones [179]. Electrolysis conditions: platinized platinum anode; anolyte; 2 N NaOH containing 0.4% cationic surfactants (long-chain tetraalkyl or arylalkyl trialkyl ammonium salts); addition of different solvents for substrate (examples given for benzene); current density, 0.5 to 6 mA/cm^2; 80°C; substrate solution finely dispersed by stirring.

Benzhydrol	$\longrightarrow$ benzophenone	23
Benzyl alcohol	$\longrightarrow$ benzaldehyde	18
α-Methylbenzyl alcohol	$\longrightarrow$ acetophenone	21
p-Methylbenzyl alcohol	$\longrightarrow$ p-tolualdehyde	29
p-Nitrobenzyl alcohol	$\longrightarrow$ p-nitrobenzaldehyde	23
Diphenylacetonitrile	$\longrightarrow$ tetraphenylsuccinonitrile	35

[180] electrolysis conditions as given above [179]

Diethyl malonate	$\longrightarrow$ tetraethyl ethanetetracarboxylate	28

[180, 181] electrolysisis conditions as given above, but electrolyte 2 N Na$_2$CO$_3$ containing bromide anions

Cyanation of aromatic compounds [183, 184]. Electrolysis conditions: platinum anode; anolyte; 3 M NaCN; addition of 10 to 20 mmole/liter cationic surfactant (tetrabutylammonium sulfate or tricaprylmethylammonium chloride); current density, 15 mA/cm^2; 20°C; substrate dissolved in methylene chloride; organic phase finely dispersed by vigorous stirring.

Table 2.14 Continued

	Conversion		Yield (%)
Naphthalene	⟶	α-naphthonitrile	60
Anisole	⟶	p-methoxybenzonitrile	35
		o-methoxybenzonitrile	22
1,2-Dimethoxybenzene	⟶	o-methoxybenzonitrile	26
		2,3-dimethoxybenzonitrile	11
1,3-Dimethoxybenzene	⟶	2,4-dimethoxybenzonitrile	70
1,4-Dimethoxybenzene	⟶	p-methoxybenzonitrile	40
1,2,3-Trimethoxybenzene	⟶	2,6-dimethoxybenzonitrile	28
1,2,4-Trimethoxybenzene	⟶	2,4-dimethoxybenzonitrile	38
1,3,5-Trimethoxybenzene	⟶	2,4,6-trimethoxybenzonitrile	56
Dimenthylbenzylamine	⟶	N,N-dimethylphenylglycinonitrile	22
		N-benzyl-N-methylglycinonitrile	24

Franklin and co-workers described anodic oxidation of micelle-solubilized depolarizers [179-181]. They did not clearly distinguish between direct anodic oxidation of micelle-solubilized substances and anodic oxidation mediated by electrogenerated oxidants (such as OCl^- and OBr^-) for which micelle-solubilizing agents may act as phase transfer catalysts (compare Ref. 182). In the anodic oxidation of micelle-solubilized aromatic alcohols to aldehydes and ketones, cationic surfactants, used as solubilizing agents, caused a marked increase in current yields while anionic and neutral surfactants had only small effects. The authors concluded that this was due partially to formation of a surfactant film on the anode by which oxygen evolution was inhibited. It is not clear whether the surfactant counter ion (chloride) mediates the anodic oxidation by hypochlorite formation. In a second paper, Franklin and Honda described the anodic dimerization of diethyl malonate [180, 181] and of diphenylacetonitrile. In the anodic diethylmalonate dimerization, the role of anodically generated bromine or hypobromite is acknowledged by the authors; however, their results for dimerization of diphenylacetonitrile were not conclusive.

Table 2.14 includes the interesting work of Eberson and Helgée, who performed the anodic cyanation of easily oxidizable arenes in aqueous emulsions of methylene chloride. Tetrabutylammonium ion was used as phase transfer agent. Eberson and Helgée believed that the cyanation of naphthalene and anisole [183, 184] which proceeds in these two-phase emulsion electrolytes with remarkable selectivity, is due to CN^- anion transfer into the organic phase. They assumed that the arenes were oxidized at the electrode/organic phase interface, since if charged transfer at the electrode/aqueous phase interface did occur, hydroxylation of the arene compounds would be expected. These results do not, however, provide sufficient information as to charge transfer, detailed reaction mechanisms, and any estimation concerning optimal process parameters and performance for this type of electrolysis technique.

Pilot and Industrial Application of Two-Phase-Electrolyte-Electrolysis

General Considerations

In spite of the many investigations reported in the literature on two-phase emulsion electrolysis, there remain few processes that have been developed to the pilot-plant stage. The number of processes that have reached the stage of fully operating production are even fewer. At the time of this writing, rumor exists that several small electrochemical processes either are in the pilot-plant stage or are running at normal production. It is our opinion that two-phase electrolysis has a chance of being employed—especially for batch-processing of various small-scale products. This last section describes process techniques for liquid/liquid emulsion electrolysis, the use of liquid/gaseous dispersion electrolysis, and the electrochemical extraction process within the PUREX cycle in which plutonium is separated from uranium by simultaneous charge and phase transfer.

Synthesis of Benzoquinone by Anodic Oxidation of Benzene

Two reports have been published on a benzoquinone process based on the oxidation of benzene dispersed in sulfuric acid solutions at lead dioxide anodes [11, 13]. In both publications it is claimed that the process was developed to the pilot-plant stage and run successfully long enough to allow estimation of large-scale performance and cost. The most detailed process description is given by Fremery and co-workers [11]. The flow sheet of this process is given in Fig. 2.20 and shows that the work up equipment is much more important than the electrolysis cell with respect to investment cost and process operation. The problem of mass transfer and substrate supply was solved by using turbulent flow of the benzene emulsion in the aqueous electrolyte and by using some sort of Venetian-blind electrodes (Fig. 2.21) to increase their surface area and to promote turbulent flow. A very important feature of the process is that the benzoquinone concentration in the dispersed benzene must not exceed 3% and that benzoquinone is continuously reduced at the cathode to hydroquinone. There it is reextracted into the aqueous catholyte, where is is not allowed to accumulate beyond a concentration of 7%. It is separated

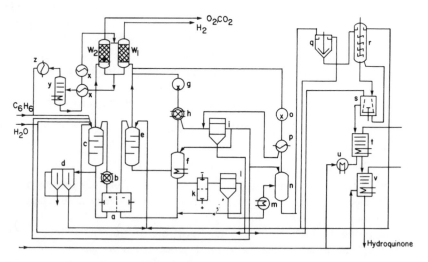

Fig. 2.20. Complete flowsheet of anodic hydroquinone-synthesis process [11]. (*a*) Electrolyzer, (*b*) heat exchanger for electrolyte, (*c, e*) degaser, (*d, i, l*) phase separators (*f*) flash evaporator, (*g, o*) pumps, (*h, p, x, z*) condensers, (*k*) electrolyzer for postreduction (*m, u*) heaters, (*n*) vacuum crystallizer, (*g*) decanter, (*r*) washer, (*s*) centrifuge, (*t, v*) drying columns.

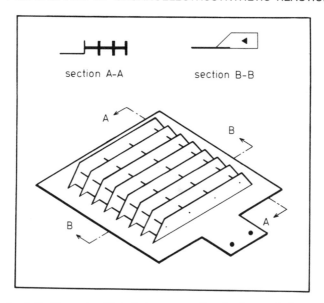

section A-A   section B-B

Fig. 2.21.   Lead dioxide anode with turbulence promoting fins for anodic benzoquinone production from benzene [11].

from the electrolyte after evaporation and is crystallized out by solvent cooling. A thorough cleaning of all waste gases recovers all benzene, which may not be lost into the environment both because of its toxicity and because of its cost.

Synthesis of Benzaldehydes by Mediated Toluene-Oxidation by Two-Phase-Electrolyte-Electrolysis

Ibl and co-workers investigated the anodic oxidation of toluene and substituted toluenes mediated by Ce^{IV} and Mn^{IV} very thoroughly [202]. They analyzed the process-technique and identified Mn^{III} in perchlorate-solutions as the most appropriate mediator. The authors gave a careful cost-estimation for this process and stated that as a result of continually rising costs for the raw-materials (toluene and other alkyl-aromates) the electrochemical route is more profitable than the conventional chemical route to benzaldehyde.

Anthraquinone Production by Mediated Anodic Anthracene Oxidation

Although the mediated anodic oxidation of anthracene has been reported [26, 28, 32, 33], a one-step process is still to be demonstrated. Instead, the classical anthraquinone process is still conducted in a batchwise oxidation of anthracene by H_2SO_4/dichromate. The reaction is performed in stirred tank reactors that have a volume of more than 20 m^3. In such reactors 3 tons of anthracene are converted by successive addition of the oxidant within 30 to 36 hr at 100 to 105°C. The spent oxidant is not recycled, but is worked up to yield Cr^{III} salts [185]. The total cost and potential use of chromic salts will determine whether a two-step-mediated anodic oxidation will eventually be advantageous. Yet anodic dichromate regeneration is a well-established process [186].

Benzidine Synthesis by Cathodic Reduction of Dispersed Nitrobenzene in Alkaline Solutions.

SODIUM-AMALGAM-MEDIATED REDUCTION OF NITROBENZENE

The reduction of nitrobenzene in alkaline solutions with sodium amalgam (which was produced in a chloralkali-mercury cell) was performed years ago on an industrial scale [187, 188]. Figure 2.22 demonstrates the process. The sodium amalgam leaves the amalgam cell and enters the stirred tank reactor in which the reduction of nitrobenzene to hydrazobenzene proceeds according to (2.31). The spent amalgam then reenters the electrolysis cell.

$$2PhNO_2 + 6H_2O + 10Na(Hg) \longrightarrow Ph-NH-NH-Ph + 10NaOH \quad (2.31)$$

Direct Cathodic Reduction of Nitrobenzene

Sekine [132–137], Dey et al. [116] and, more recently, Udupa et al. [168] worked on the direct reduction of dispersed nitrobenzene at a cathode covered by spongy metal deposits (preferentially lead). In 1969, Udupa and co-workers published the design for a 1000-Å cell, suitable for small- to medium-scale production of benzidine [189]. As shown in Fig. 2.23, they used a massive rotating rod cathode surrounded by three anodes encased in diaphragm-enclosed chambers to prevent reoxidation of the product to azobenzene. The complete flowsheet of the process given by Dey, Govindachari, and Rajagopalan is shown in Fig. 2.24. Except for the usual workup equipment, a xylene dosimeter is included for the addition of controlled amounts of xylene to the electrolyte to avoid cathode clogging due to hydrazobenzene precipitation [116].

Aminophenol Synthesis by Cathodic Reduction of Nitrobenzene Being Dispersed in Acidic Solutions

According to a study of a modified "Gattermann-procedure," Dey and co-workers came to the conclusion that aminophenol production by reduction of nitrobenzene emulsions in moderately acidic solution (20 to 30% H_2SO_4) might be economically attractive, although approximately 30% of the nitrobenzene was converted to aniline [115]. Later Monastyrskii and co-workers [154] published a report on their experience with the same process. The flowsheet is given schematically in Fig. 2.25. The process was said to operate at current densities of 20mA/cm^2 and 73% mass yield.

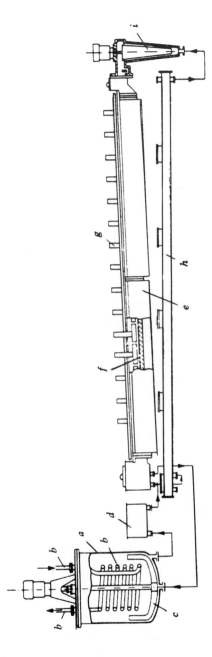

Fig. 2.22. Amalgam process for sodium-amalgam-mediated reductions of organic compounds. (*a*) Stirred tank reactor for performance of mediated conversion, (*b*) cooler, (*c*) stirrer, (*d*) phase separator, (*e*) electrolyzer, (*f*) mercury cathode, (*g*) graphite anode, (*h*) amalgam pile, (*i*) mercury pump [188].

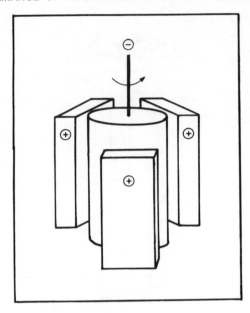

Fig. 2.23. Rotating cathode for hydrazobenzene production [189]. Cathode is surrounded by three diaphragm-walled pockets that contain the anodes.

Production of Propylene Chlorohydrin and Propylene Oxide from Propylene and Electrochemically Generated Hypochloric Acid

In contrast to the industrial chemical production of ethylene oxide, the conversion of propylene to propylene oxide cannot be performed by heterogeneously catalyzed gas-phase oxidation. Instead, propylene oxide synthesis is based primarily on propylene chlorohydrin production in the first step, with subsequent hydrolysis of the chlorohydrin to the desired epoxide:

$$2NaCl + 2H_2O \longrightarrow 2NaOH + H_2 + Cl_2 \longrightarrow NaOCl + NaCl + H_2O \tag{2.32a}$$

$$CH_3\text{-}CH = CH_2 + HOCl \longrightarrow CH_3\text{-}CHOH\text{-}CH_2Cl \tag{2.32b}$$

$$CH_3\text{-}CHOH\text{-}CH_2Cl + NaOH \longrightarrow CH_3\text{-}CH\text{-}CH_2 + H_2O + NaCl \atop \diagdown\diagup \atop O \tag{2.32c}$$

This reaction sequence has some disadvantages with respect to environmental pollution, for example, the base added for epoxide formation is lime so that it was necessary to dispose of large amounts of $CaCl_2$.

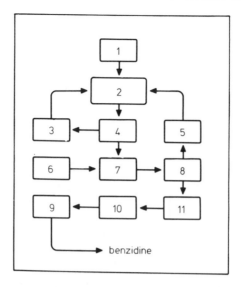

Fig. 2.24. Schematic flowsheet for benzidine production [98, 116, 154]. (*1*) Nitrobenzene tank, (*2*) electrolyzer, (*3*) alkali tank, (*4*) phase separator, (*5*) xylene dosimeter, (*6*) sulfuric acid tank, (*7*) mixing tank, (*8*) centrifuge, (*9*) vacuum distillation unit, (*10*) neutralizer, (*11*) benzidine sulfate collector.

Electrochemical chlorohydrin formation and saponification in a single-stage reactor, that is in a chloralkali-electrolysis cell, were thus envisaged. As a result consumption and co-production of sodium (or alkali) chloride are internally compensated. But the problem with respect to process technique is how to react gaseous propylene with dissolved hypochloric acid to achieve acceptable mass and current efficiencies.

Ibl and Selvig [190] in 1970 published some results on chlorohydrin formation in the anode chamber of a diaphragm-divided cell and found the optimum conditions (current density, flow rate, volumetric ratio propylene/electrolyte) for obtaining nearly 100% current yield. Propylene was introduced into the cell, which had an anode–diaphragm distance of 1 cm, by means of a frit that was inserted into the bottom of the cell. Since propylene feed was always a factor of 4 to 7 in excess of its consumption, a gas/liquid dispersion was processed in the anode chamber (Fig. 2.26).

Later Simmrock [191] reinvestigated the proposed combined process. He found that, because of the need for very precise pH control for the chlorohydrin formation and its subsequent saponification, it was advisable to perform

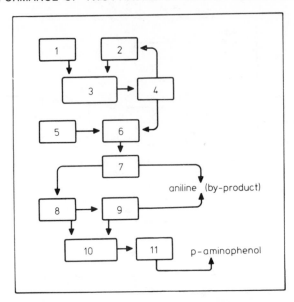

Fig. 2.25. Schematic flowsheet for p-aminophenol production [98, 115]. (1) sulfuric acid dosimeter, (2) nitrobenzene dosimeter, (3) electrolyzer, (4) steam distiller, (5) sodium carbonate dosimeter, (6) neutralizer, (7) steam distiller, (8) settler, (9) flash evaporator, (10) filter, (11) vacuum drier.

the electrolysis and the subsequent chemical reactions separately. The optimal process design is depicted in the flowsheet of Fig. 2.27. Propylene enters the process *after* the electrolysis cell in the second reactor where chlorohydrin is formed. Simmrock used membrane chloralkali electrolysis, which with the newly developed and commercially available DuPont Nafion® membranes, gave current efficiencies for Cl_2 production close to 100%. Conversion of the chlorohydrin to propylene oxide is 83% with respect to total propylene consumption because of careful process control in the chemical steps. The process cost according to Simmrock compares favorably to other processes currently in use for propylene oxide production.

Fluorination of Gaseous Organic Compounds

Fluorination of organic compounds is performed electrochemically by anodic substitution according to the Simons process. Fluorination is achieved usually in hydrogen fluoride/potassium fluoride electrolytes, and the substrate is dissolved in this electrolyte [192]. Suitable organic substrates for this process usually possess satisfactory solubility in the electrolyte because of hydrogen bonding of HF to polar groups such as, $-COOH$, $-SO_3H$, and

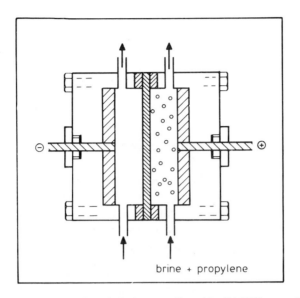

brine + propylene

Fig. 2.26. Schematic presentation of diaphragm cell used by Ibl [190] to produce propylene chlorohydrin. Propylene was introduced through a fritte at the bottom of the electrolysis cell.

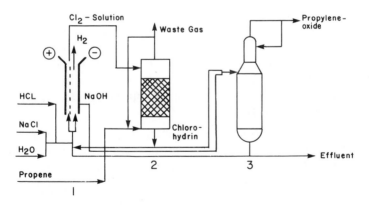

Fig. 2.27. Schematic flowsheet for propylene production according to Simmrock [191]. (*1*) Electrolyzer, (*2*) chlorohydrin formation, (*3*) propylene oxide formation.

$-NH_2$. Fluorination is performed at a current density of 20 mA/cm^2 at nickel anodes.

Nonpolar substrates of higher volatility, however, often do not dissolve in the electrolyte and must be processed in gas/liquid dispersions or in special electrolyzers. The involved mass transfer problems are considerable.

Schmeisser and Sartori [193] proposed inserting a gas disperser below the electrode stack in the Simons electrolyzer (see Fig. 2.8 in section 3 of this chapter).

About the same time a new process technique was developed and piloted using porous carbon anodes.

Electrochemical Extraction: An Electrochemical Step in the PUREX Process*

The separation of uranium and plutonium is an important step in the workup procedure of spent nuclear fuels. It is brought about by reduction of Pu^{IV}, which together with U^{VI} is dissolved in complexed form in a tributyl phosphate/kerosene solvent. Pu^{III}, in contrast to Pu^{IV}, is soluble perferentially in water and is thus extracted into the aqueous phase [196–198]:

$$(Pu^{IV})_{organic} + reductant \longrightarrow (Pu^{III})_{aqueous} \qquad (2.33)$$

Whereas the reduction process utilized ferrous ions at the outset, today U^{IV} ions, generated electrochemically, are employed as reducing agents. However, as Baumgärtner and co-workers showed [199], this reductive extraction process can be performed without any dissolved and/or mediating reductant by direct cathodic reduction of the complexed Pu^{IV} species. Detailed investigations of this cathodic extraction of plutonium showed that the rate-determining step for the reduction is the phase transfer of the Pu(IV) complex (which possesses finite, but low, solubility in water) from the nonaqueous into the aqueous phase [200, 201]. Thus it seems very likely that the reduction of Pu^{IV} to Pu^{III} proceeds without any further complication at the aqueous phase/electrode interface.

The cathodic extraction of Pu^{III} into the aqueous phase is performed in two completely different devices:

1.   As shown in Fig. 2.28, mixer/settler batteries made of titanium are used. The mixer walls serve as cathodes. Reduction of Pu^{IV} is conducted by stationary supply of the two-phase mixture to the mixer section. Subsequent stationary phase separation is carried out in the much larger settler compartment.

2.   The same reaction may be performed in a pulsed perforated plate col-

---

*PUREX: *P*lutonium—*U*ranium separation by *ex*traction.

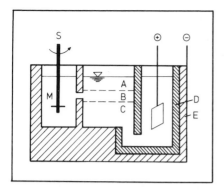

Fig. 2.28. Mixer electrolyzer/settler separator for cathodic extraction of Pu^{III} into the aqueous phase [198]. (A) Organic phase, (B) settling two-phase mixture, (C) aqueous phase, (D) insulating wall, (E) titanium wall (cathode), (F) anode.

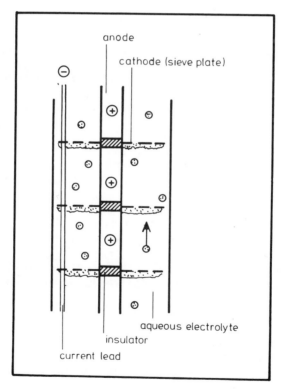

Fig. 2.29. Continuously operated cathodic extractor for Pu^{III} extraction into coherent aqueous phase (pulsed column). Cathodes: perforated plates: anodes: central shaft [198].

umn as shown in Fig. 2.29. Organic and aqueous phase flow is countercurrent. The plates of the column serve as cathodes, whereas the anode is the common shaft that supports the plates.

The cathodic reduction of Pu^{IV} and U^{VI} which has previously been extracted into the aqueous phase proceeds at those parts of the perforated metallic plate cathodes which are not covered by the non-aqueous TBP-containing phase [203].

References for Section 4

1. T. Kempf, German Patent 117, 251 (1899); *Chem. Zentr.*, **1901/I**, 348.
2. F. Fichter and R. Stocker, *Ber. Dtsch. Chem. Ges.*, **47**, 2003 (1914).
3. F. Fichter, *Trans. Electrochem. Soc.*, **45**, 107 (1925).
4. F. Fichter, *J. Soc. Chem. Ind.*, **48**, 325 (1929).
5. K. S. Udupa, G. S. Subramanian, and H. V. K. Udupa, *Bull. Acad. Polon. Sci.*, *Sér. Sci. Chim.* **9**, 45 (1961) [*Chem. Abstr.*, **58**, 13443 (1963)].
6. J. S. Clarke, R. E. Ehigamusoe, and A. T. Kuhn, *J. Electroanal. Chem.* **70**, 33 (1976).
7. Y. Isomura, *J. Chem. Soc. Jap.*, **62**, 1167 (1941) [*Chem. Abstr.*, **41**, 3070f (1947)].
8. H. Inoue and M. Shikata, *J. Chem. Ind. (Jap.)*, **24**, 567 (1921) [*Chem. Abstr.*, **16**, 1046 (1922)].
9. S. Horrobin and R. G. A. New, U. S. Patent 2,285,858 [*Chem. Abstr.*, **36**, 6922 (1942)].
10. J. C. Ghosh, S. K. Bhattacharyya, M. S. Muthanna, and C. R. Mitra, *J. Sci. Ind. Res. (India)*, **11B**, 356 (1952) [*Chem. Abstr.*, **47**, 2064e (1953)]; *Curr. Sci.*, **16**, 87 (1947) [*Chem. Abstr.*, **41**, 4725c (1947).
11. M. Fremery, H. Höver, and G. Schwarzlose, *Chem. Ing. Tech.*, **46**, 635 (1974)].
12. A. Seyewitz and G. Miodon, *Bull. Soc. Chim. Fr.*, **33**, 449 (1923).
13. J. P. Millington, *Chem. Ind. (Lond.)*, **1975**, 780.
14. C. B. Medinski, *Chem. Zentr.*, **1934/II**, 2523.
15. M. Yokoyama, *Bull. Chem. Soc. Jap.*, **8**, 71 (1933) [*Chem. Abstr.*, **27**, 3917 (1933)].
16. L. Gattermann and F. Friedrichs, *Chem. Ber.*, **27**, 1942 (1894).
17. F. T. Kitchen, U.S. Patent 1,322,580 (1919) [*Chem. Abstr.*, **14**, 287 (1920)].
18. A. Renard, *C. R.*, **91**, 175 (1880).
19. R. M. Archibald, *Trans. R. Soc. Can.*, **26/III**, 69 (1932).
20. A. Wacher, German Patent 614,041 (1932).
21. S. Kikuchi and H. Masago, *J. Soc. Sci. Photogr. Jap.*, **14**, 11 (1951).
22. A. P. de Bottens, *Z. Elektrochem.*, **8**, 673 (1902).
23. Hoechst AG, German Patent 152,063 (1902); *Chem. Zentr.*, **1904/II**, 71; *Z. Elektrochem.*, **11**, 278 (1905).
24. G. Schetty, Thesis, University of Basel, (Switzerland), 1925.
25. E. G. White and A. Lowy, *Trans. Electrochem. Soc.*, **62**, 223 (1932).

26. R. A. C. Rennie (ICI), German Patent 2,301,803 (1973) [*Chem. Abstr.*, **79,** 115357v (1973)]; German Patent 1,804,727 (1969); German Patent 1,804,728 (1969).

27. L. A. Joo and L. A. Bryan, French Patent 1,558,481 (1969).

28. J. P. Millington, German Patent 2,201,017 (1972) [*Chem. Abstr.*, **77,** P121554g (1972)]; British Patents 1,377,681 and 1,373,611 (1975) [*Chem. Abstr.*, **82,** P125099p (1975)].

29. C. Liebermann, *Chem. Ber.*, **7,** 805 (1874); W. H. Perkin, Jr., *J. Chem. Soc.*, **59,** 1012 (1891); M. S. Newman, *J. Am. Chem. Soc.*, **64,** 2324 (1942).

30. C. H. Rasch and A. Lowy, *Trans. Electrochem. Soc.*, **56,** 477 (1929).

31. F. M. Perkin and A. Fontana, *Z. Elektrochem.*, **10,** 441 (1904).

32. K. Shirai and K. Sugino, *Denki Kagaku*, **25,** 284 (1957) [*Chem. Abstr.*, **52,** 8794g (1958)].

33. S. Krishnan, V. A. Vyas, M. S. Venkatachalapathy, and H. V. K. Udupa, *J. Electrochem. Soc. India*, **14,** 32 (1965) [*Chem. Abstr.*, **64,** 1618h (1966)].

34. P. A. Florenski and K. T. Metelkin, Anilinokras. Prom., **1931** (8) 9 and (9/10) 7; *Chem. Zentr.* **1932/II,** 1689.

35. J. Babic, *Arh. Rudarstvo Technol.*, **6,** 49 (1968) [*Chem. Abstr.*, **70,** 83576m (1969)].

36. C. J. Thatcher, U.S. Patent 1,397,562 (1922) [*Chem. Abstr.*, **16,** 878 (1922)].

37. K. Niki, T. Sekine, and K. Sugino, *J. Electrochem. Soc. Jap. (Overseas Ed.)*, **37,** 74 (1969) [*Chem. Abstr.*, **72,** 42517r].

38. N. S. Drosdow and S. S. Drosdow, *Chimitscheski J. Ser. B, J. Prikl. Chim.*, **6,** 897 (1935); *Chem Zentr.*, **1935/I,** 230.

39. A. Renard, *C. R.*, **92,** 965 (1880).

40. K. Elbs, *Z. Elektrochem.*, **2,** 522 (1896).

41. K. Puls, *Chem. Z.*, **25,** 263 (1901).

42. W. Lang, U.S. Patent 808,095 (1905); German Patent 189, 178 (1902).

43. H. D. Law and F. M. Perkin, *Trans. Faraday Soc.*, **1,** 31,251 (1905).

44. F. Fichter, *Z. Elektrochem.*, **19,** 781 (1913).

45. F. Fichter, and E. Uhl, *Helv. Chim. Acta,* **3,** 22 (1920).

46. C. F. Cullis and J. W. Ladbury, *J. Chem. Soc.*, **1955,** 555.

47. W. S. Trahanovsky and L. B. Young, *J. Org. Chem.*, **31,** 2033 (1966).

48. P. S. R. Murti and S. C. Pati, *Chem. Ind. (Lond.)*, **1967,** 702.

49. P. J. Andrulis, Jr., M. J. S. Dewar, R. Dietz, and R. L. Hunt, *J. Am. Chem. Soc.*, **88,** 5473 (1966).

50. T. A. Cooper and W. A. Waters, *J. Chem. Soc. (B)*, **1967,** 687.

51. E. J. Heiba, R. M. Dessau, and W. J. Koehl, Jr., *J. Am. Chem. Soc.*, **91,** 6830 (1969).

52. M. S. Venkatachalapathy, R. Ramaswamy, and H. V. K. Udupa, *Bull. Acad. Pol. Sci., Sér. Sci. Chim.*, **6,** 487 (1958) [*Chem. Abstr.*, **53,** 3853d (1959)].

53. J. Mizuguchi and S. Matsumoto, *J. Pharm. Soc. Jap.*, **71,** 737 (1951) [*Chem. Abstr.*, **45,** 10099d (1951)].

54. J. Kato and M. Sakuma, Japanese Patent 1,625 (1952) [*Chem. Abstr.*, **47,** 5826i (1953)].

55. C. A. Mann and P. M. Paulson, *Trans. Electrochem. Soc.*, **47,** 101 (1925).

56. G. Kawada, *J. Pharm. Soc. Jap.*, **521**, 628 (1925) [*Chem. Abstr.*, **20**, 339 (1926)].

57. R. W. Mitchell, *Trans. Electrochem. Soc.*, **56**, 495 (1925).

58. B. M. Dey and R. K. Maller, *J. Sci. Ind. Res. (India)*, **12B**, 255 (1953) [*Chem. Abstr.*, **47**, 12056g (1953)].

59. M. S. Venkatachalapathy, R. Ramaswamy, and H. V. K. Udupa, *Bull. Acad. Polon. Sci.*, *Sér. Sci. Chim.*, **8**, 361 (1960) [*Chem. Abstr.*, **56**, 12788a (1962)]; India Patent 62,379 (1959) [*Chem. Abstr.*, **53**, 7384b (1959)]; Indian Patent 62,426 (1960) [*Chem. Abstr.*, **54**, 7654a (1960)].

60. F. Fichter and D. Müller, *Helv. Chim. Acta*, **18**, 831 (1935).

61. F. M. Perkin, *Practical Methods of Electrochemistry*, Lomgmans, Green, London, 1905, p. 279.

62. J. H. James, *J. Am. Chem. Soc.*, **21**, 889 (1899).

63. A. Merzbacher and E. F. Smith, *J. Am. Chem. Soc.*, **22**, 723 (1900).

64. F. Fichter and M. Rinderspacher, *Helv. Chim. Acta*, **10**, 40 (1927).

65. F. Fichter and J. Meyer, *Helv. Chim. Acta*, **8**, 74 (1925).

66. F. Fichter and G. Grisard, *Helv. Chim. Acta*, **4**, 928 (1921).

67. F. Fichter and M. Rinderspacher, *Helv. Chim. Acta*, **9**, 1097 (1926).

68. F. Fichter and J. Meyer, *Helv. Chim. Acta*, **8**, 285 (1925).

69. F. Fichter and G. Schetty, *Helv. Chim. Acta*, **20**, 150 (1937).

70. F. Fichter and G. Schetty, *Helv. Chim. Acta*, **20**, 563 (1937).

71. F. Fichter and M. Adler, *Helv. Chim. Acta*, **9**, 279 (1926).

72. R. F. Dunbrook and A. Lowy, *Trans. Electrochem. Soc.*, **45**, 81 (1924).

73. F. Fichter and P. Lotter, *Helv. Chim. Acta*, **8**, 438 (1925).

74. F. Fichter and G. Bonhôte, *Helv. Chim. Acta*, **3**, 395 (1920).

75. F. Sachs and R. Kempf, *Ber. Dtsch. Chem. Ges.*, **35**, 2704 (1902).

76. O. W. Brown and A. E. Brown, *Trans. Electrochem. Soc.*, **75**, 393 (1939).

77. F. Fichter and H. Ris, *Helv. Chim. Acta*, **7**, 803 (1924).

78. F. Fichter and F. Ackerman, *Helv. Chim. Acta*, **2**, 583 (1919).

79. K. Ono, *Helv. Chim. Acta*, **10**, 45 (1927).

80. F. Fichter and A. Christen, *Helv. Chim. Acta*, **8**, 332 (1925).

81. F. Fichter and W. Dietrich, *Helv. Chim. Acta*, **7**, 131 (1924).

82. F. Fichter and R. Senti, *Chem. Zentr.*, **1927/II**, 54.

83. R. H. McKee and J. R. Heard, *Trans. Electrochem. Soc.*, **65**, 301 (1934).

84. F. Fichter, H. E. Suenderhauf, and A. Goldach, *Helv. Chim. Acta*, **14**, 249 (1931).

85. H. Tryller, German Patent 100,417, 1897.

86. D. F. Calhane and C. C. Wilson, *Trans. Electrochem. Soc.*, **63**, 247 (1933).

87. F. Fichter and E. Plüss, *Helv. Chim. Acta*, **15**, 236 (1932).

88. J. C. Ghosh, S. K. Bhattacharyya, M. R. A. Rao, M. S. Muthanna, and R. B. Patnaik, *J. Sci. Ind. Res. (India)*, **11B**, 361 (1952) [*Chem. Abstr.*, **47**, 2064e (1953)].

89. C. W. Croco and A. Lowy, *Trans. Electrochem. Soc.*, **50**, (1926) [*Chem. Abstr.*, **20**, 3396 (1926)].

90. G. Bionda and M. Civera, *Ann. Chim.*, **41**, 814 (1951) [*Chem. Abstr.*, **46**, 7908f (1952)].

91. G. Bionda and M. Civera, *Ann. Chim.*, **43**, 11 (1953) [*Chem. Abstr.*, **47**, 5821b (1953)].

92. I. A. Atanasiu and C. Belcot, *Bull. Sect. Sci. Acad. Roum.*, **19**, 101, 106 (1937) [*Chem. Abstr.*, **32**, 7830 (1938).

93. R. Tomat and A. Rigo, *J. Electroanal. Chem.*, **75**, 629 (1977).

94. D. Pletcher and S. J. D. Tait, *J. Chem. Soc. Perkin Trans.*, **2**, **1979**, 788.

95. H. Lund, "Cathodic Reduction of Nitro Compounds", *Organic Electrochemistry*, in M. Baizer, Ed., Dekker, New York, 1973, p. 315.

96. F. G. Thoma and K. G. Boto, The Electrochemistry of Azoxy, Azo, and Hydrazo Compounds, in *The Chemistry of Functional Groups—The Chemistry of the Hydrazo, Azo and Azoxy Compounds Part 1*, S. Patai Ed., Wiley, New York, 1975.

97. F. Beck, *Elektroorganische Chemie*, Verlag Chemie, Weinheim 1974, p. 169.

98. A. P. Tomilov, editor, *The Electrochemistry of Organic Compounds*, Leningradskoe Otdelenie, Leningrad, 1968; Israel Program for Scientific Translations, Jerusalem, London, 1972, pp. 239, 257, 263.

99. Landolt-Börnstein, Zahlen Werte und Funktionen, Vol II, Part IIb, 6th ed., Springer-Verlag, Berlin, Göttingen, Heidelberg 1962.

100. H. E. Vermillion, B. Werbel, J. H. Saylor, and P. M. Gross, *J. Am. Chem. Soc.*, **63**, 1346 (1941).

101. F. Haber, *Z. Elektrochem.*, **4**, 506, 577 (1898).

102. M. Heyrovsky, B. Kastening, S. Vavricka, and L. Holleck, *J. Electroanal. Chem.*, **26**, 399 (1970).

103. M. Heyrovsky, S. Vavricka, and L. Holleck, *Coll. Czech. Chem. Commun.*, **36**, 971 (1971).

104. L. Holleck, S. Vavricka, and M. Heyrovsky, *Electrochim. Acta*, **15**, 645 (1970).

105. M. Heyrovsky and S. Vavricka, *J. Electroanal. Chem.*, **28**, 411 (1970).

106. G. Pezzantini and R. Guidelli, *J. Electroanal. Chem.*, **102**, 205 (1979).

107. F. Haber and C. Schmidt, *Z. Phys. Chem.*, **32**, 271 (1900).

108. K. Brand, *Ber. Dtsch. Chem. Ges.*, **38**, 3077 (1905).

109. Y. Ogata, M. Tsushida, and Y. Takagi, *J. Am. Chem. Soc.*, **79**, 3397 (1957).

110. G. A. Russell and E. J. Geels, *J. Am. Chem. Soc.*, **87**, 122 (1965).

111. L. Meites and P. Zumen, *Electrochemical Data*, Part 1, Vol A, Wiley-Interscience, New York, 1974.

112. A. Bellefontaine and J. Repplinger, "Anilin," in *Ullmanns Enzyklopädie der Technischen Chemie*, Vol 7, 4th ed., Verlag Chemie, Weinheim 1972, p. 566.

113. C. J. Thatcher, *Trans. Electrochem. Soc.*, **48**, 175 (1927).

114. H. J. Schwenecke, "Benzidin", in *Ullmanns Enzyklopädie der Technischen Chemie*, Vol 8, 4th ed., Verlag Chemie, Weinheim, 1972, p. 352.

115. B. B. Dey, T. R. Govindachari, and S. C. Rajagopalan, *J. Sci. Ind. Res. (India)*, **4**, 574 (1946) [*Chem. Abstr.*, **40**, 4965 (1946)].

116. B. B. Dey, T. R. Govindachari, and S. C. Rajagopalan, *J. Sci. Ind. Res. (India)*, **4**, 559 (1946) [*Chem. Abstr.*, **40**, 4965 (1946)]; *ibid;* 569 [*Chem. Abstr.*, **40**, 4965 (1946)].

117. B. B. Dey, T. R. Govindachari, and S. C. Rajagopalan, Indian Patent 34,756 (1948) [*Chem. Abstr.*, **44**, 6886c (1950)].

118. B. B. Dey, T. R. Govindachari, and S. C. Rajagopalan, Indian Patent 34,757 (1948) [*Chem. Abstr.*, **44**, 6886d (1950)].

119. B. B. Dey, R. K. Maller, and B. R. Pay, *J. Sci. Ind. Res. (India)*, **7B**, 198 (1948) [*Chem. Abstr.*, **43**, 4150g (1949)]; Indian Patent 40,261 (1950) [*Chem. Abstr.*, **44**, 9984c (1950)].

120. B. B. Dey, T. R. Govindachari, and S. C. Rajagopalan, *J. Sci. Ind. Res. (India)*, **5B**, 75 (1946) [*Chem. Abstr.*, **41**, 4046f (1947)].

121. B. B. Dey, T. R. Govindachari, and S. C. Rajagopalan, *J. Sci. Ind. Res. (India)*, **5B**, 77 (1946) [*Chem. Abstr.*, **41**, 4046h (1947)].

122. B. B. Dey, T. R. Govindachari, and S. C. Rajagopalan, *J. Sci. Ind. Res. (India)*, **4**, 637 (1946) [*Chem. Abstr.*, **40**, 6347 (1946)].

123. B. B. Dey, T. R. Govindachari, and S. C. Rajagopalan, *J. Sci. Ind. Res. (India)*, **4**, 642 (1946) [*Chem. Abstr.*, **40**, 6347 (1946)].

124. B. B. Dey, T. R. Govindachari, and S. C. Rajagopalan, *J. Sci. Ind. Res. (India)*, **4**, 645 (1946) [*Chem. Abstr.*, **40**, 6347 (1946)].

125. B. B. Dey, R. K. Maller, and B. R. Pay, *J. Sci. Ind. Res. (India)*, **7B**, 107 (1948) [*Chem. Abstr.*, **43**, 3729g (1949)]; Indian Patent 39,428 (1950) [*Chem. Abstr.*, **44**, 9278h (1950)].

126. B. B. Dey, R. K. Maller, and B. R. Pay, *J. Sci. Ind. Res. (India)*, **7B**, 113 (1948) [*Chem. Abstr.*, **43**, 3730f (1949)]; Indian Patent 39,429 (1950) [*Chem. Abstr.*, **44**, 9278h (1950)].

127. B. B. Dey, R. K. Maller, and B. R. Pay, *J. Sci. Ind. Res. (India)*, *7B85 71 (1948)* [*Chem. Abstr.*, **42**, 7643f (1948)]; Indian Patent 39,427 (1950) [*Chem. Abstr.*, **44**, 9279a (1950)].

128. B. B. Dey, R. K. Maller, and B. R. Pay, *J. Sci. Ind. Res. (India)*, **8B**, 206 (1949) [*Chem. Abstr.*, **44**, 2869c (1950)].

129. B. B. Dey and H. V. Udupa, *J. Sci. Ind. Res. (India)*, **6B**, 83 (1947) [*Chem. Abstr.*, **42**, 3681h (1948)].

130. B. B. Dey, T. R. Govindachari, and H. V. K. Udupa, *Curr. Sci.*, **15**, 163 (1946) [*Chem. Abstr.*, **40**, 7012 (1946)].

131. G. S. Krishnamurthy, H. V. K. Udupa, and B. B. Dey, *J. Sci. Ind. Res. (India)*, **15B**, 47 (1956).

132. T. Sekine, *Denki Kagaku*, **28(8)** (1960) [*Chem. Abstr.*, **61**, 14187g (1964)].

133. T. Sekine, and K. Sugino, *J. Electrochem. Soc. Jap.*, **21**, 383 (1953) [*Chem. Abstr.*, **48**, 13485h (1954)].

134. T. Sekine, *J. Chem. Soc. Jap., Pure Chem. Sect.*, **77**, 67 (1956) [*Chem. Abstr.*, **50**, 9181c 1956)].

135. T. Sekine, *Kogyo Kagaku Zasshi*, **60**, 918 (1957) [*Chem. Abstr.*, **53**, 8883f (1959)].

136. T. Sekine, *J. Electrochem. Soc. Jap. (Overseas Ed.)*, **26**, 145 (1958) [*Chem. Abstr.*, **55**, 16220c (1961)].

137. K. Sugino and T. Sekine, *J. Electrochem. Soc.*, **104**, 497 (1957) [*Chem. Abstr.*, **51**, 15307d (1957)].

138. K. Sugino, K. Odo, T. Sekine, K. Shirai, and E. Ichikawa, *J. Electrochem. Soc. Jap.*, **30**, E167 (1962) [*Chem. Abstr.*, **63**, 1474g (1965)].

139. K. Sugino, Japanese Patent 5,717 (1956) [*Chem. Abstr.*, **52**, 8805g (1958)].

140. K. Sugino and T. Sekine, Japanese Patent 4,962 (1959) *Chem. Abstr.*, **54,** 5298i (1960)].
141. H. E. Heller, E. D. Hughes, and C. K. Ingold, *Nature,* **168,** 909 (1951).
142. L. Gattermann, *Ber. Dtsch. Chem. Ges.,* **26,** 1844 (1893).
143. A. S. McDaniel, L. Schneider, and L. Ballard, *Trans. Electrochem. Soc.,* **39,** 441 (1932).
144. F. Darmstädter, German Patent 150,800 (1901); German Patent 154,086 (1903); *Z. Elektrochem.,* **10,** 198 (1904); *ibid.,* **11,** 274 (1905).
145. F. M. Brigham and H. S. Lukens, *Trans. Electrochem. Soc.,* **61,** 281 (1932).
146. O. Imray, British Patent 18,081 (1915).
147. O. Imray, *J. Soc. Chem. Ind.,* **36,** 129 (1917).
148. O. Imray, *J. Chem. Soc. Abstr.,* **112,** 197 (1917).
149. T. Shoji, *J. Chem. Ind. Tokyo,* **21,** 117 (1918).
150. T. Shoji, *J. Soc. Chem. Ind.,* **37,** 439A (1918) [*Chem. Abstr.,* **12,** 1878 (1918)].
151. J. Thatcher, U.S. Patent 1,501,472 (1924) [*Chem. Abstr.,* **18,** 2715 (1924)].
152. R. H. McKee and B. G. Gerapostolou, *Trans. Electrochem. Soc.,* **68,** 329 (1935).
153. C. A. Mann, R. E. Montonna, and M. G. Larian, *Trans. Electrochem. Soc.,* **69,** 367 (1936).
154. L. M. Monastyrskii, L. V. Armenskaya, Z. S. Smolyan, and E. N. Lysenko, *Vestn. Tek. Ekon. Inf.,* **11,** 19 (1963).
155. C. L. Wilson and H. V. Udupa, *J. Electrochem. Soc.,* **99,** 289 (1952).
156. N. M. Winslow, *Trans. Electrochem. Soc.,* **88,** 81 (1945).
157. Meister Lucius und Brüning, German Patent 127,727 (1927).
158. W. Löb and J. Schmitt, *Z. Elektrochem.,* **10,** 756 (1904).
159. W. Löb, German Patent 116,467 (1900); *Z. Elektrochem.,* **7,** 627 (1901).
160. W. Löb, *Z. Elektrochem.,* **7,** 627 (1901).
161. C. Kerns, *Trans. Electrochem. Soc.,* **62,** 183 (1932).
162. Weiler-ter-Mer, German Patent 140,613 (1900).
163. K. Elbs, *Chem. Z.* **17,** 209 (1893).
164. Bayer & Co, German Patent 121, 899 (1900).
165. F. Darmstädter, German Patent 181,116 (1904).
166. Gesellschaft für Chemische Industrie, German Patent 297,019 (1917).
167. L. E. Ter-Minasyan, *Izv. Akat. Nauk. Arm. SSR, Chem. Sci. Ser.,* **10,** 173 (1957).
168. K. S. Udupa, G. S. Subramanian, and H. V. K. Udupa, *J. Electrochem. Soc.,* **108,** 373 (1961).
169. Boehringer & Söhne, German Patent 116,942 (1899); German Patent 117,007 (1899); *Angew. Chem.,* **1901,** 64, 142, 380.
170. W. E. Bradt and A. W. Erickson, *Trans. Electrochem. Soc.,* **75,** 401 (1939).
171. W. E. Bradt and H. B. Linford, *Trans. Electrochem. Soc.,* **69,** 353 (1936).
172. A. Chilesotti, *Z. Elektrochem.,* **7,** 768 (1901).
173. C. N. Otin, *Z. Elektrochem.,* **16,** 674 (1910).
174. P. N. Anantharaman and H. V. K. Udupa, Extended Abstracts, The Electrochemical Society, Spring Meeting, Washington, DC May 2-7, 1976.
175. W. Löb, *Ber. Dtsch. Chem. Ges.,* **29,** 1894 (1896).

176. R. H. McKee and C. J. Brockman, *Trans. Electrochem. Soc.*, **62**, 203 (1932).
177. R. H. McKee and J. R. Heard, *Trans. Electrochem. Soc.*, **65**, 135, 161 (1934).
178. K. L. Mittal and P. Mukerjee, "The Wide World of Micelles, in *Micellization, Solubilization and Microemulsions*, K. L. Mittal," Ed., Vol. 1, Plenum Press, New York, 1977, p. 1.
179. T. C. Franklin and L. Sidarous, *J. Electrochem. Soc.*, **124**, 65 (1977).
180. T. C. Franklin and T. Honda, "The Use of Phase Transfer Catalysts with Emulsions and Micelle Systems, in *Electroorganic Synthesis,*" in *Micellization, Solubilization and Microemulsions*, K. L. Mittal Ed., Vol. 2, Plenum Press, New York, 1977, p. 617.
181. T. C. Franklin and T. Honda, *Electrochim. Acta*, **23**, 439 (1978).
182. D. Pletcher and N. Tomov, *J. Appl. Electrochem.*, **7**, 501 (1977).
183. L. Eberson and B. Helgée, *Chem. Scr.*, **5**, 47 (1974).
184. L. Eberson and B. Helgée, *Acta Chem. Scand.*, **B29**, 451 (1975).
185. Ullmanns *Enzyklopädie der Technischen Chemie*, Vol. 3, 3rd ed., Verlag Chemie, Weinheim 1961, p. 660.
186. M. Käppel, *Chem. Ing. Tech.*, **35**, 386 (1963).
187. C. Kellner, German Patent 947,361 (1896).
188. W. Funke, *Chem. Ing. Tech.*, **35**, 336 (1963).
189. T. D. Balakrishnan, K. S. Udupa, G. S. Subramanian, and H. V. K. Udupa, *Chem. Ing. Tech.*, **41**, 776 (1969).
190. N. Ibl and A. Selvig, *Chem. Ing. Tech.*, **42**, 180 (1970).
191. K. H. Simmrock, *Chem. Ing. Tech.*, **48**, 1085 (1976).
192. J. Burdon and J. C. Tatlow, *Advances in Fluorine Chemistry* Vol. 1, Butterworth Scientific, London, 1966, p. 129.
193. M. Schmeisser and P. Sartori, *Chem. Ing. Tech.*, **36**, 9 (1964).
194. H. M. Fox, F. N. Ruehlen, and W. V. Childs, *J. Electrochem. Soc.*, **118**, 1246 (1971).
195. Philips Petroleum Company., U.S. Patent 3,511,760 (1967); U.S. Patent 3,686,082 (1970).
196. G. Koch, *Chem. Z.*, **101**, 64 (1977).
197. F. Baumgärtner, *Chem. Ing. Tech.*, **49**, 756 (1977).
198. F. Baumgärtner and H. Schmieder, *Radiochim. Acta*, **25**, 191 (1978).
199. F. Baumgärtner, E. Schwind. and P. Schlosser, German Patent 1,905,519 (1970).
200. F. Baumgärtner and L. Finsterwalder, *Solvent Extraction Research*, Wiley, New York, 1969, p. 313.
201. F. Baumgärtner and L. Finsterwalder, *J. Phys. Chem.*, **74**, 108 (1970).
202. K. Kramer, P. M. Robertson, and N. Ibl, *J. Appl. Electrochem.*, **10**, 29 (1980).
203. H. Feess and H. Wendt, *Chem. Ing. Tech.*, **53** (1981) in press.

Acknowledgment

The authors are indebted to the AIF (Arbeitsgemeinschaft Industrieller forschungsvereinigungen) for financial support for their work on fundamentals of two-phase-electrolyte electrolysis, which stimulated the writing of this chapter.

List of Symbols

U	electrode potential
i	current density
i_d	diffusion limited current density
z, n	number of electrons transferred
m	number of mediating molecules
F	Faraday constant
γ	current efficiency
a	volume-specific electrode area
η	space time yield
k_m	mass transfer coefficient
$\overline{k}_m$	mean mass transfer coefficient
D	diffusion coefficient
ν	kinematic viscosity
ω	velocity
Re	Reynolds number
t	time
T	temperature
δ_N	Nernstian diffusion layer thickness
δ_{Pr}	Prandtl-layer thickness
y	distance, coordinate vertical to electrode surface
$\overline{x}$	mean diffusion distance
c	concentration
c^∞	bulk concentration
c^{sat}	saturation concentration
$c_{y=0}$	concentration at zero distance
$\dot{c}$	variation of concentration with time
$\dot{n}$	passage of moles per unit area and unit time
r_{drop}	droplet radius
r_{crit}	critical droplet radius
ϑ	surface tension
$\Delta\delta$	density difference
d_p	pore diameter
g	gravitation constant
$\dot{v}$	surface specific volumetric flow rate

Chapter III

DESIGN OF A MULTIPURPOSE,
MODULARIZED ELECTROCHEMICAL CELL

L. Carlsson,[1] H. Holmberg,[2] B. Johansson,[2] and A. Nilsson[2]

[1]Swedish National Development Company,
Åkersberga, Sweden
[2]Foundation for Industrial Organic Electrochemistry,
Chemical Center,
University of Lund,
Lund, Sweden

1 INTRODUCTION

Through the years a great deal of published information about synthetic and mechanistic aspects of organic electrochemistry has accumulated. Nevertheless, only a limited number of reactions have been commercialized [1]. A large number of the investigated reactions fail for economic reasons, but a number of interesting reactions still remain as possible candidates for electrochemical manufacture. One may wonder why a changeover from chemical to electrochemical production technology on a commercial scale should be such a sluggish process in most companies. There are, of course, many contributing reasons and these vary from one company to another. In some cases, lack of experience with the new technology probably constitutes a retarding factor. The economic resources available for risk ventures are also limited in a situation characterized by tough competition. As a result, convincing process results from laboratory experiments do not constitute sufficient source material for reaching decisions on production on a commercial scale in accordance with the new technology. Reproducibility on a pilot scale is usually required to shed light on the process economy.

Too often the daring researcher who sets out to push a process through this unknown territory is left alone to make his own mistakes and gain his own experience. To make things worse, the cost of mistakes increases rapidly with increasing production scale. Careful planning eases the burden a little, but is more than offset by the irritating fact that one cannot buy parts for this increasingly expensive equipment. A practical cell, the heart and most crucial part of any electrochemical process, has not been commercially available until now. One had to start by constructing a cell, a time-consuming and expensive task, which puts new and alien demands on the frustrated organic electrochemist.

2   DEMANDS ON A GENERAL PURPOSE PILOT CELL

These problems were deeply felt in the organic electrochemistry group at the University of Lund, so a decision was made to try to narrow the gap between laboratory experiments and practical applications. In any such undertaking the equipment problem must be resolved. One must construct and manufacture a cell for pilot-plant studies. The ideal would be a cell that could be used for almost any electrochemical process. The cell should give reliable and useful information about the process and be reasonably inexpensive. Such a cell, jointly developed by the electrochemical group at the chemical center and the Swedish National Development Company, is described below.

The first decision we reached was to aim for a modularized cell design made up of identical plate and frame units, because this design solves all the problems of scale-up effects encountered when the scale is stepped up to manufacturing level. Chief among the problems is the transport of substances to and from the electrodes. Mass transport is dependent in a complicated way on the flow pattern and the geometric design of the electrolyte chambers. To convert one set of data for a given arrangement to another, larger design is a cumbersome and unreliable procedure. If the pilot plant runs, however, are carried out in a unit that is identical to the one making up the manufacturing plant, this conversion need not be made at all. The information and results obtained in the pilot-plant studies will be directly applicable to the full-size plant.

The best known design using repeating units is the parallel-plate-and-frame type. To avoid difficulties, this arrangement was chosen for the cell. The next step in the development program was to select the other criteria that the cell had to fulfill. Since the cell was intended for use in any type of process, we decided to put the emphasis on flexibility and on a concept that could be implemented either as divided or undivided—we intentionally limited ourselves to two different media, thereby cutting ourselves off from more complex reaction schemes, which require three or more compartments. This decision led to a demand for replaceable electrodes. The next consideration was the construction material. Of course the cell had to be resistant to the most common solvents, that is, water of any pH, acetic acid, and the lower aliphatic alcohols, but a number of more exclusive solvents such as toluene, DMF, DMSO, methylene chloride, and acetonitrile were considered to be of interest too.

The size of the cell was not specified but we aimed at a unit suitable for a production of between 10 and 10,000 tons/year and preferably between 50 and 1000 tons/year. These limits reflect the range of production volumes for a list of potential candidates for electrochemical manufacturing. A production volume in excess of 10,000 tons/year makes the creation of specific cell concepts, which reduce capital and operating costs for that particular process, worth considering. Bearing in mind other demands that various organic electrosyntheses would put on the cell, we concluded that the cell ought to have (a) a current density range up to 40 A/dm^2, (b) a maximum allowable potential gradient in the electrodes of 0.1 V, and (c) an operating temperature between -40 and 100°C. Most organic electrosyntheses fall within these limits. On the other hand, we decided to exclude all high-pressure processes because of the different engineering problems involved in such designs.

Guaranteeing a high and uniform mass transport to the electrodes was very important. A turbulent plug flow appeared to be the best way to achieve this and to offer accurate reproducibility of the flow pattern from one run to another. When using organic solvents with low conductivity it is very important that the distance between the electrodes be kept as small as possible to avoid high ohmic losses. Such losses entail high energy costs and require increased cooling capacity. But small interelectrode distances presented us with another problem. If the membranes were allowed to come in contact with the electrodes, they were badly damaged under certain operating conditions.

Another major problem was to avoid any leakage of gases or liquids from the cell. This was even more important, since some of the possible solvents are highly flammable or poisonous and often have a high vapor pressure. Last, but not least, came the necessity of obtaining a piece of equipment that was practical to handle. The stacks had to be easy to disassemble and reassemble. All of the parts had to be easily accessible and reasonably light in weight, and maintenance had to be kept to a minimum. One important point was to try to avoid gaskets, since they tend to swell in organic solvents and are difficult to reposition when the stack is disassembled. All these considerations led to the following "demand profile" for the cell concept:

1. The cell should be built in a parallel-plate arrangement with repeating identical units.

2. The cell should be capable of operating in a divided or undivided manner.

3. The electrodes should be replaceable.

4. The cell components should be resistant to water of any pH, CH_3COOH, lower aliphatic alcohols, toluene, DMF, DMSO, CH_2Cl_2 and CH_3CN.

5. It should be possible to use the cell for producing chemicals with a production volume of 10 to 10,000 tons/year. The optimum range should be 50 to 1000 tons/year.

6. The cell should manage a current density between 1 and 40 A/dm².

7. The maximum allowable potential gradient in the electrodes should be 0.1 V.

8. It should be possible to work at temperatures between −40 and 100°C.

9. The cell should operate at atmospheric pressure.

10. A high and even mass transport to the electrodes must be guaranteed.

11. The electrode gap should be minimized.

12. The cell must be leakproof for both gases and liquids.

13. No contact should be allowed between electrodes and membranes.

14. The cell must be easy to handle.

3 DEVELOPMENT OF THE CELL DESIGN

Many of the important functions included in the design objectives for the cell are of such a nature that they must be linked directly to the design of the plastic frames. Uniform and turbulent plug flow, leakproof construction, membrane sealing, electrode fixation, and so on are examples of functions of this type. Several of these functions require more or less complicated design solutions as well as high dimensional accuracy.

Selecting the injection molding method for manufacturing the frame components meant that these requirements could be met at the same time as production could be simplified and rationalized, keeping the waste minimal. Plastic materials such as poly (vinylidene fluoride) plastic, which are also attractive from the chemical resistance viewpoint, could be used without great expense.

An important task at the beginning of the cell development work was to reach an optimum size and design for the electrodes. A monopolar cell design was chosen to avoid problems with parasitic currents. An optimization study was carried out aiming for a standardized electrode element for use in the whole range from laboratory and pilot scale up to an industrial scale of 50 to 1000 tons/year. Important parameters for the study were to meet the demands on current densities, potential gradients, and voltage drop for a flexible choice of electrode materials.

The final electrode design (8 dm^2 calculated with double-sided load) with two current collectors of composite construction on each electrode means that the electric requirements can be met for most electrode materials. The maximum current must, however, be limited for graphite electrodes, since considerably larger potential gradients than those aimed at in the specification of requirements will otherwise occur in the electrode because of the relatively poor conductivity of graphite.

When designing the plastic frames that hold the electrodes in position our aim was to include as many of the necessary functions, such as flow distribution, turbulence formation, seals, and spacing, in as few frame components as possible. The aim was also to achieve a design that made it possible to keep the electrode and frame components together in a single unit so as to simplify handling when assembling the cell. The four different components included in an electrode element are shown in Fig. 3.1 and assembled in Fig. 3.2.

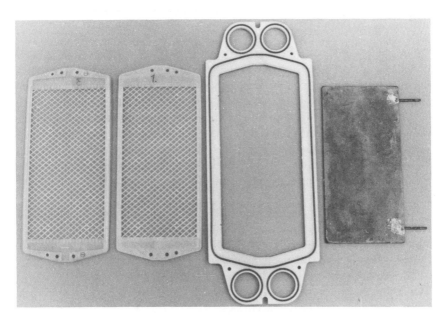

Fig. 3.1.   Disassembled electrode element.

Fig. 3.2.   Assembled electrode element.

The components consist of an outer frame, two inner frames, and an electrode. The outer frame distributes the electrolyte to the anode and cathode cell compartments. The outer frame has a groove that permits sealing by means of an O-ring to prevent leakage from the cell. It also has a specially developed groove sealing system for sealing and fixing the membrane so that catholyte leakage to the anolyte and vice versa are avoided.

The inner frames have a barrier system at the top and bottom for distributing the electrolyte and a grid-shaped turbulence generator between these that also acts as a separating element to prevent contact between the electrodes and the membrane.

Flow studies, with a colored electrolyte, were carried out as a basis for the present design of these components. These studies made it possible to optimize the design so as to achieve the turbulent plug flow sought. They also included adding gas to the incoming electrolyte flow to adapt the design to facilitate gas release from the cell compartment and to avoid pockets of gas. This is important since considerable gas development occurs in most processes, at least at one of the electrodes.

Laboratory modules can be constructed in a size range of 0.04 (one-sided load) to 1.04 m^2 for the electrode surface by alternating cathode and anode elements with membranes (if necessary). The number of modules required for the process under consideration is added for pilot- and industrial-scale applications.

The electrodes are connected in parallel within each module, while the modules are connected in series to each other. Electrolyte is supplied in parallel to the various modules. Figure 3.3 shows the principles for the electric connection and flow distribution in the system.

Figure 3.4 shows the construction of a seven-module section of a larger production unit, as well as the various components of a module.

A pilot cell with six modules with a total electrode area of 3.84 m^2 consisting of Pb/PbO_2 electrodes has been produced and installed at the Foundation for Industrial Organic Electrochemistry at the University of Lund, Sweden. The cell, which is used for process studies, can be seen in Fig. 3.5.

4 POSSIBLE APPLICATIONS FOR THE CELL

Thanks to its highly flexible design, the cell can be used in several areas of electrochemistry. The most obvious is, of course, inorganic electrochemistry. Since inorganic electrochemical applications involve aqueous media, it does not give rise to any serious problems. One of the most common inorganic applications is local production of chlorine or hypochlorite for swimming pools or small industries. Equipped with dimensionally stable anodes and steel cathodes, the cell is ideally suited for this reaction. Chlorine and caustic can be produced with membranes, and hypochlorite without membranes.

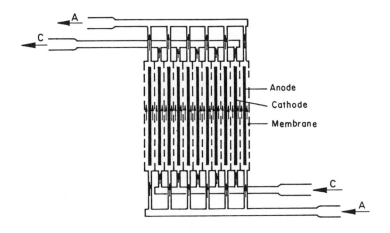

Fig. 3.3. Schematic flow distribution and electrical connections.

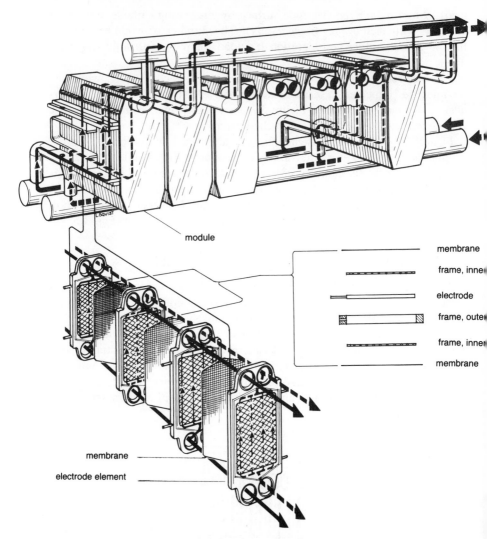

module

membrane

frame, inner

electrode

frame, outer

frame, inner

membrane

membrane

electrode element

Fig. 3.4.   The sketch above shows a seven-module section of a larger production cell.

Another simple but very useful case is the oxidation or reduction of metal salts. The cell can be put to use in regenerating the oxidant in electrochemical machining, chromium plating, and copper etching. All applications involving the use of metal salts to oxidize or reduce a substrate and then regenerate the active form of metal electrochemically belongs to this group. The ultimate in metal reduction is electroplating, where the material is retrieved

Fig. 3.5.   A six-module pilot cell on a 3.84-m^2 electrode area.

in solid form on the cathode, a practical way to solve an effluent problem and save the value of the metal [2–4]. These applications mostly consist of fairly small operations working under conditions that rarely put any excessive demands on the cell.

Solving environmental problems is an area where electrochemistry has created a niche of its own. The retrieval of heavy metals from the plating industry is mentioned earlier, but the cyanides, which constitute another major problem in that industry, can also be tackled. They can be oxidized to form harmless cyanates either directly [5] or by means of electrochemically generated chlorine [6]. Chlorine can be a problem too, at least in the form of chlorinated hydrocarbons. However, this chlorine can be removed electrochemically and the hydrocarbon retrieved [7, 8]. Perhaps the most general and useful application of electrochemistry in effluent treatment is in oxidizing organic impurities in low concentrations to carbon dioxide at ambient temperatures [2–4]. Though this is not electrosynthesis, the reactions are of the same type and the same kind of equipment is also used in many cases.

A borderline case between organic and inorganic electrochemistry, which we have studied, is the oxidation of bromide to bromine in spent brominating solutions. The pharmaceutical and fine chemicals industry carries out quite a lot of bromination of organic substances. This procedure consumes a large quantity of rather expensive bromine, which is only used to the extent of 50%, since the other half goes to bromide ion. Bromine is cumbersome to handle and the spent bromide solution often contains organic impurities, making it an effluent problem. However, the spent bromide solution can be oxidized on graphite anodes in a divided cell to regenerate bromine with satisfactory yields (see Fig. 3.6). The bromine can then be recovered by solvent extraction or blowing with air.

Quinones can be made by anodic oxidation of phenols. The phenols are dissolved in mixtures of acetonitrile and dilute sulfuric acid and are oxidized on a lead dioxide anode in a divided cell. Because of the low solubility of the phenols in these mixtures (the proportion of acetonitrile should not exceed about 50%) the current density cannot be too high. Reactants tested so far include 2,3,4,6-tetramethylphenol, 2,6-dimethylphenol, and 2,6-di-t-butylphenol.

Indirect Oxidation with Ceric Ion

One field that appears to show great promise for electrosynthesis is the generation of oxidizing and reducing agents. Unless regeneration is applied, few of these agents can be used without giving rise to major disposal problems. For the more expensive agents, even a shortsighted economic evaluation would suggest the necessity for regeneration.

Considerable interest has been shown in the use [9, 10] and regeneration [11, 12] of the ceric ion, mainly because the ceric ion can be used for oxidizing toluene and substituted toluenes to aldehydes with satisfactory yields.

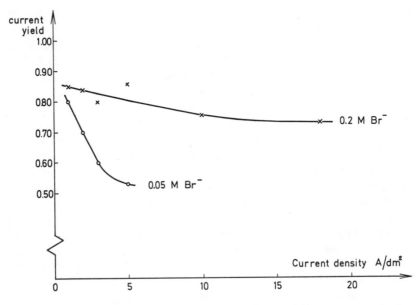

Fig. 3.6. Oxidation of spent brominating solution. Current yield versus current density.

The reaction between the ceric ion and the toluene is usually carried out in a two-phase system, consisting of an aqueous acid solution of the oxidant and usually a mixture of the toluene and an inert solvent, such as hexane and methylene chloride. Without the added solvents, this gives a high number of dimers and other by-products with most substrates.

The oxidizing power of the ceric ion depends mainly on the counterion involved. A weak complexing agent such as perchlorate gives a strong oxidant, which can be used for toluene itself and weakly deactivated toluenes. Other counterions such as sulfate, are used to attack easily oxidized substrates such as p-methoxytoluene.

One must also consider the solubility of the ceric and cerous ions and the cost of regeneration when choosing the aqueous system. We have chosen to use ceric sulfate in dilute sulfuric acid in our investigations, mainly to make use of inexpensive lead anodes possible. Then low solubility of the cerous ions permits use of low current densities and the low oxidizing strength restricts the choice of substrates.

The regeneration causes some problems. To avoid the reduction of ceric ion on the cathode one should use a divided cell, but available membranes are not very selective to protons relative to ceric and cerous ions. The total concentration of ceric and cerous ions in the anolyte will not, however, change very much because of this shortcoming of the membranes, since they also allow a small amount of water to pass from anolyte to catholyte. If the catholyte is made up of saturated cerous sulfate and the electrolyte volumes are balanced by pumping catholyte to the anolyte, the concentration of ceric and cerous ions is maintained in the electrolytes. However, then the sulfuric acid concentration changes so that the acid is enriched in the anolyte. This problem can be dealt with by choosing a higher concentration of acid in the anolyte and also by pumping anolyte to the catholyte.

Although the regeneration takes place at rather low current densities, lead anodes give rise to problems. The layer of lead dioxide developed is not stable on pure lead and particles of the dioxide soon flow around and cause trouble with an increasing pressure drop across the cell stacks. A small amount of silver in the lead reduces this problem [13].

The main problem associated with regeneration of metal ions is the avoidance of organic material in the anolyte. Small amounts of toluene or aldehyde may cause large losses in efficiency, either by consuming electrons or by fouling of the anode. Efficient cleaning of the spent oxidant solution is, consequently, essential before the solution is returned to the anolyte.

A study of the economic aspects of regeneration soon shows that capital costs are high compared to the costs for electrical energy. Obtaining a high *effective current density* (the product of current density and current efficiency) is, therefore, more important than obtaining a high current efficiency. Provided sufficient experimental data are available, a relation can easily be derived between the relative change of the current density and the relative change of the current efficiency as a basis for determining the economic optimum.

We have made a rough estimate of the most economic current density at different flow rates and cerous ion concentrations for the oxidation of cerous sulfate in sulfuric acid in the SU Electro Syn Cell (Fig. 7). At the most economic alternative the current efficiency is only about 50%.

The anolyte flow has a considerable influence on the effective current density. This is raised by at least 40% if the flow is increased from 0.2 to 0.3 m/sec (10 to 15 dm^3/min, frame) and the current efficiency is kept constant (by increasing the current density).

Another important determining factor for the regeneration efficiency is the ratio of ceric to cerous ion in the anolyte. A high proportion of ceric ion gives a high conversion of toluene in the chemical reactor per unit volume of aqueous solution passed, but it also gives a poor effect in the cell. Nevertheless, we

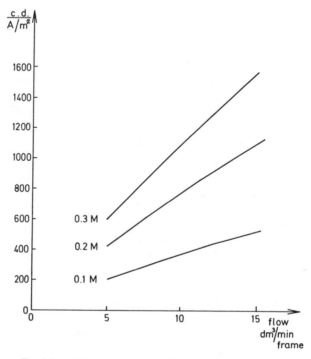

Fig. 3.7.   Oxidation of cerous sulfate at different flow rates.

consider that the high costs for cleaning spent oxidant require a fairly high concentration of ceric ion.

Figure 3.8 shows a pilot plant for the production of p-methoxybenzaldehyde. The plant consists of a six-module SU Electro Syn Cell for the regeneration of the oxidant, a reactor in which toluene is oxidized and a similar unit for cleaning spent oxidant. The chemical reaction and cleaning are carried out batchwise. When the cleaning operation was optimized the current efficiency was fairly good, although not so good as with uncontaminated electrolyte.

Fig. 3.8. The pilot-cell installation including external electrolyte system, heat exchanger, and so on.

Reduction of Oxalic Acid

At the time of our study general opinion was optimistic about the competitive possibilities of the electrochemical approach to glyoxylic acid from oxalic acid compared with the traditional approach. Environmental restrictions, it was assumed, would sooner or later force producers to use a cleaner process and the cathodic reduction of oxalic acid would then be the choice. In spite of considerable work by electrochemists [14] and despite the attractive features of this reduction, it does not look like a winner today. Nevertheless, we describe our experience to demonstrate the effectiveness of the SU Electro Syn Cell.

In the electrochemical approach the reduction was run continuously in a divided cell with lead cathodes. The recirculating catholyte was an aqueous solution of glyoxylic acid saturated with oxalic acid. Oxalic acid and water were supplied and a comparable quantity of catholyte was drawn off. Most of the oxalic acid in this outflow was recovered after evaporation and crystallization.

On the anodic side oxygen was produced from a dilute sulfuric acid solution on a dimensionally stable anode. The electrolyses could be run at a high current density with high material and current yields. The losses mainly were due to the simultaneous reduction of glyoxylic acid to glycollic acid and the evolution of hydrogen.

The formation of the glycollic acid increased with higher glyoxylic acid concentration and higher temperature in the catholyte. It was, however, independent of the current density. The rate of formation might be controlled by the transformation of glyoxylic acid hydrate to the anhydrous form. Figure 3.9 shows some experimental values of the ratio (α) between the concentrations of glyoxylic acid and glycollic acid in the catholyte from electrolyses with the same glyoxylic acid concentration. The ratio naturally increased with a higher current density and it was found to increase exponentially with the inversed absolute temperature.

The ratio of evolved hydrogen was fairly independent of the parameters discussed above. As long as the mass transfer in the cell was good and the catholyte was saturated with oxalic acid, the current density was increased to at least 4000 A/m^2 without causing a pronounced increase in the relative amount of hydrogen formed. The rate of hydrogen evolution was, however, sensitive to metallic deposits on the cathodes. Contact had to be avoided between the electrolytes and metallic construction materials. This meant that, for example, cooling was a problem.

We made a number of runs in a module of the SU Electro Syn Cell* in the

---

*The SU Electro Syn Cell is available from the Swedish National Development Company or in the United States and Canada from The Electrosynthesis Co., P.O. Box 16, East Amherst, NY 14051.

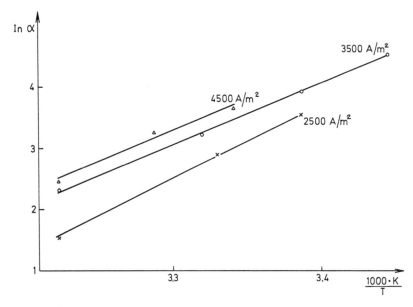

Fig. 3.9. The temperature dependence of the ratio between the concentrations of glyoxylic and glycollic acid.

manner suggested above. At the beginning of the electrolyses the catholyte was a pure solution of oxalic acid. The concentration of glyoxylic acid increased toward an equilibrium value and was dependent on the current density and the amount of water and oxalic acid added. The lead anodes used evolved some ozone. The ozone may have diffused into the catholyte and caused some oxidation.

The catholyte flow was about 0.5 m/sec, the temperature in the catholyte 15 to 16°C, and the current density was 1350 or 2700 A/m². The current yields were above 80% in both cases. With the lower current density the equilibrium concentration of glyoxylic acid was 1.2 moles/dm³, with the higher 1.4 moles/dm³. The ratio between glyoxylic acid and glycollic acid was about 40 in both cases, which means that the material yield was nearly 98%.

References

1. M. M. Baizer, *J. Appl. Electrochem.*, **10**, 285 (1980).
2. H. Tamura and T. Arikado, *Kagaku (Kyoto)*, **30** (9), 736 (1975).
3. G. Pini, *Chem.-Tech. (Heidelberg)*, **4** (7), 257 (1975).
4. T. Kaczmarek and R. Dylewski, *Chemik (Gliwice)*, **28** (11), 3 (1975).
5. J. P. Divn, *Trait. Surf.*, **14** (124), 35; (125) 46; (126) 59; (127) 48 (1973).
6. M. Beevers, *Met. Finish. J.*, **18** (211), 232 (1972).

7. P. K. Mitskevich and L. D. Guturova, *Ukr. Khim.*, 2h **39** (1), 82 (1973).
8. J. W. Sease and R. C. Reed, *Tetrahedon Lett.*, **6**, 393 (1975).
9. T. L. Ho, *Synthesis* **1973**, 347.
10. L. Syper, *Tetrahedon Lett.*, **1966**, 4493.
11. G. Smith, G. Frank and A. E. Kott, *Ind. Eng. Chem. Anal. Ed.*, **1940**, 12, 268.
12. R. Ramaswamy, M. S. Venkatachalapathy, and H. V. K. Udupa, *Bull. Chem. Soc., Jap.*, **35**, 175 (1962).
13. K. Scott, Ph.D. Thesis, University of Newcastle, 1977.
14. U. C. Tainton, A. G. Taylor and H. P. Erlinger, *Am. Inst. Mining Met. Eng. Tech. Publ.*, **221**, 12 (1929).

Chapter **IV**

ELECTRODE MATERIALS

B. V. Tilak

Hooker Research Center,
Long Road,
Grand Island, New York

S. Sarangapani,

Union Carbide, Parma Technical Center
Cleveland, Ohio

and

N. L. Weinberg

Electrosynthesis Company,
East Amherst, New York

---

1   INTRODUCTION

Of all the electrochemistry-related variables, such as electrode material and current density, it is the electrode material that continues to play a dominant role in electroorganic chemistry for a variety of reasons. Its primary role is achieving and/or imparting selectivity for a given course of reaction at high faradaic efficiencies. Interfacial electrochemistry deals with "events" occurring at the metal/solution interface that are complex not only by themselves, but are made much more complex by the influence of adsorption of ions and/or neutral species, surface oxide layers, and several other factors. The rewards for successful understanding of the interfacial phenomena are significant in the wake of energy conservation in electrolytic industries forced by escalating power costs.

One of the ways of accomplishing energy savings is by reducing the power consumption per unit of product. Power consumption is directly related to the cell voltage and it is indeed possible to achieve considerable voltage savings by appropriate choice of cell design. Cell voltage is the additive sum of several distinctive components (see Chapter I). Two of these components relevant to the present discussion are the cathodic and anodic overvoltages, which are dependent on the nature of electrode materials and could force selectivity and specificity, which are characteristic of electrochemical processes as is evident in aluminum and other electrowinning operations and the chloralkali industry.

The choice of electrode materials for a given process should be judicious, based on cost, chemical stability, electrochemical performance characteristics, and the nature of the process involved. As such, the basis for choice should be carefully developed and what is applicable for one electrolytic process need not necessarily be "effective" for another process, as is illustrated in the following sections.

The purpose of this chapter is to illustrate the complexities involved in electrode material selection in electroorganic reactions and to rationalize some of the earlier excellent discussions [1–5] on the criteria for selection of materials.

2 NATURE AND CLASSIFICATION OF ELECTROORGANIC REACTIONS

Generally, electroorganic reactions are complex in the sense that most of the reactions, following the initial electron-transfer step [see (4.1)] where the electrode materials play a major role, proceed further in several steps either chemically or electrochemically—the electrochemical steps sometimes competing with the initial step by means of surface coverage and the chemical steps competing in a multitude of pathways with the reactants, products, and/or intermediates formed during the electrolysis.

$$\text{Molecule} \xrightarrow{+e} (\text{radical ion}) \xrightarrow[\text{fast}]{\text{reaction}} \text{intermediate} \xrightarrow[\text{fast}]{+e} \text{product} \qquad (4.1)$$

The manner in which the electron-transfer step can occur can be *direct* or *indirect* as shown in Table 4.1. Examples of indirect routes are presented in Table 4.2.

Thus, following the first electron-transfer reaction, the electrochemically produced radicals, radical cations, or anions react further, resulting in the desired product(s) (see Figs. 4.1 and 4.2). The subsequent chemical or electrochemical reactions are manifold and should be made selective through a detailed knowledge (or at least an intuitive understanding) of the kinetics of different competing consecutive reactions.

3 ELECTRODE MATERIALS USED IN ELECTROORGANIC REACTIONS

Many thousands of electroorganic synthetic studies of an exploratory nature have been conducted [7, 8] over the years. The electrode materials employed in these studies are summarized in Table 4.3. Table 4.4 illustrates materials used in other industrial electro-chemical operations.

The commonly used anode materials are Pt, C, and PbO_2, which may be employed for oxidation of organic substrates such as olefins, aromatics, and saturated hydrocarbons. Hg, Ag, Au, Cu, Fe, and Ni form a secondary group of useful anodes for easily oxidizable organic compounds. Hg itself is too readily oxidized to be useful, but has found application to organics with very low oxidation potential (i.e., those that are oxidized more readily than Hg). Fe and Ni are especially useful in aqueous alkaline media, where they are relatively stable to corrosion. Ni and Monel are excellent for electrofluorination reactions. With several metals organometallics may be the major products when certain organics are used. (e.g., Hg, Pb, and Al for organo-

Table 4.1.   Classification of Electroorganic Reactions

1. Direct: Charge transfer is the primary act with the organic substrate of interest
   a. Cation radical formation: $R \longrightarrow R^{+\cdot} + e$
   b. Anion radical formation: $RX \longrightarrow RX^{-\cdot} - e$
      (where X may be halogen, H or other functional group)
   c. Carbonium ion formation: $RH \longrightarrow R^+ + H^+ + 2e$

   $(e.g., CH_3\text{-}COO^- \xrightarrow[\text{anode}]{\text{carbon}} CH_3^+ + CO_2 + 2e)$

   d. Carbanion formation: $RX + 2e \longrightarrow R^- + X^-$
   e. Reduction of carbonium ions or oxidation of anions:

product

2. Indirect: Charge transfer occurs with some other species, which then reacts with the substrate of interest
   a. With electroregenerated redox species (e.g., Cr^{6+}, Ce^{4+}, Ag^{2+}, Br_2, etc.)
   b. With adsorbed intermediates generated during the course of a reaction (e.g., Cl_{ads}, CO_2^-, HO_2^- (ads), H^+ or OH^-)

metallics). Other useful alloys for anodic reactions include: Pb/Sb, Monel, ferrosilicon, Duriron, WC, and Pt/Ir.

Among the cathode materials Hg, Pb, Al, Ag, Zn, Ni, Fe, Cu, Sn, Cd, C, and Pt (in nonaqueous media) have found the widest use, and Au, Mg, Ru, and Pd have been used to a lesser extent. Useful solid mercury cathodes include amalgams of Pb, Ni, Zn, and Cu; organometallics are readily formed with Hg, Sn, and Pb under certain conditions. The use of Hg as a cathode material has declined in recent years for industrial processes because of pollution problems.

Table 4.2.   Indirect Electrolysis

Redox Couple	Electrochemical Conversion
Ti^{4+}/Ti^{3+}	Nitroaromatics $\longrightarrow$ anilines,
	quinone $\longrightarrow$ hydroquinone
Fe^{3+}/Fe^{2+}	Acrylonitrile polymerization
$Fe(CN)_6^{-3}/Fe(CN)_6^{-4}$	Benzene oxidation
MnO_4^-/MnO_4^{-2}	Oxidation of aromatics
Ni^{3+}/NiF_6^{-2}	Electrofluorination
Tl^{3+}/Tl^+	1-Butene to methyl ethyl ketone
Co^{3+}/Co^{2+}	Oxidation of aromatics
Sn^{4+}/Sn^{2+}	Reduction of nitrocompounds
Ce^{2+}/Ce^{3+}	Anthracene to anthraquinone
Cu^{2+}/Cu^+	Hydroxylation of aromatics
VO_3^-/VO^{2+}	Oxidation of aromatics
HIO_4/HIO_3	Dialdehyde starch process
$NaHg/Hg$	Hydrodimerization
$NaOCl/NaCl$ or	Propylene oxide from propylene, oxidation of sugars
$NaOBr/NaBr$	
$OsO_4/[OsO_2(OH)_4]^{2-}$	Olefins to glycols
Br_2/Br^-	Alkoxylation of furans
I_2/I^-	Halofunctionalization; Prevost reaction

4 INFLUENCE OF ELECTRODE MATERIALS ON THE SELECTIVITY OF ELECTROORGANIC REACTIONS

Electroorganic synthesis offers many examples wherein electrode materials play a significant role in altering the course of the reaction and result in the formation of different products. Some examples (see Figs. 4.3 to 4.6) involving the influence of electrode material are discussed here to illustrate the complexities involved in electroorganic reactions that involve not only the initial charge-transfer step, but also complex "follow-up" chemistry and/or electrochemistry.

Depending mainly on the choice of the cathode material, acetone can be converted [10] to isopropyl alcohol, pinacol, propane, or diisopropyl mercury (see Fig. 4.3). While mechanisms of formation of isopropyl alcohol and pinacol seem straightforward enough, factors involved in propane synthesis are not as clear. Diisopropyl mercury derives from electrogenerated radical species that react with the mercury cathode by metal-atom abstraction.

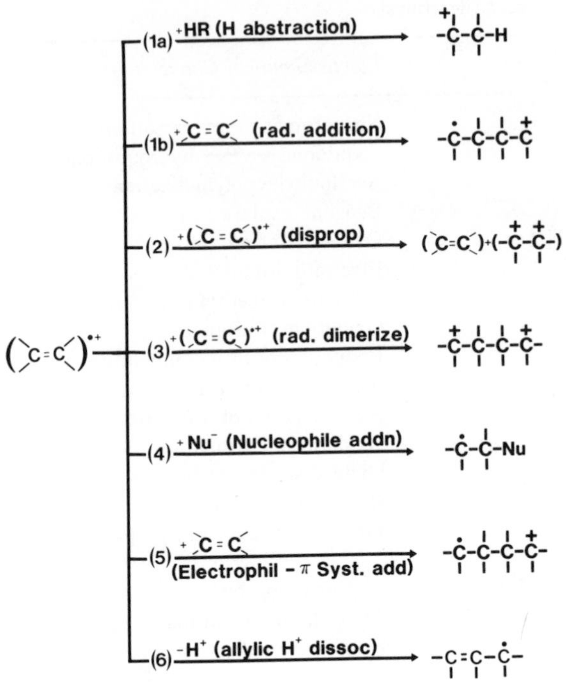

Fig. 4.1. Reaction routes [6] for radical cations obtained from unsaturated hydrocarbons: (*1a-3*) radical reactions; (*4-6*) cationic reactions. Reproduced with permission from the authors and Plenum Press.

Electrofluorination of organics occurs [11] at only a few kinds of anode materials (see Fig. 4.4). These include nickel in HF solution, porous carbon in molten HF/KF, and Pt in certain organic solvents containing fluoride ion. The Ni anode (3M Process) leads to nonspecific fluorination apparently involving a high-valent nickel fluoride (NiF_6^{2-} ?) species as fluorinating agent. This high-valent species is formed as an insoluble, continuously renewable anode surface coating (redox agent?) that attacks the organic. The porous carbon anode (Phillips ECF Process) also leads to nonspecific fluorinations, but the mechanism probably involves electrogeneration of fluorine atoms, which may be the fluorinating species. It is possible that the carbon surface may be protected from corrosion by stable, conducting C/F surface layers. The porous carbon anode is a hollow structure, providing a "conduit" for both the incoming feed and the outgoing fluorinated product. Fluorination occurs within the pores in the Phillips ECF Process. In contrast, Pt behaves

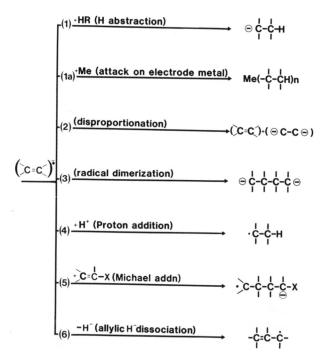

Fig. 4.2. Reaction routes [6] for radical anions from unsaturated hydrocarbons (e.g., vinyl compounds): (*1-3*) radical reactions; (*4-6*) anionic reactions. Reproduced with permission from the authors and Plenum Press.

completely differently. The organic is believed to be directly oxidized to a cationic species, which then undergoes nucleophilic attack by fluoride ion. Other materials such as Cu, Ag, and Fe, undergo severe passivation or corrosion that results in zero efficiency for electrofluorination. The Phillips ECF Process is reviewed in Chapter VII.

Hydrodimerization of acrylonitrile provides a further example of the role of electrode material in organic electrosynthesis (see Fig. 4.5). Thus, whereas acrylonitrile is reduced to propionitrile at Pt or Ni cathodes by adsorbed H atoms, Pb in strong acid solutions yields allylamine, Sn cathodes give the organometallic (Hg does not), and Hg, Pb, and C cathodes lead to the formation of adiponitrile and propionitrile. The Monsanto Adiponitrile Process is described in Chapter VI.

The diversity of products generated during electroreduction [13] of nitrobenzene (see Fig. 4.6) depends largely on the nature of the cathode

Table 4.3. Electrode Materials For Electroorganic Synthesis[a]

Electrode Materials	Oxidations	Reductions
Pt	Saturated hydrocarbons Aromatics Olefins Acetylenes Alcohols, phenols Enols Esters Carboxylates Anhydrides Aldehydes, ketones Ethers Alkylacetates Ketone/bisulfite complexes Amines Amides, lactams Amidines Azo and hydrazo compounds Hydrazines N-Heterocycles Aromatic amine polymerization Alkyl and aryl halides Grignard reagents Mercaptans Disulfonates Xanthates Alkyl and aryl sulfides Sulfoxides Thioureas Sulfonic acids Nitroalkanes Halofunctionalization of olefins Halogenation of aromatics Specific electrofluorination CO, HCN, $(NH_4)_2CO_3$ Ammonium acetate	Aromatics Olefins Acetylenes Nitriles Nitro compounds Aldehydes, ketones $>C=N$-compounds Organometallics Olefin electroinitiated poly- merizations
Carbon/ graphite	Saturated bicyclic hydro- carbons Aromatics, alkylaromatics	Nitro compounds, alkyl halides Olefin electroinitiated poly- merization

Table 4.3. (Continued)

Electrode Materials	Oxidations	Reductions
	Olefins	
	Alcohols, phenols	
	Aldehydes	
	Carboxylates	
	Ethers	
	Nitroalkane salts	
	Alkyl isocyanides	
	Amines	
	Hydrazines	
	N-Heterocycles	
	Alkyl and aryl sulfides	
	Disulfides	
	Sulfoxides	
	Halofunctionalization of olefins	
	Electrofluorination	
PbO$_2$	Saturated hydrocarbons	—
	Aromatics, alkylaromatics	
	Electroinitiated polymerization	
	of aromatics	
	Olefins	
	Acetylenes	
	Aryl halides	
	Alcohols, phenols	
	Lactones	
	Aldehydes, ketones	
	Ethers	
	Amides, lactams	
	N-Heterocyclics	
	Sulfoxides	
	Sulfonic acids	
	Disulfonates	
Hg	Olefins (to organometallics)	Aromatics
	Pinacols	Olefins
	N-Heterocycles	Acetylenes
	Hydrazines	Aldehydes, ketones
	Hydrazides	Nitro compounds
	Hydroxylamines	Alkyl halides
	Grignard reagents	$>$C=N-compounds
		Nitriles
		$>$C=S (hydrogenolysis)

Table 4.3. *(Continued)*

Electrode Materials	Oxidations	Reductions
		–S–S-cleavage
		N-Heterocycles
		Onium Salts
		Organometallics
		Olefin electroinitiated poly-
		merization
Ag	Nitroalkane salts	Aromatics
	Aldehydes	Acetylenes
	Alcohols	Aldehydes, ketones
	Amines	
	Hydrazines	
	Oximes	
Ni	Alcohols	Olefins
	Aldehydes, ketones	Acetylenes
	Nitroalkanes	Aldehydes, ketones
	Electrofluorination	Nitriles
		Nitro compounds
Sn	—	Aromatics
		Aldehydes, ketones
		Nitro compounds
		Alkyl halides (to R_4Sn)
Cu	Acetylenes	Olefins
	Alcohols, phenols	Acetylenes
	CO	Aldehydes, ketones
	N-Heterocycles	Nitro compounds
		$>$C=S (hydrogenolysis)
Pb	Alcohols, phenols	Aromatics
	Grignard reagents (to R_4Pb)	Aldehydes, ketones
		Alkyl halides (to R_4Pb)
		Nitro compounds
		$>$C=N-compounds
		Nitriles
		N-Heterocycles
		$>$C=S (hydrogenolysis)
		–S–S-cleavage
		Onium salts
Fe	Olefins	Aldehydes, ketones
	Alcohols	Nitro compounds
	Ketones	
	Nitroalkanes	
	Anilines	
	Quinolinium salts	

Table 4.3. (Continued)

Electrode Materials	Oxidations	Reductions
Au	Olefins	Nitro compounds
	Acetylenes	Amides
	Alkylacetates	
	Hydrazines	
Zn	Phenol polymerization	Nitro compounds
		Alkyl halides
		Amides
Mn	Alcohols	—
Cr	Alcohols	—
Al	Alkyl halides	Aromatics
	(to organometallics)	Olefins
	Organometallics	Nitro compounds
	(to R_3Al)	N-Heterocycles
Mg	—	Olefins
Ru	—	Aromatics
Pd	Olefins	Nitriles
Cd	—	Aromatics
		Aldehydes, ketones
		N-Heterocycles
Pb/Hg	—	Glucose, amides, nitro compounds
Ni/Hg	—	Nitriles
Zn/Hg	—	Aldehydes, ketones
Cu/Hg	—	Salicylic acid, organometallics
Pb/Sb	Electrolysis	Aldehydes, ketones
Pb/Sn	—	Aldehydes, ketones
Pb/Ag	—	Salicylic acid
Cu/Ag	—	Acetylenes
Monel	Electrofluorination	Nitro compounds
	Alcohols	
Ferrosilicon	Nitration of aromatics	—
Durion	Nitration of aromatics	—
Ni/Pb	—	Nitriles
Cd/Bi	—	Aldehydes, ketones
Phosphor Bronze	—	Nitro compounds
WC	Aldehydes	—
Pd/Ni	—	Nitriles
Pt/Ir	Phenols, carboxylates	—

[a]The data in Table 4.3 were obtained from the tabulations of reactions in Parts I and II of this series [7, 8].

Table 4.4. Electrode Materials Used in Electrolytic Industries

Material	Anode/Cathode		Electrolytic Process
Pb	+	−	Electrolysis of solutions containing H_2SO_4
	+	+	Organic compounds
Fe	+	+	Water electrolysis
	+	+	Organic compounds in alkaline solutions
	−	+	Chloralkali electrolysis
	−	+	ClO_3^-, ClO_4^- and persalts
	−	+	Fused-salt electrolysis (Na, Li, Be, Ca)
Graphite	+	−	Chloralkali electrolysis
	+	+	Hypochlorite production
	+	−	ClO_3^- production
	+	+	Fused salt electrolysis (Na, Li, Be, Ca)
	+	−	Organic compounds
	+	−	Al electrolysis
Fe_3O_4	+	−	Chloralkali and chlorate
Nickel	+	−	Water electrolysis
	+	+	Fe (III) cyanide and permanganate
	+	+	Organic compounds
	+	+	Fused salt electrolysis (Na)
Platinum	+	+	Organic compounds in solutions containing chloride
	+	−	ClO_3^-, ClO_4^-, persalts, hypochlorite
Mercury	−	+	Chloralkali electrolysis
	−	+	Amalgam electrolysis (Cd, Tl, Zn)
	−	+	Organic compounds
Dimensionally stable anodes (Mixed anodes of Ti and Ru or other noble metals)			Chloralkali electrolysis, hypochlorite, ClO_3^- production, electrowinning and cathodic protection operations
Ta or Ti/Pt			Persulfate production, cathodic protection, hypochlorite production and electrodialysis
Ti/Pt-Ir			ClO_3^- and hypochlorite production
Ti/PbO$_2$			ClO_3^- production, organic compounds in acid media

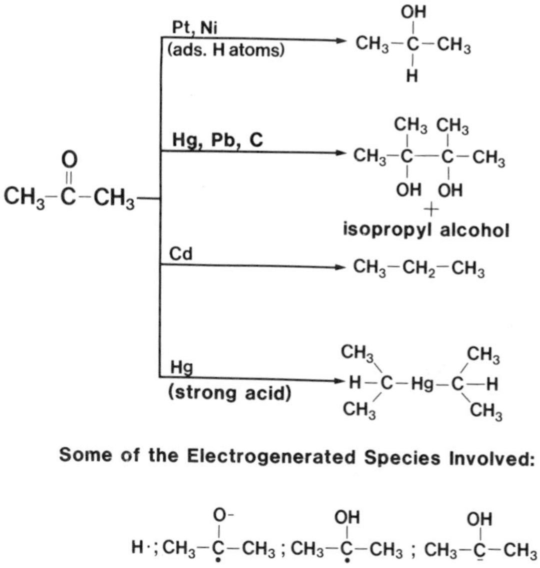

Some of the Electrogenerated Species Involved:

$$H \cdot \;;\; CH_3-\overset{O^-}{\underset{\cdot}{C}}-CH_3 \;;\; CH_3-\overset{OH}{\underset{\cdot}{C}}-CH_3 \;;\; CH_3-\overset{OH}{\underset{|}{C}}-CH_3$$

Fig. 4.3.   Influence of electrode material on the electrochemical reduction of acetone.

material, the solvent and supporting electrolyte, and other factors such as temperature and reactant concentration. Only the major product is shown and it is often accompanied by unwanted by-products.

Several of these major products and by-products may also be influenced by the magnitude of the electrode surface coverage by adsorbed hydrogen and the orientation of the adsorbed reactant or intermediate, which is dependent on the surface charge.

Thus the selectivity, as is noted earlier, depends on the nature and stability of intermediates and the rates at which the consecutive reactions proceed either in the bulk or at the interface. These factors dictate the appropriate choice of hydrodynamic regimes in the cell or appropriate electrode struc- tures that allow the reactions to proceed at significant rates, providing selec- tivity at the same time. It may be noted that the penalties associated with low selectivity and hence low yields are enormous, as the overall process economics would be adversely affected by the high energy requirements and the costs involved in separation and purification of the desired product.

5   CRITERIA FOR SELECTION OF ELECTRODE MATERIALS

The types of materials used and the nature of the selectivity achieved in electroorganic reactions are discussed in earlier sections. Although extensive

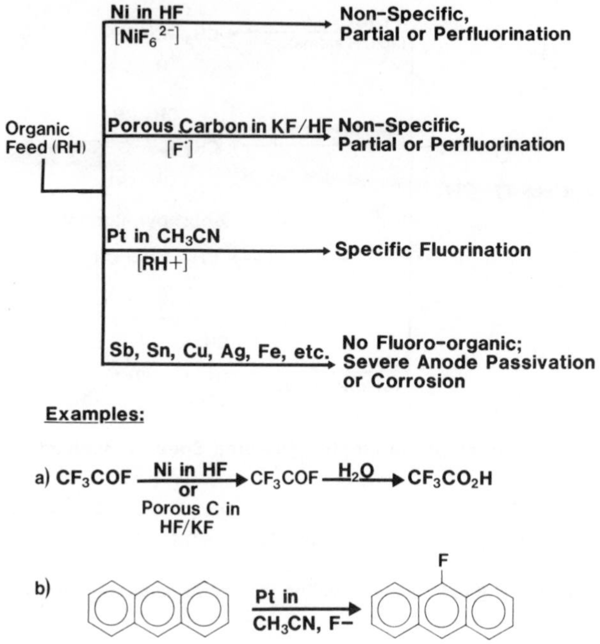

Ni in HF
[NiF$_6^{2-}$] → Non-Specific, Partial or Perfluorination

Organic Feed (RH)

Porous Carbon in KF/HF
[F'] → Non-Specific, Partial or Perfluorination

Pt in CH$_3$CN
[RH+] → Specific Fluorination

Sb, Sn, Cu, Ag, Fe, etc. → No Fluoro-organic; Severe Anode Passivation or Corrosion

Examples:

a) $CF_3COF \xrightarrow[\substack{or \\ Porous\ C\ in \\ HF/KF}]{Ni\ in\ HF} CF_3COF \xrightarrow{H_2O} CF_3CO_2H$

b)

Fig. 4.4. Influence of electrode material during electrofluorination.

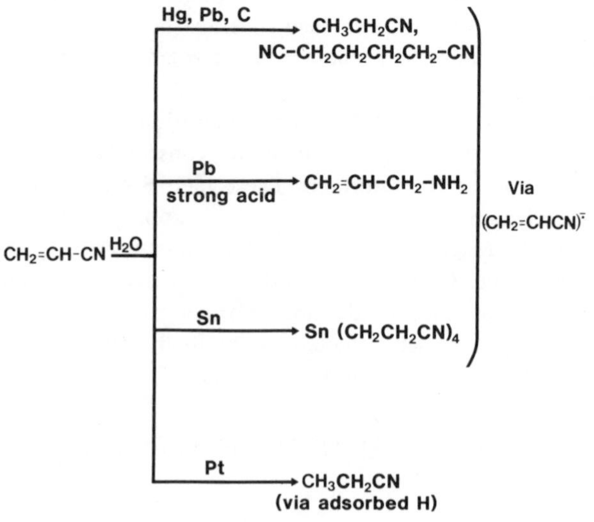

Hg, Pb, C → CH_3CH_2CN, $NC-CH_2CH_2CH_2CH_2-CN$

Pb strong acid → $CH_2=CH-CH_2-NH_2$

Sn → $Sn(CH_2CH_2CN)_4$

$CH_2=CH-CN \xrightarrow{H_2O}$

Pt → CH_3CH_2CN (via adsorbed H)

Via $(CH_2=CHCN)^{\bar{}}$

Fig. 4.5. Influence of electrode material during hydrodimerization of acrylonitrile.

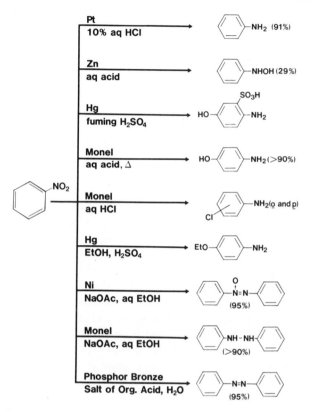

Fig. 4.6. Influence of electrode material during electrochemical reduction of nitrobenzene. Reproduced with permission from American Chemical Society.

investigations were carried out, most of these studies were only exploratory and few or no kinetic and electrocatalytic data characterizing these reactions were obtained. Lack of understanding of the electrode kinetics of electroorganic reactions is one of the hurdles in preventing a rational choice of electrode materials.

The purpose of this section is to discuss the basic criteria required of any material so that it will have low energy requirements and a high degree of selectivity, and to provide a basis on which to make a skillful choice of material for a given reaction based on experiences in other areas of industrial electrochemical operations (e.g., chloralkali industry, water electrolysis, and metal winning).

The main criteria relevant to choosing an electrode material—be it an anode or cathode—include the following:

1. Cost

2. Stability toward oxidation and/or attack by the reactants, products, and/or intermediates formed during electrolysis.

3. Conductivity of the bulk and surface layers formed at potentials where the products are generated. Note that the general tendency is either metal or metal oxide dissolution or loss of activity through the formation of insulating surface films or hydrides.

4. Electrocatalysis to achieve low overvoltage and high product selectivity.

Cost

The cost of an electrode material is an integral part of the overall process economics and a company's DCF/ROI (discounted cash flow/return on investment) requirements and is very much dependent on the type of electrode configuration, material, and coating technology (if any) used. Since the material costs, especially those of noble metals are escalating rapidly, no attempt is made here to provide these costs. However, one can "crudely" estimate these costs from the data published weekly in *Chemical Marketing Reporter* by Schnell Publishing Company, Inc. Other factors in arriving at costs of proprietary electrodes purchased from electrode manufacturers such as Electrode Corporation, Engelhard Industries include the type of negotiations regarding front-end fee and, running royalty—the details of which are not in the public domain.

Stability

Electrode stability is usually one of those factors requiring constant evaluation, especially in the case of anodes. Some general rules based on melting point, electronegativity, and so on have been proposed [14, 15] for stability. However, these are not applicable to electrochemical systems, since in electrochemical operations, the metal is immersed in a conducting electrolyte and a finite potential difference is established at the interface. These medium and polarization effects impose severe restrictions on the choice of electrode materials, especially for chlorine evolution, which gives rise to strongly oxidizing and corrosive conditions. Thus the material (be it a metal or metal oxide) used as an anode not only should exhibit stability toward dissolution, but also should not form insulating layers.

Stability can be judged from three factors: (a) thermodynamic considerations, (b) structural factors associated with substrate/coatings), and (c) reactivity of the electrode material with the intermediates, products, and/or reactants.

Thermodynamic Considerations

When a metal is immersed in an aqueous solution not containing its own ions, the metal either assumes a corrosion potential setup by the kinetically favored reaction(s) (e.g., Fe/1 M H_2SO_4) or an equilibrium potential established by the redox species in the solution. If such an electrode is polarized in the anodic direction, as is the case during Cl^- ion discharge, its stability can be approximately predicted, based on potential/pH diagrams [16]. Such diagrams are valid only for metal/H_2O systems in which the activity of the metal is assumed to be unity and for which equilibrium conditions are assumed. However, if the activity of the metal is different from unity, then it is, of course, possible to extend the *thermodynamic stability* of the metal on the anodic side. Thus, by alloying and/or incorporating the metal atom into a simple or mixed oxide lattice, the region of thermodynamic stability can be extended with the result that the "new" oxide structure may behave like a noble metal in so far as its natural tendency to remain in the metallic state is concerned. However, this "imparted" stability to the metal atom may be only transient because of unfavorable kinetics and/or the difficulties in achieving stability through incorporation of a sufficiently low concentration of the metal ions into a host lattice.

Thus, for example, the conducting sodium tungsten bronze system Na_xWO_3 (where $0.5 < x < 1$), is initially stable [17] at potentials where Cl_2 is generated. However, during electrolysis, the Na^+ ions are gradually leached out leaving a WO_3 matrix that is an insulator.

Another example of interest is $La_{0.5}Sr_{0.5}CoO_3$, which is suggested [18] as an electrocatalyst for O_2 reduction. During O_2 reduction, on most surfaces, O_2 chemisorbs in an end-on position resulting in HO_2^- species. The reduction of HO_2^- to OH^-, which involves splitting of the O–O bond, does not proceed at the same rate as the first reaction and hence H_2O_2 accumulates, promoting irreversibility. However, if O_2 is adsorbed in a side-on position, splitting of O_2 and further reduction to OH^- are facilitated rapidly by a series of proton switches accompanied by electron movement through the catalyst. Electron jumping between Co^{3+} and Co^{4+} by double exchange by way of intermediate O_2 ions provides conductivity and magnetic properties as well. This electrode, when used as an anode for chlorine generation exhibited passivity, and the loss of conductivity appears to arise from preferential dissolution of La and/or Sr much in the same way as in sodium bronzes, where passivity sets in as a result of "surface depletion of Na^+ ions."

NaCl electrolysis is carried out in aqueous solutions at large potentials positive to the hydrogen electrode; hence there is a tendency for most nonoxidic metallic compounds, be they sulfides or phosphides, to become converted into oxides, since the oxygen discharge reaction starts prior to the chlorine evolution reaction. It is from this viewpoint that simple or mixed oxides that exhibit (or approach) metallic conductivity are chosen as anode coatings. If the metal oxide is not in its highest valence state, then the tendency is for the metal of the oxide to be further oxidized to this state. Thus mixed oxides existing in their highest valence states and forming solid solutions should exhibit remarkable stability in environments such as those encountered during chlorine evolution. This may be the reason for the success of Electrode Corp. DSA* anodes for chlorine generation.

Structural Factors Associated with Substrate/Coating(s)

It is assumed so far that the coating formed on a given substrate is compact, porefree, and devoid of any imperfections or problems associated with incompatibility, as in the case of RuO_2 + TiO_2 coatings on Ti, which are "perfectly matched" in terms of lattice parameters, structure, and lattice volume (62 Å^3) (see Table 4.5). However, if imperfections arise as a result of improper methods of preparation of the surface and the coatings, one should anticipate failure of the coatings. Thus the performance losses [19] of Ti-supported anodes in NaCl seem to depend on the anode coating composition (see Fig. 4.7); compositions containing > 30 mole % RuO_2 in thick layers exhibit long life, while electrodes containing < 30 mole % RuO_2 can fail. This is attributed [20, 21] to poor doping and hence poor conductivity of the layers between Ti and the catalytic layer, resulting in an additional potential drop, varying with time, across the TiO_2 layer. Even with coatings contain-

Table 4.5. Some Crystallographic Features of RuO_2, TiO_2, Ti, and Ru

Property	TiO_2	RuO_2
Symmetry group	$P4_{2/mnm}$	$P4_{2/mnm}$
Crystal habit	Tetragonal	Tetragonal
Crystal structure	Rutile	Rutile
a_0	4.594 Å	4.4902 Å
c_0	2.958 Å	3.1059 Å
V	62.4280 Å	62.6202 Å
	Ti	Ru
Size	0.68 Å	0.65 Å
Electronegativity	1.5	1.8
Valence	+4	+4

*DSA is a registered trademark of Diamond Shamrock Technologies, S.A.

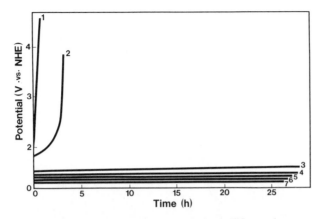

Fig. 4.7. Variation of potential [19] with time using RuO_2 + TiO_2 anodes at a current density of 1 A/cm² and various anode compositions in 290 g/liter NaCl (pH ~ 2 and temperature ~ 353°K). (*1*) 5% RuO_2 + 95% TiO_2; (*2*) 10% RuO_2 + 90% TiO_2; (*3*) 20% RuO_2 + 80% TiO_2; (*4*) 94% TiO_2 + 6% TiO_2 in the first layer and 10% RuO_2 + 90% TiO_2 in the layer next to the solution; (*5*) 42% RuO_2 + 58% TiO_2; (*6*) 20% RuO_2 + 80% TiO_2 in the layer close to the solution with an initial layer (i.e., adjacent to the Ti substrate) containing 94% TiO_2 + 6% RuO_2; (*7*) 100% RuO_2. Reproduced with permission from Plenum Publishing Corp.

ing > 30% RuO_2, degradation was observed and attributed to Ru dissolution [22] and O_2 evolution [23], the dissolution rate being higher initially and during power interruption.

Reactivity of the Electrode Material with the Intermediates, Products, and/or Reactants

Potential/pH diagrams are extremely useful for selecting materials. However, they should not be misused since the stability domains in Fig. 4.8 are subject to changes depending on the medium, complexing species in the electrolyte, temperature, and so on; in such instances, construction of specific potential/pH diagrams is necessary. Thus graphite can severely corrode because of intercalation by certain ions resulting in exfoliation or expansion of lattice planes. These ions include SO_4^- and ClO_4^- at anodes [24] and alkali metal, and NH_4^+ ions at the cathode [25]. The corrosion arising from the intercalating ion depends on the solvent medium, nature of the graphite material, and the concentration of the intercalating species. Electrogenerated oxidizing agents such as HO_2^-, $Cl\cdot$, and inorganic species can destroy various carbons. Yet it is surprising to note that electrofluorination of organics [10] occurs quite readily at certain carbon and nickel electrodes without any significant electrode degradation.

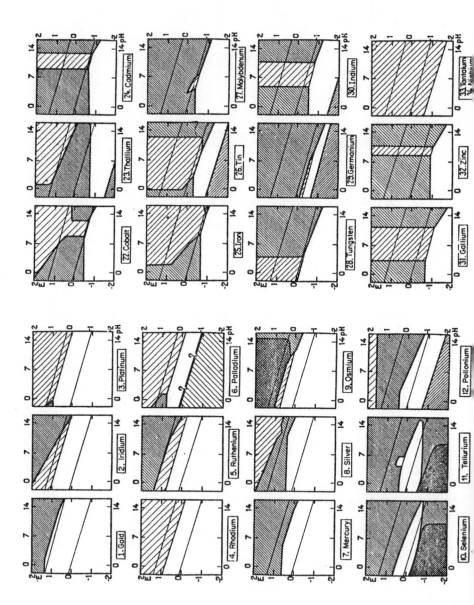

214

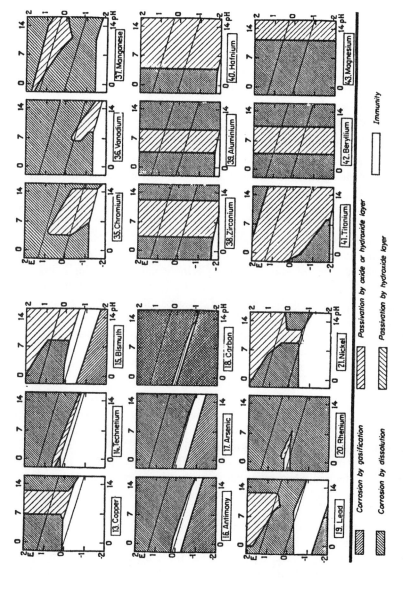

Fig. 4.8. Corrosion, immunity, and passivation domains of metals. Reproduced with permission from Pergamon Press.

215

Corrosion of metals can also occur by means of metal abstraction involving electrode material [M] and the discharged species $R^{\cdot}$:

$$R^{\cdot} + [M] \longrightarrow [RM^{\cdot}] \longrightarrow etc. \qquad (4.2)$$

Such reactions may occur at both cathodes and anodes and may involve a very broad range of substrates, such as alkyl and aryl halides, organometallic reagents, aldehydes, ketones, onium salts, aromatics, and olefins. Often this kind of degradation goes undetected in short-term experiments, but it may become apparent in long-term studies, especially with Hg, Zn, Cd, and Sn electrodes. (It should be noted that formation of organometallics may often be the desired process [26].)

Open-Circuit Potential and Its Significance

A measurement that is particularly useful in judging the stability of an electrode, is the open-circuit potential of the electrode (E_{ocp}) in the given medium. If E_{ocp} is different from the reversible potential (E_r) of the given reaction, then one can infer events occurring at the electrode that are responsible for the departure of the E_{ocp} from E_r. Thus iron in NaOH solutions (cathode in the chloralkali cell) exhibits E_{ocp} values negative to the hydrogen electrode in the same solution. This corrosion potential may be attributed to the anodic dissolution of iron as $HFeO_2^-$—the conjugate reaction being H^+, OCl^-, or ClO_3^- reduction schemes. If the latter two are responsible for the mixed potential set up at the Fe/NaOH aqueous interface, then metals such as Ni and Co, should also exhibit degradation or dissolution in the chloralkali catholyte especially during shutdowns when the concentration of hypochlorite is significant in the cathode compartment.

Let us now turn our attention to the cathodic behavior [27] of high-surface-area nickel in 20 to 25% KOH solutions in water electrolysis cells. Nickel exhibits negative open-circuit potentials (with respect to a reversible hydrogen electrode in the same solution) that appear to arise from Ni/"NiH"-type equilibria. While significant corrosion is not anticipated because of "NiH" formation, one may expect progressive buildup of "NiH" resulting in a change in the mechanism of H_2 evolution and/or blocking up of the electrode surface. It should be noted that this effect is accentuated by the adsorption of impurities and it is difficult to delineate the role of impurities vis-à-vis NiH formation.

Thus observation and interpretation of open-circuit potentials is very important, since this not only helps in understanding the behavior of electrodes, but also facilitates appropriate choice of electrode materials. It may be noted that generally, in industrial operations, the number of regular and unscheduled shutdowns may vary anywhere from 3 to > 12 (?) per year depending on the efficiency of plant operations, and hence consideration of chemistry and electrochemistry under open-circuit conditions is a key factor.

Conductivity Characteristics of Electrodes

Consideration of conductivity is important from the viewpoint of minimizing (a) voltage losses associated with ohmic drop across the coating and (b) formation of insulating layers. Hence we briefly discuss electrical properties of metals and metal oxides. Usually, under anodic conditions, the electrode is covered with a surface oxide through which electronic conduction occurs. The cathode, on the other hand, may or may not be covered with a surface oxide.

The commercial success of DSA electrodes has attracted the attention of electrochemists who focused on understanding the electrical conduction characteristics of metallic oxides in order to synthesize alternative and better electrocatalysts for chlorine evolution. Several binary and ternary oxides containing platinum group metal and nonplatinum metal double oxides have been synthesized [28] and their electrical conduction characteristics have been examined in various laboratories not only to elucidate the electrical transport properties, but also to find applications in other industrial processes. The subject of electrical properties in oxides, fluorides, and sulfides has been thoroughly discussed in the literature [29–36] and hence is not discussed in detail in this chapter. However, an attempt is made here to outline and highlight some of the developments that are of great importance in predicting the electrical conductivity of materials that may find application as electrode materials in electrolytic industries.

Most pure crystalline materials may be classified as either metals or insulators. Metals are characterized by low resistivity (10^{-2} to 10^{-6} $\Omega \cdot cm$ at room temperature) and a linear increase in resistivity with increasing temperature. On the other hand, insulators have resistivities of the order of 10^3 to 10^{17} $\Omega \cdot cm$, with resistivity decreasing exponentially with increasing temperature. These criteria would permit one to classify NiO, CoO, MnO, Fe_2O_3, Cr_2O_3, MnS, FeS_2, and MnS_2 as insulators and TiO, CrO_2, NbO, RuO_2, ReO_3, TiS, CoS_2, and CuS_2 as metals. However, there are some compounds, such as FeO, that exhibit semiconductivity and have a negative temperature coefficient and compounds such as MnO_2 whose conductivity is in between that of a nearly degenerate semiconductor and that of an impure metal.

There are two limiting descriptions [36] of the atomic outer electrons that contribute to electrical conduction in solids—band theory and localized electron theory. Band theory considers a solid as a collection of atoms and electrons that move more or less independently in a self-consistent field potential setup by the ions and other electrons, resulting in an appreciable overlap of electron orbitals of an atom. When both the ionic size and electronegativity of anion and cation are considerably different (as in transition metal oxides), the outer S and P orbitals form a filled valence band and an empty conduction band separated by a large energy gap of ~ 5 to 10 eV in oxides. If, when filled following Fermi-Dirac statistics, the Fermi level is in the middle of allowed states, the resulting solid is a metal and if the Fermi level lies well inside a forbidden gap, it is an insulator or a semiconductor. While this theory explains the experimental observations in the cases of ReO_3, TiO, and VO, it fails to unify the conductivity characteristics of transition metal oxides such as TiO, VO, MnO, FeO, CoO, and NiO. All these oxides have (a) identical structures (rock-salt structure), (b) similar lattice constants (ranging from 4.09 Å for VO to 4.43 Å for MnO), and (c) similar melting points (varying from 1420°C for FeO to 1990°C for NiO). Nevertheless, their conductivity characteristics are completely different—TiO with a room temperature conductivity of 4.7 × 10^{-3} $\Omega^{-1} \cdot cm^{-1}$, and MnO with a conductivity of ~ 10^{-15} $\Omega^{-1} \cdot cm^{-1}$—the other monoxides being insulators. According to elementary band theory, all these compounds should be conducting except FeO. (There are many examples cited in the literature showing the discrepancy between experiment and the predictions of band theory; the reader is referred to Ref. 36 for details.)

Several models [20 to 36] have been proposed to describe the conductivity and magnetic and optical properties of a wide range of materials (e.g., simple and complex oxides, sulfides, and phosphides) and of these, one generally accepted model is that of Goodenough.

Goodenough [33] based his approach on the nature of chemical bonding and crystal structure and developed "empirical" criteria to generalize the magnetic and electrical properties of oxides, bronzes, perovskites, and spinels. This he did by invoking the "critical" distance concept for the overlap of cation–cation (c-c) and cation–anion–cation (c-a-c) orbitals as determined by the crystal structures—the c-c and c-a-c interactions describing the behavior of d electrons whether they be localized or collective, and the location of the Fermi surface "distinguishing" these extremes. Two classes of metallic oxides are distinguished: (1) those having large b (where b refers to spin-independent transfer energy between nearest-neighbour cations) because of large cation–cation interactions as a result of a small cation–cation separation (Class I oxides); and (2) those having large b because of cation–anion–cation interaction due to large covalent mixing of anionic p orbitals into the cationic d orbitals (class II oxides). Mixed oxides of Class I-II, where both mechanisms prevail, were also considered by Goodenough.

To illustrate Goodenough's procedure, two case studies are presented to explain the electrical properties of TiO_2, an insulator, and ReO_3, a metal. TiO_2 crystallizes in a rutile lattice and in this lattice, each oxygen anion is surrounded by three Ti^{4+} cations and each Ti^{4+} cation by six oxygen anions. The distorted octahedral arrangement (see Fig. 4.9) around each cation generates a cubic crystal field that lowers the energy of three of the d orbitals

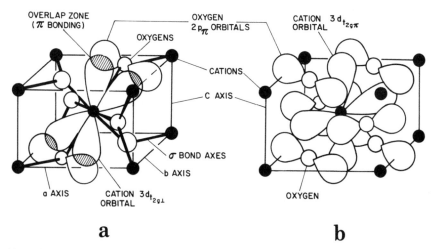

Fig. 4.9. (*A* and *B*) Crystal structure and bonding and (*C*) schematic band structure (*c*) diagram for TiO_2. Reproduced with permission from Plenum Press. (*a*) Overlap of only one of the $t_{2g} \perp$ orbitals with the two oxygen $2P_{\pi}$ orbitals with which *it* forms π-bonds. An additional cation $t_{2g} \perp$ oxygen $2P_{\pi}$ overlap occurs with the oxygen ions at the upper left and lower right with the $t_{2g} \perp$ orbital plane perpendicular to the one shown. (The second set is now shown for clarity.) Cations are filled circles at corners and center, and oxygens are open circles. Heavy solid lines denote axes of cation–oxygen σ-bonds. Note that π-bonds extend only along a chain parallel to the *C*-axis. (*b*) Locations of all the oxygen $2P_{\pi}$ orbitals are shown along with the cation $t_{2g} \parallel$ orbital. Note that no overlap occurs between this cation orbital and any of the surrounding oxygen $2P_{\pi}$ orbitals.

(d_{xy}, d_{xz}, d_{yz}) relative to the other two ($d_{x^2-y^2}$, d_{z^2}). The former three are designated as t_{2g} and the latter two as e_g orbitals. The two e_g, one 4S and the three 4P orbitals of the cation form strong σ bonds with appropriate orbitals of the anion (i.e., 6 sp^2 orbitals), and six additional σ* antibonding states. [Note here that occupancy of antibonding states reduces cohesiveness, whereas occupation of valence bonds (σ and π bands) contributes to cohesivity.] Two of the three $3t_{2g}$ orbitals combine with the remaining oxygen $2P_{\pi}$ orbitals to form π bonds—the third t_{2g} orbital being positioned such that two lobes lie along the C-axis and combine with another $3t_{2g}$ orbital of another Ti^{4+} cation on the C-axis. This is shown in Fig. 4.9a and 9b. The σ band and π band formed from mixing of cation d^2sp^3/oxygen sp^2 and cation $t_{2g} \perp$/oxygen P_{π} states are shown crosshatched in Fig. 4.10 to indicate that these bands are filled in TiO_2. The sum of the valence electrons for Ti (4) and two oxygens (12) is sufficient to fill the σ and π valence bands, with the result that the high-energy conduction bands remain empty, offering high electrical resistivity to TiO_2.

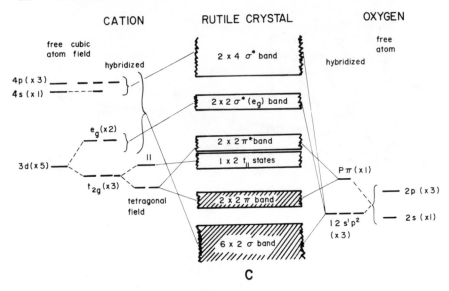

Fig. 4.10. Schematic band structure showing filled valence bands and empty conduction bands. Reproduced with permission from Plenum Press.

Conductivity can be imparted to TiO_2, for example, by doping with an element having one or more additional valence electrons, which will occupy the conduction band and offer electrical conductivity—sometimes equivalent to that of a metal. (This may, however, reduce the stability to oxidation generally.)

ReO_3 behaves like a metal and belongs to the perovskite class of structure. Figure 4.11a shows the atomic structure of the perovskite-like unit cell of ReO_3 and Fig. 4.11c gives the listing of atomic levels corresponding to valence shells for Re (on the left) and O (on the right). There are 7 valence electrons in Re and 18 valence electrons from the three O atoms in ReO_3. The splitting of the $5d$ Re orbitals and the mixing of the e_g, S, and P orbitals of Re in the direction of the anions, and the S and P_σ orbitals of the anions in the direction of cations is shown in Fig. 4.11b. The two e_g, 1S and 3P orbitals of the cation combine with $S^1P^1\sigma$ orbitals of O to form a σ band accommodating 12 electrons and generating σ^* and σ^* (e_g) antibonding bands. The three t_{2g} orbitals combine with $2P_\pi$ orbitals to form the π band accommodating 12 electrons. Thus, of the 25 valence electrons, 24 are located in the π and σ valence bands, leaving the extra electron in the next higher band of primarily t_{2g} character. This explains [35] the metallic conductivity exhibited by ReO_3.

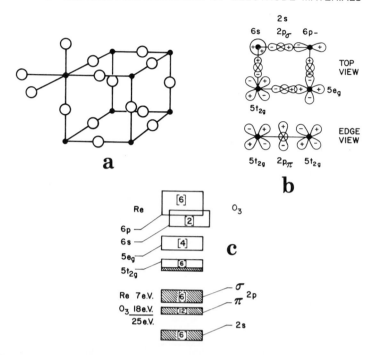

Fig. 4.11. (*a*) Crystal structure of ReO_3. (*b*) Band-structure diagram, based on orbitals for ReO_3. Atomic orbitals corresponding to valence shells for Re (left-hand side) and for O (right-hand side). Reproduced with permission from Marcel Dekker, Inc. (*c*) Types of bonding in ReO_3 [35].

Since the anode used for Cl_2 generation is a Ti substrate coated with noble metal compounds that are good electrocatalysts for chlorine evolution, it is relevant at this stage to examine the conductivity behavior of metal dioxides. Figure 4.12 shows the electrical conduction characteristics of various metal dioxides and Fig. 4.13 depicts the predicted Fermi levels for cation electronic configurations in various crystal structures along with some selected examples. Thus, from structural and electronic configuration of the oxides, it is indeed possible to predict and synthesize materials (see Table 4.6 for resistivities of common materials) having good electronic conductivity and chemical stability.

One of the materials used extensively in electroorganic syntheses is PbO_2, which exhibits metallic conductivity and is a better conductor than Hg or graphite. PbO_2 has found considerable use [37] in such processes as oxidation of organics, hydrodimerization of acrylonitrile, and chlorate, hypochlorite, iodate, periodate, and bromate production and can be readily

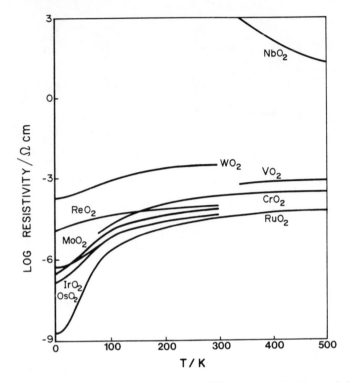

Fig. 4.12. Resistivity of some rutile structure oxides [34]. Reproduced with permission from Marcel Dekker, Inc.

electrodeposited on graphite, Ti, Zr, and so on. While PbO_2 electrodes should be kept at positive potentials to avoid corrosion, caution should be exercised, especially under open-circuit conditions in electrolytes containing complexing species.

Electrocatalysis

An understanding of electrocatalysis is very important since it would provide insight into the "characteristic parameters" required for a rational choice of materials for electroorganic processes. Judicious selection based on electrocatalytic considerations would (a) permit choice of materials exhibiting low overvoltage (η) and on which the interference from parasitic reactions (e.g., H_2, O_2 evolution) with the main reaction is minimal or negligible and (b) offer opportunities to synthesize "tailor-made" electrodes for electroorganic reactions, which would not only enhance selectivity, but might also improve the overall process economics (by reduction in the total

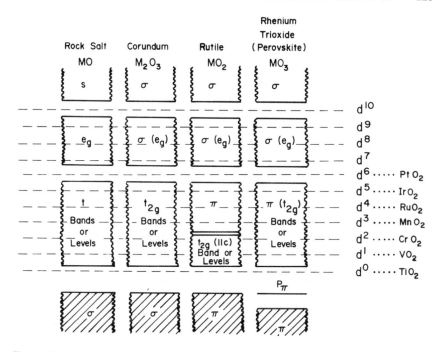

Fig. 4.13. Predicted [34] Fermi levels for cation electronic configurations in various crystal structures with examples from oxides having rutile structure. Reproduced with permission from Marcel Dekker, Inc.

number of unit operations). An example of interest is the case of propylene oxide, which is generally prepared [41] in two stages: chlorination of propylene to form propylene chlorhydrin and subsequent hydrolysis. However, in the presence of copper or cobalt acetate, propylene can be directly oxidized [42] to propylene oxide (on noble metal electrodes) in high yields, thus reducing the total number of unit operations.

In addition, a knowledge of the magnitude of exchange current density, i_0, for a given reaction would facilitate appropriate choice of electrode structure and cell geometry that would permit the reaction to proceed at an optimal rate. Mechanistic features involved in electrochemical reactions are reflected in the magnitude of the Tafel slope, b^{-1}, and hence overvoltage. Under Tafel conditions overvoltage is related to the exchange current density according to:

$$i = i_0 \exp(b\eta) \qquad (4.3)$$

Table 4.6. Resistivities of Common Materials at 25°C

Material	Resistivity ($\Omega \cdot$cm)	Reference
Al	2.826×10^{-6} (20°C)	38
Au	2.44×10^{-6}	38
Ag	1.6×10^{-6}	38
Cu	1.72×10^{-6}	38
Fe	9.2×10^{-6}	1
Hg	95.78×10^{-6}	38
Ir	6.1×10^{-6} (20°C)	38
Nb	13.2×10^{-6}	1
Ni	7.8×10^{-6} (20°C)	38
Pd	11×10^{-6}	38
Pt	10×10^{-6}	38
Sn	11.5×10^{-6}	38
Ta	15.5×10^{-6} (20°C)	38
Ti	55×10^{-6}	1
W	5.5×10^{-6} (20°C)	38
Zn	5.8×10^{-6}	38
Graphite	0.5 to 3×10^{-3}	1
Fe_3O_4	3.96×10^{-3}	1
PbO_2	9.08×10^{-5}	1
Rh_2O_3	130	39
RuO_2	2×10^{-5}	39
IrO_2	3×10^{-5}	39
Pt_3O_4	4×10^3	39
α-PtO_2	1×10^6	39
β-PtO_2	$\sim 6 \times 10^{-4}$	39
MnO_2	10	36
SiO_2	3×10^4 (27°C)	40
SnO_2	4×10^6 (20°C)	40
WO_3	2×10^5	40
TiO	$\sim 8 \times 10^{-3}$	34
TiO_2	$> 10^{13}$ (20°C)	40

The value of i_0, for a given reaction, is dependent on the nature of the electrode material, and this relative ability of metallic or nonmetallic surfaces in influencing the rate of the reaction may be termed [43] "electrocatalysis."

It is important to understand why the rate should be dependent on the nature of the electrode material. Let us consider the H_2 evolution reaction [44-46] as a case study. In general, the elementary steps in this reaction are

$$M + H^+ + e \longrightarrow MH \qquad \text{(discharge)} \qquad (4.4)$$

$$2MH \longrightarrow 2M + H_2 \qquad \text{(recombination)} \qquad (4.5)$$

$$MH + H^+ + e \longrightarrow M + H_2 \qquad \text{(electrochemical desorption)} \qquad (4.6)$$

The rate equations for (4.4) and (4.5) are of the form:

$$i = \frac{k \, \exp\,(-\Delta G_H/2)}{1 + K \exp\,(-\Delta G_H)} \qquad (4.7)$$

which indicates log i is an ascending and descending linear function of free energy of adsorption of H (ΔG_H), with a maximum at $\Delta G_H = 0$ at constant ϕ, pH, and $\{H^+\}$ (see Fig. 4.14). Under conditions when the Temkin isotherm applies, the maximum is not well defined and a plateau-like behavior is observed around $\Delta G_H = 0$. This behavior was given the name "volcano plot" by Balandin [47], who discussed catalytic cases. It can now be argued that when ΔG_H is negative, M-H is too strongly bound, resulting in a decrease of overall reaction rate. On the other hand, when ΔG_H is positive, M-H is weakly bound, resulting in a decrease in the rate of the reaction. These theoretical predications may now be compared with experimental

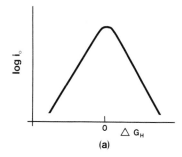

(a)

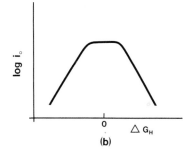

(b)

Fig. 4.14. Log i_o variation with standard free energy of adsorption of H_2 on the electrode surface assuming that adsorbed atoms obey (*a*) Langmuir adsorption isotherm and (*b*) Temkin's logarithmic adsorption isotherm.

observations plotted against ΔH_{ads} (see Fig. 4.15). Since plotting against ΔH would correspond *only* to the loss of three translational degrees of freedom (which implies that standard entropy of adsorption is metal independent), the agreement appears satisfactory. It may be noted that the mechanism of H_2 discharge on either side of the curve is different. Discharge is rate controlling (followed by electrochemical desportion) on *sp* metals and electrochemical desorption is slow on *d* metals except at low overpotentials (preceded by fast discharge).

An interesting analysis of volcano curves was made by Parsons [48], addressing the question, "Is there a possibility of enhancing the overall rate by using an electrode that has two types of surfaces with different adsorption properties?". Reactions (4.8) and (4.9) are

$$A^+ + e \longrightarrow B(ads) \tag{4.8}$$

$$B(ads) + e \longrightarrow C^- \tag{4.9}$$

assumed to occur on both surfaces and the possibility of enhancement of rate arises from the transfer of the intermediate B between sites of different adsorption energy. Assuming noninteracting sites and constant transfer coefficient, which implies i_{0-1}/i_{0-2} is independent of the free energy of adsorption, Parsons arrived at the conclusion that it is almost impossible to produce a good electrocatalyst from two poor ones when only one adsorbed intermediate is involved in the overall reaction. While this appears to be generally true in the case of the H_2 evolution reaction, by understanding the structural and compositional features of the intermediates, it is indeed possi-

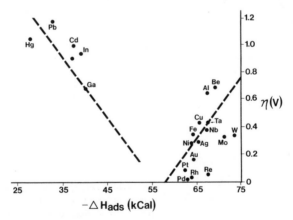

Fig. 4.15. Volcano plots [46] for the hydrogen evolution reaction in acid solutions. Reproduced with permission from Plenum Press.

ble to "manipulate the surfaces" to achieve not only a low overvoltage, but also a high degree of selectivity for a given reaction.

Kinetic studies related to electrocatalytic factors involved in electroorganic reactions are scarce except in a few instances where some attempts have been made. One such example [98] includes oxidation of sodium formate on Au/Pd alloys, where maximum electrocatalytic activity was noticed with electrodes containing 50 atom % Au. Another example of interest is the oxidation of formamide [99] in acid media, where maximum activity was observed with Pt electrodes containing 35 atom % Au. While such investigations involving electroorganic reactions are still in their infancy, the nature of intermediates formed during the course of electroinorganic reactions (e.g., H_2, O_2, Cl_2 evolution, O_2 reduction) and their reactivity with "surface orbitals" has been extensively investigated. The role of the intermediates present in reactions of practical interest and the present status in these areas is *briefly* discussed in the following section. (The reader is referred to Refs. 27, 28, 46, and 49 to 51 for details on the mechanistic aspects of these reactions and to Refs. 50 and 52 for electrocatalytic factors involved in fuel cell reactions involving organic substances.)

As is noted in the introduction, the total cell voltage is determined by the electrode potentials of both the anode and the cathode. In a few cases, it is possible to carry out reactions at both the cathode and the anode simultaneously, thereby significantly reducing the power consumption per unit of product. However, in most industrial processes, this is not possible, with the result that sustaining reactions such as hydrogen and oxygen evolution take place at the counterelectrode. These gas evolution reactions have very high E^0 values to start with and in addition are irreversible on most of the commonly used electrode materials [26]. While it is possible to minimize the overvoltage by a proper choice of electrode material, a more expeditious way to conserve power would be to replace the gas evolution reactions with the corresponding ionization or reduction reactions. The later reactions have a thermodynamic advantage of several hundred millivolts; in addition, the absence of void-fraction effects results in substantial voltage savings. Electrode poisoning due to organic substances may pose a problem, but is not insurmountable in view of developments in membrane technology. With this in mind, a brief discussion of gas evolution, ionization or reduction reactions are given below together with a compilation of kinetic parameters to provide guidelines in the selection of electrode materials.

Hydrogen Evolution Reaction

The elementary steps involved in the H_2 evolution reaction [see (4.4) to (4.6)] and the origin of the dependence of i_0 on the nature of the metal are discussed in earlier sections in terms of M–H bond strength. Kinetic parameters for this reaction in acid and alkaline media are presented in Table 4.7.

Noble metals and noble metal based compositions are the best catalysts in acid solutions and Ni and Ni based materials are promising in alkaline solutions. However, one of the problems associated with Ni based compositions in alkaline solutions appears to arise from the loss of activity [27] due to either adsorption of impurities or to the formation of hydrides.

It is of interest to note here that low Tafel slopes arise from the potential dependence of coverage by adsorbed intermediates. However, coverage by the intermediates may be deleterious [55] with metals such as Fe, Ni, and Ti, which undergo hydrogen embrittlement as a result of the diffusion of adsorbed H atoms. While these effects can be minimized by appropriate doping of the surface to prevent H_2 permeation, low overpotentials can be achieved by using high-surface-area electrodes on which discharge is the slow step.

Two examples are presented here to show how the "surface characteristics" can be altered to develop electrodes with characteristics desired for a given operation. Jalan et al. [56] examined the Cr^{2+}/Cr^{3+} couple in acid solutions for application in redox battery systems. The system requirements are that $Cr^{2+} \rightleftharpoons Cr^{3+}$ oxidation and reduction be favored without interference from H_2 evolution during reduction. The hydrogen evolution reaction was suppressed by use in a carbon electrode coated with 100 μg of Au (in HCl solutions containing 10 mg Pb^{2+}), promoting the oxidation of Cr^{2+} and reduction of Cr^{3+} (see Figs. 4.16 and 4.17). While one may rationalize this behavior in terms of low η on Au for Cr^{2+} oxidation and high η on Pb for H_2 evolution to promote Cr^{3+} reduction, the character of the "active nature of the surface" and the kinetic factors involved in this reaction remain to be understood. Another example relates to the enhanced activity of the H_2 evolution reaction on Raney nickel in the presence of nitrates of Pb or Cd adatoms. This is attributed [57, 58] to the influence of Pb or Cd adatoms on the bond energy of adsorbed H atoms on the electrode surface resulting in a change in the H_2 evolution mechanism from electrochemical desorption to recombination as exhibited by a Tafel slope of ~ 30 mV. It should, however, be borne in mind that a low Tafel slope may adversely influence the long-term performance of these surface modified electrodes by the diffusing adsorbed atomic hydrogen on the surface.

Oxygen Evolution Reaction [46, 49]

Studies on the electrocatalytic aspects of the O_2 evolution reaction are limited by the restricted choice of materials because of complications arising from anodic dissolution and the formation of surface oxide film (often having

Table 4.7. Kinetic Parameters for the H_2 Evolution Reaction from Acid and Alkaline Media (Compiled from Refs. 27, 53, and 54)

Material	Medium (Temp.)[a]	Acid i_0 (A/cm²)	Tafel slope (mV)	Medium (Temp.)	Alkaline i_0 (A/cm²)	Tafel slope (mV)
Ag	5 N HCl	5×10^{-6}	120	1 N NaOH (30°C)	3.16×10^{-7}	120
Au	2 N H$_2$SO$_4$ (20°C)	4.36×10^{-6}	116	0.1 N NaOH (25°C)	10^{-5} to 10^{-7}	71–120
C	pH ~ 0	2.56×10^{-5} to 0.3×10^{-6}		40% NaOH (40°C)	2.95×10^{-5}	148
Co				6 N NaOH (25°C)	5×10^{-5}	140
Cu	0.1 N HCl (25°C)	1.45×10^{-7}	114	0.1 N NaOH (25°C)	10^{-7}	120
Cd	0.5 N H$_2$SO$_4$	1.7×10^{-11}	135	6 N NaOH (25°C)	3.98×10^{-7}	160
Cr	pH ~ 0	3.98×10^{-6} to 10^{-7}		6 N NaOH (25°C)	10^{-7}	120
Fe	0.5 N HCl	6.6×10^{-6}	133	0.1 N NaOH (25°C)	1.58×10^{-6}	120
Hg	0.1 N HCl	7.76×10^{-13}	116	0.1 N NaOH (25°C)	3×10^{-15}	~120
Mo	0.1 N HCl	3.98×10^{-7} to 5×10^{-8}	80–104	0.1 N NaOH (25°C)	5×10^{-7} to 5×10^{-8}	80–116
Nb	1 N HCl	10^{-9}	110	1 N NaOH (25°C)	3.16×10^{-8}	140
Ni	0.5 N H$_2$SO$_4$	6.02×10^{-6}	124	0.5 N NaOH	7.94×10^{-7}	96
Pb	0.1 N H$_2$SO$_4$	1.26×10^{-13}	118	0.5 N NaOH	3.16×10^{-7}	130
Pd	2 N H$_2$SO$_4$	1.99×10^{-3}	29	0.1 N NaOH	10^{-5}	125
Pt	1 N HCl	6.92×10^{-4}	30	0.1 N NaOH	6.76×10^{-5}	114
Ir	pH ~ 0	1.99×10^{-3}		0.1 N NaOH	5.5×10^{-4}	125
Rh	1 M H$_2$SO$_4$	10^{-3}	27.5	0.1 N NaOH	10^{-4}	118
Sn	pH ~ 0	10^{-9}		6 N NaOH	3.16×10^{-7}	150
Ti	1 N H$_2$SO$_4$	8.5×10^{-7}	135	6 N NaOH	10^{-6}	140
W	0.1 N HCl	3.16×10^{-6} to 7.76×10^{-7}	130–135	0.5 N NaOH	2.5×10^{-7} to 3.16×10^{-8}	80–100
Zn	1 N H$_2$SO$_4$	1.58×10^{-11}	120	6 N NaOH	3.98×10^{-7}	210

[a]Values of i_0 refer to at room temperature unless otherwise specified.

229

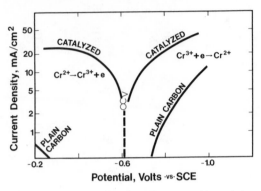

Fig. 4.16. Cr^{2+}/Cr^{3+} Redox characteristics of Au/Pb on graphite rod electrode in 1 N HCl + 0.09 M Cr^{2+} + 0.09 Cr^{3+} solutions. Reproduced from Ref. 56 with permission from the authors.

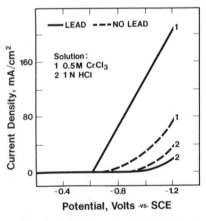

Fig. 4.17. H_2 evolution and Cr^{3+} reduction on a carbon felt electrode containing 25 μg Au/cm^2; (with and without 100 μg Pb/cm^2). Reproduced from Ref. 56 with permission from the authors.

poor electronic conductivity). The most generally accepted mechanism involves the following steps:

$$M + H_2O \longrightarrow MOH + H^+ + e \qquad (4.10)$$

$$2MOH \longrightarrow MO + H_2O \qquad (4.11)$$

$$2MO \longrightarrow 2M + O_2 \qquad (4.12)$$

Since OH_{ads} is the intermediate formed, the rate of O_2 evolution should be dependent on M–OH bond strength [59] as shown in Fig. 4.18. (See Table 4.8 for kinetic data in acid and alkaline media.)

In acid solutions, noble metal based compositions, PbO_2, and MnO_2 are the most promising catalysts, whereas in alkaline media, sintered Ni electrodes impregnated with "nickel oxide," NiOOH, high-surface-area $NiCo_2O_4$, and 50 Ni/50 Fe alloy appear [27] to be the optimized electrode compositions.

Hydrogen Ionization Reaction

The mechanism of hydrogen ionization is not necessarily the reverse of the H_2 evolution reaction. This is because H_2 evolution occurs, for instance, on most transition metals in regions of potentials, where these metals are covered with at least a monolayer of adsorbed H atoms, whereas in the region of oxidation reaction, the coverage with adsorbed H is less than a monolayer. On Pt, which is most extensively studied in H_3PO_4 solutions, dual-site dissociation of H_2 appears to be the slow step [46, 50] (see Table 4.9 for kinetic data in acid alkaline solutions).

In work on acid solutions most efforts so far have been directed to developing electrodes insensitive to impurities such as CO and S; CoP_3, WC, CoPS, FeP_2 were reported [72] to be promising. In alkaline media, Pt, Ni, and Ni alloys are potential catalysts [72]. However, titanium containing Raney Ni showed [73] superior performance with respect to electrocatalytic and stability considerations.

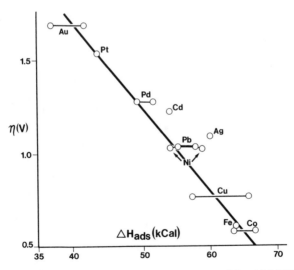

Fig. 4.18. Correlation [59] between oxygen evolution overpotential and M–OH bond strength. Reproduced with permission from the *Journal of Chemical Physics*.

Table 4.8. Kinetic Parameters for the O_2 Evolution Reaction from Acidic and Alkaline Solutions

Material	Acid				Alkaline			
	Medium (Temp.)[a]	i_0 (A/cm²) (activation energy, kcal/mole)[b]	Tafel slope (mV)	Ref.	Medium (Temp.)	i_0 (A/cm²) (activation energy, kcal/mole)	Tafel slope (mV)	Ref.
Pt	1 M H₂SO₄ (80°C)	1.3×10^{-11}	90	60	30 wt % KOH (80°C)	1.2×10^{-9}	46	66
Ir	1 M H₂SO₄ (80°C)	1.5×10^{-7}	85	60	30 wt % KOH (80°C)	5×10^{-8}	55	66
Rh	85% H₃PO₄ (75°C)	8.5×10^{-10}	98	61	30 wt % KOH (80°C)	3.8×10^{-7}	67	66
Ru	1 M H₂SO₄ (80°C)	5.1×10^{-9}	41	60	30 wt % KOH (80°C)	3×10^{-8}	67	66
RuO₂ + TiO₂	1 M H₂SO₄ (80°C)	1.3×10^{-8}	66	60				
Ti/RuO₂	1 M HClO₄ (30°C)	(3.9)	45	62	1 M KOH (30°C)	(10.9, 11.2)	46, 43	68
Ti/Rh₂O₃	0.5 M H₂SO₄ (30°C)	(4.4, 3.8)	64, 65	62	1 M KOH (30°C)	(13.2, 13.6)	50	68
Ti/IrO₂	0.5 M H₂SO₄ (30°C)	(2.9)	61	62	1 M KOH (30°C)	(8.5, 9.2)	53, 46	68
Ti/PtO₂	0.5 M H₂SO₄ (30°C)	(14.6, 13.5)	118, 126	62	1 M KOH (30°C)	(Low η: 19.1; high η: 14.0)	Low η: 75; high η: 114	68
Pt/β-MnO₂	0.5 M H₂SO₄ (30°C)	(4.5, 3.8)	82, 80	62	1 M KOH (30°C)	(13.9)	75	68
Ti/Co₃O₄	0.5 M H₂SO₄ (30°C)	(4.2)	71	62	1 M KOH (30°C)	(12.0, 12.5)	64, 63	68
α-PbO₂	H₂SO₄	1.7×10^{-16}	50	63				
β-PbO₂	H₂SO₄	6×10^{-10}	120	63				

Pt/Rh	$HClO_4$ (pH ~ 0)	3×10^{-11}	Low η: 65–70 high η: 120	64	KOH (pH ~ 14)	Low η: 10^{-12}; high η: 10^{-6}	Low η: 40; high η: 120	64
Au	$HClO_4$ (pH ~ 0)	1.6×10^{-22}	45	65	KOH (pH ~ 14)	high η: 1.1×10^{-8}	244	65
Pd	$HClO_4$ (pH ~ 0)	4×10^{-11}	100	65	KOH (25°C) (pH ~ 14)	3.1×10^{-11}	113	65
					30 wt % KOH (80°C)	1.9×10^{-7}	67	66
Fe					30 wt % KOH (80°C)	1.7×10^{-5}	191	66
Co					30 wt % KOH (80°C)	3.3×10^{-6}	126	66
Ni					30 wt % KOH (80°C)	2.3×10^{-7}	62	66
Ag					1 M KOH	10^{-10}	120	67
$NiCo_2O_4$					30 wt % KOH (80°C)	2.9×10^{-8}	42	69

[a] Values of i_0 refer to room temperature unless otherwise specified.
[b] Activation energy at a given overpotential defined as $E_a(\eta) = -\,2.303\,R\,[\partial \log i / \partial(1/T\eta)]$.

233

Table 4.9. Kinetic Parameters for the H_2 Ionization Reaction from Acid and Alkaline Media

Material	Acid				Alkaline			
	Medium (Temp.)[a]	i_0 (A/cm^2)	Tafel slope (mV)	Ref.	Medium (Temp.)	i_0 (A/cm^2)	Tafel slope (mV)	Ref.
Pt	1 N H_2SO_4 (25°C)	10^{-3}	32	54	0.1 N NaOH (25°C)	3.98×10^{-4}	105	54
Ir	1 N H_2SO_4	3.98×10^{-4}	50	54	0.1 N NaOH (25°C)	1×10^{-4}	120	54
Rh	1 N H_2SO_4	3.98×10^{-4}	38	54	0.1 N NaOH (25°C)	2.5×10^{-4}	105	54
Pd	1 N H_2SO_4	1.28×10^{-3}	100	54	0.1 N NaOH (25°C)	1.28×10^{-6}	125	54
Au	1 N H_2SO_4	7×10^{-6}	105	54	0.25 N NaOH (25°C)	$\sim 10^{-4}$	60	54
Fe	1 N H_2SO_4				0.1-1 N NaOH	10^{-6}	25	54
Ni	1 N H_2SO_4	10^{-8}						
MnP_3	1 M H_2SO_4 (22°C)	3.5×10^{-5}		70				
FeP_3	1 M H_2SO_4 (22°C)	5×10^{-5}		70				
CoP_3	1 M H_2SO_4 (22°C)	1.5×10^{-5}		70				
NiP_3	1 M H_2SO_4 (22°C)			70				
Raney Ni					6 M KOH (22°C)	0.45×10^{-6}		71
Raney Ni + 2% Ti					6 M KOH (22°C)	1.75×10^{-6}		71
Raney Ni − 7.5% Mo					6 M KOH (22°C)	5.4×10^{-6}		71
Raney Ni + 21.3% Mo					6 M KOH (22°C)	9.1×10^{-6}		71

[a] Activation energy values refer to room temperature unless otherwise specified.

234

Oxygen Reduction Reaction

Several complexities are involved in studying the kinetics of oxygen reduction, including (1) instability of most metals and alloys in the region of potentials where the reaction occurs in acid and alkaline solutions; (2) relatively slow reaction rates ($i_o \sim 10^{-10}$ A/cm^2); (3) competing reactions such as oxide formation, and (4) generation of H_2O_2 as an intermediate or in parallel reaction in many cases.

Pathways [51, 74, 75] involved in O_2 reduction are complex and some generally accepted schemes are shown below.

1. *Peroxide pathway* (active carbon, graphite, Au in alkaline solutions).

$$HOH + O_2 \xrightarrow{\ 2e\ } HO_2^- \text{ (ads)} + OH^- \tag{4.13}$$

$$\searrow \ 2e$$

$$HO_2^- + OH^- \tag{4.14}$$

   a. *Peroxide reduction* (Au).

$$HOH + HO_2^-$$

$$\xrightarrow{2e}$$

$$3OH^- \tag{4.15}$$

$$\xrightarrow{2e}$$

$$HOH + HO_2^- \text{ (ads)}$$

   b. *Peroxide catalytic decomposition.*

$$2HO_2^-$$

$$2OH^- + O_2 \tag{4.16}$$

$$2HO_2^- \text{ (ads)}$$

2. *Direct four-electron pathway* (clean Pt, macrocyclics, e.g., adsorbed Fe tetrasulfonated phthalocyanine, dicobalt porphyrin on graphite in acid solutions).

$$2M + O_2 \longrightarrow 2MO \tag{4.17}$$

$$MO + H^+ + e \longrightarrow MOH \tag{4.18}$$

$$MOH + H^+ + e \longrightarrow M + H_2O \tag{4.19}$$

Studies on the O_2 reduction reaction are extensive and kinetic data reported on some substrates are presented in Table 4.10. (See also Fig. 4.19 for electrocatalytic correlations involved in O_2 reduction reaction.) Several materials were examined in acid media, including metal chelates, sulfides, thiospinels, and macrocyclic complexes; however, Pt and Pt/Au alloys continue to be used [72]. Finding O_2 electrodes for alkaline media is easier than finding them for acidic environments. Activated carbon, (Bi, Ni, Ti) doped Ag

Table 4.10. Kinetic Parameters for the Oxygen Reduction Reaction from Acid and Alkaline Solutions

	Acid				Alkaline			
Material	Medium (Temp.)[a]	i_0 (A/cm²)	Tafel slope (mV)	Ref.	Medium (Temp.)	i_0 (A/cm²)	Tafel slope (mV)	Ref.
Pt	HClO₄ (pH ~1.0) (70°C)	3×10^{-10}	60	76	NaOH (pH ~13)	10^{-11}	65	76
Ir	HClO₄ (pH ~0.2)	10^{-11}	105	64	KOH (pH ~13)	10^{-11}	100	64
Rh	HClO₄ (pH ~0.2)	10^{-11}	65	64	KOH (pH ~13)	1.2×10^{-12}	55	64
Au	0.1 N HClO₄	1.5×10^{-11}	120	77	0.1 M NaOH	1.8×10^{-4} (for O₂/HO₂⁻ couple)	120	119
Pd	HClO₄ (pH ~1.0)	3×10^{-11}	60	77	NaOH (pH ~13)	1.2×10^{-11}	55	78
Ag					NaOH (pH ~13)	1.5×10^{-9}	110	78
Fe					NaOH (pH ~13)	4.3×10^{-10}	120	78
Ni					NaOH (pH ~13)	4.3×10^{-10}	120	78
Pt black	50% H₃PO₄ (70°C)	4×10^{-10}	60	79	30% KOH (70°C)	4×10^{-10}	50	79
Pt	85% H₃PO₄ (25°C)	1.2×10^{-11}	125 (10)[c]	80				
Active Pt	85% H₃PO₄ (22°C)	2.7×10^{-10}	106 (13.1)[c]	46				
Pt crystallites (on Pt)					0.1 M NaOH	4.55×10^{-7}	120	120
Pt (bulk)	85% H₃PO₄ (130°C)	1.7×10^{-8}	120	121				

		i_0		
Pt crystallites (on graphite)	85% H_3PO_4 (130°C)	2.4×10^{-7}	120 (12.5)[c]	121
Au (prereduced)	85% H_3PO_4 (22°C)	1.9×10^{-15}	107 (23.6)[c]	46
Ir (prereduced)	85% H_3PO_4 (22°C)	1.1×10^{-11}	112 (12.6)[c]	46
Ru (prereduced)	85% H_3PO_4 (22°C)	6.2×10^{-12}	115 (11.7)[c]	46
Rh (prereduced)	85% H_3PO_4 (22°C)	4.3×10^{-15}	60 (22)[c]	46
Pd (oxide free)	85% H_3PO_4 (22°C)	1.9×10^{-13}	63 (22.4)[c]	46
Ag (oxide free)	85% H_3PO_4 (22°C)	2.4×10^{-15}	119 (26.2)[c]	46
Colloidal Pt	97% H_3PO_4 (450°K)		105 (20)[b]	81
$HfPt_4$, $ZrPt_4$	97% H_3PO_4 (450°K)		120 (110)[b]	81
VPt_3	97% H_3PO_4 (450°K)		115 (75)[b]	81
$TaPt_5$, $NbPt_5$	97% H_3PO_4 (450°K)		120 (100)[b]	81

[a] Values of i_0 refer to room temperature unless otherwise specified.
[b] The values in parentheses refer to specific activity in $\mu A/cm^2$ at 0.9 V.
[c] The values in parentheses refer to activation energy in kcal/mole.

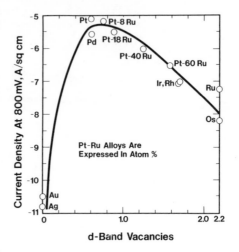

Fig. 4.19. Plot of i at $\eta = -460$ mV at 25°C in 85% H_3PO_4 during O_2 reduction as a function of d-orbital vacancy values [46]. Reproduced with permission from Plenum Press.

containing 11.5% Hg, $NiCo_2O_4$, Ag, and Pt are reported [82] to be most promising in alkaline solutions. It is of interest to note that while prereduced Au and oxide-free Ag show poor performance in acid solutions, they are catalytically active in alkaline media [83].

One way to enhance the rate of O_2 reduction is by "modification of the electrode surface." On Au electrodes in alkaline solutions, O_2 is reduced to OH^- by way of HO_2^- by a well-separated two-electron process. However, in the presence of underpotentially deposited Tl (see Fig. 4.20 and 4.21), the O_2 reduction reaction proceeds [84] reversibly (4e process) at the O_2/HO_2^- reversible potential. This catalytic effect is a result of rapid surface-catalyzed decomposition of HO_2^- and may be attributed to enhanced adsorption of O and/or O_2^- radicals participating in regenerative chain reaction and/or rapid redox cycles facilitated by low desolvation energies of heavy metal ions adsorbed in the double layer.

Chlorine Evolution Reaction

Kinetics of the chlorine evolution reaction have been extensively studied in recent years, since the advent of dimensionally stable anodes and the kinetic data for this reaction are presented in Table 4.11. The reaction pathways involved during chlorine evolution are similar [28] to those proposed for the hydrogen evolution reaction and are as follows:

$$M + Cl^- \longrightarrow MCl^{\cdot} + e \qquad (4.20)$$

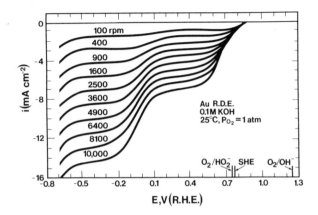

Fig. 4.20. O$_2$ reduction at an Au RDE in 0.1 M KOH [84]. Reproduced with permission from the *Journal of the Electrochemical Society.*

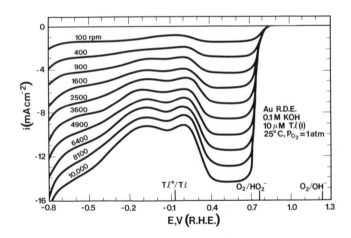

Fig. 4.21. Electrocatalysis [84] of O$_2$ reduction by upd T(0) on an Au electrode in 0.1 M KOH; T(I) = 10 μm. Reproduced with permission from *Journal of the Electrochemical Society.*

$$2MCl^{\cdot} \longrightarrow Cl_2 + 2M \qquad (4.21)$$

$$MCl^{\cdot} + Cl^- \longrightarrow M + Cl_2 + e \qquad (4.22)$$

Some $MCl^{\cdot}$ is involved during Cl_2 evolution; i_o appears to be function [92] of M–Cl bond strength as shown in Fig. 4.22. However, caution should be exercised since Cl_2 evolution, like O_2 evolution, occurs on oxide or oxide-covered surfaces.

One of the parasitic reactions occurring during the discharge of Cl^- ions is the O_2 evolution reaction. Selectivity is achieved [28] in this case by doping the surface with oxides such as SnO_2 to suppress the O_2 evolution reaction or by lowering the effective surface area.

Bromine and Fluorine Evolution Reactions

These reactions have not been studied as extensively as the other reactions noted above and the limited kinetic data available in the literature is pre-

Table 4.11.   Kinetic Parameters for the Chlorine Evolution Reaction

Material	Medium (Temp.)	i_0 (A/cm^2)	Tafel slope (mV)	Ref.
Pt	0.2 N HCl + 2NH$_2$SO$_4$ (25°C)	3.97×10^{-3}	120	85
	3 M NaCl (30°C)	6.7×10^{-3}	40	86
Ir	0.2 N NCl + 2NH$_2$SO$_4$ (25°C)	0.19×10^{-3}	120	85
Rh	0.2 N HCl + 2NH$_2$SO$_4$ (25°C)	4×10^{-5}	30–40	85
TiO$_2$	5 M NaCl (20°C)	3.9×10^{-5}	30–120 (at high η)	87
RuO$_2$	5 M NaCl (20°C)	6.3×10^{-3}	30 at low η and at high η	87
	5 M NaCl (25°C)	35×10^{-6}	40	89
IrO$_2$	5 M NaCl (25°C)	39.8×10^{-3}	120	87
	5 M NaCl (25°C)	1.17×10^{-3}	40	89
γ-MnO$_2$	6 M NaCl (25°C)	6×10^{-3}	30–120	88
β-PbO$_2$	6 M NaCl (25°C)	6×10^{-6}	>120	88
Pd, PdO	3 M NaCl (30°C)		30	86
Graphite	Satd. NaCl (pH = 0.5) (50°C)	1.2×10^{-3}	40–120	90
	4 M NaCl + 1 M HCl (50°C)	0.26×10^{-3}	101	91

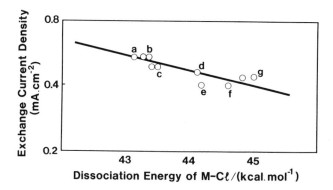

Fig. 4.22. Relationship [92] between exchange current density for Cl_2 evolution from aqueous Cl^- at various metal anodes and the metal-to-Cl bond energy as evaluated from data for corresponding metal chlorides. (a) Pt; (b) Pt/Ir 10.1%; (c) Pt/Ir 15.2%; (d) Pt/Ru 17.6%; (e) Pt/Pd 17%; (f) Pt/Pd 24%; (g) Pt/Rh 32.4%. Reproduced with permission from Pergamon Press.

sented in Table 4.12. Stability is one of the problems in HF—the only metals exhibiting [97] low corrosion rates with a conducting surface layer being Monel, Pt, and Ni. It may be noted that the high Tafel slopes observed during F_2 evolution from HF + 1 M NaF on Ni, Pt, and vitreous carbon are indicative of complex mechanisms [96] that are not yet understood.

Surface Modified Electrode Materials

Surface modified electrode materials have been the subject of a number of reviews [100–107] by various authors. Surface modification involves treating electrode materials in such a manner that surface layers of potentially catalytic character are bonded to the electrode surface. Several methods of surface modification, such as silanization, adsorption, polymer coating, and reaction with surface groups on carbon, are available. Detailed descriptions of preparative methods* and characterization of the resulting product(s) are beyond the scope of this chapter and the reader is referred to the various review articles cited above. The present discussion is restricted to the special applications of such electrodes for electroorganic reactions.

The motivation for using surface modified electrodes for electroorganic

___

*For techniques involved in coating substrates with electrocatalytic conducting compositions (e.g., simple and/or mixed oxides on Ti), see Refs. 1 and 118.

Table 4.12.   Kinetic Data for Br_2 and F_2 Evolution Reactions

Material	Medium	i_0 (A/cm^2)	Tafel slope (mV)	Ref.
		Bromine		
Pt	2 M NaClO$_4$ (25°C)	13.6×10^{-6}		93
Ir	HClO$_4$ (20°C)	36×10^{-6}		94
C	3 M ZnBr$_2$ (pH ~2; 25°C)	1.09	~120	117
		Fluorine		
Graphite	KF + 2HF (80-100°C)	3×10^{-6}	$2.303RT/\alpha F$ ($\alpha = 0.8$)	95
Carbon	KF + 2HF (80-100°C)	3×10^{-5}	$2.303RT/\alpha F$ ($\alpha = 0.8$)	95
Pt	KF + 2HF (80-100°C)	2×10^{-5}	$2.303RT/\alpha F$ ($\alpha = 0.8$)	95
Ni	HF/1 M NaF (0°C)		350	96
Pt	HF/1 M NaF (0°C)		370	96
Vitreous carbon	HF/1 M NaF (0°C)		280	96

reactions may be twofold: (a) to enhance catalytic activity of the electrode and (b) to force specificity through a proper choice of surface modification. Specific examples for each case are illustrated here.

Evans et al. [107] were the first to demonstrate the surface catalysis using a benzidine coated graphite electrode. They showed catalytic enhancement of ascorbic acid oxidation according to the following scheme [illustrated using quinone (Q)/hydroquinone (HQ) functionality]:

$$\vdash QH_2 \longrightarrow \vdash Q + 2e + 2H^+$$

$$\vdash Q + AH_2 \longrightarrow \vdash QH_2 + A$$

(4.23)

The electrocatalytic effect is attributed to the high surface coverage of the electroactive surface species and the ease with which both the proton and electron transfer steps proceed. Dautartas and Evans [108] continued this investigation and examined plasma-polymerized vinyl ferrocene film coated on pyrolytic graphite as an electrocatalyst for the oxidation of ascorbic acid.

They found the film to be very stable and catalytically effective. As the coverage increased they observed ascorbic acid transport in the film and kinetic limitations to charge transport and concluded that higher surface coverages of the catalyst may not necessarily result in higher rates.

Selective electrocatalysis appears to be achievable [109] by both electroinactive and electroactive surface modifiers; the former chemically interacts with the incoming organic substrate molecules, whereas the latter relays charge from the base material to the molecule undergoing reaction. Selective electrocatalysis may be exemplified during the reduction [109] of 1,2-dibromo-1,2-diphenylethane at a Pt/poly-p-nitrostyrene surface. Since reductions of halides could be carried out successfully on many other metals, this example should be viewed only as a demonstration of the feasibility of such reactions. The electrode deterioration was attributed to the decrease in the amount of chargeable polymers—a conclusion that could be inferred from an earlier study by Merz and Bard [110], who observed that charge transfer to a solution species by means of an electroactive polymer electrode only takes place in the region of potentials where the polymer is charged.

The importance of reversible electroactive couples in indirect oxidation and reduction reactions of organic compounds is outlined earlier (see Table 4.2). An ideal situation would be to localize these reactive centers on the electrode itself. Attachment of a reversible couple at 10^{-10} moles/cm^2 of coverage would correspond to a volume concentration of up to 1 M (depending on the thickness of the coating). Oyama and Anson [111] showed that the Ru/EDTA complex can be attached to graphite by a covalent bond. This complex can undergo a ligand substitution reaction with ease, and electron transfer between the graphite and the complex does not pose any problem. The attachment of such reversible couples to the electrode surface results in a higher rate for electron transfer (10^4/sec) since the collision number is as large as kT/h [112].

Another variation of the above technique may be to incorporate the redox reagent into a polymer matrix and coat the electrode with this combination. The polymer coating consists of several monolayers; hence the redox couple can be introduced at a much higher concentration. Increased stability has also been attributed to polymer immobilization. However, such electrodes have not yet been used to study indirect reduction and oxidation of organic molecules.

One of the advantages associated with electrode surface modification is the opportunity to exercise selectivity during synthesis. Thus when a chiral agent is used to modify the surface, one can expect stereospecific synthesis. Watkins et al. [113] examined this hypothesis by modifying a carbon electrode with (s)-phenyl alanine methyl ester and showed that this chiral surface induced asymmetry in the production of chiral alcohols from prochiral ketones.

In a subsequent study, Miller et al. [114] showed that the asymmetric chemistry takes place only on the edge surfaces and not on the basal plane of pyrolytic graphite. In the same study, these authors also demonstrated anodic asymmetric synthesis of sulfoxides using phenylalanine modified electrodes. By this process, p-tolyl methyl sulfide was anodically oxidized to form the pure sulfoxide. Chirally modified surfaces have been produced on oxide anodes for the asymmetric synthesis of sulfoxides [115]. Preferential synthesis of p-chloroanisole (over the o-isomer) has been demonstrated by Matsue et al. [116] through the use of α-cyclodextrin modified graphite substrates. The mechanism is shown in Fig. 4.23. Selectivity is achieved because the ortho position is blocked by the inner wall of the α-cyclodextrin molecule. Matsue et al. also studied the stability of the absorbed and chemically modified electrodes and found that the latter exhibit high selectivity ($p/o \sim 10:1$) even after a month. While α-cyclodextrin has been used for the selective chlorination of anisole in the homogeneous nonelectrochemical reaction, the reactant concentrations required to ensure high selectivity are very high and separation of the product becomes difficult.

Electrocatalysis of organic electrode reactions on surfaces modified by foreign adatoms has not been studied except in the context of fuel cells. Oxidation of formic acid, formaldehyde, and methanol on Pt electrodes is catalyzed in the presence of Pb^{2+} [105]. Adzic has reviewed [105] the effects of adsorbed metal ions on various other noble metal surfaces. It appears that either an underpotentially deposited metal or a surface complex can effectively interact with the adsorbed intermediate as is discussed earlier. Since intermediates play a major role in electroorganic reactions, future work

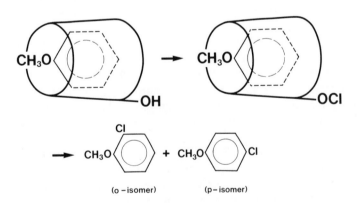

(o –isomer)   (p– isomer)

Fig. 4.23. Mechanism of anodic chlorination of anisole [116] in the presence of α-cyclodextrin. Reproduced with permission from the *Journal of the Electrochemical Society*.

should focus attention on the effect of foreign metal adatoms on the stability of the intermediate chemical species.

6 CONCLUSIONS

Three factors of significance for a rational choice of electrode materials are

1. *Stability considerations,* especially under open-circuit conditions.
2. *Conductivity characteristics* which can be evaluated.
3. *Electrocatalytic correlations* relating the kinetic parameters to the surface characteristics of the electrode materials.

Thus of all the variables in electrochemical technology, the nature of the electrode material remains the most evasive. While many fundamental studies are needed to generate relationships for the electrochemical parameters and material and/or surface characteristics, herein lie the opportunities for developing new and unusual electrode materials exhibiting low overvoltage and high selectivity, which are mandatory for successful development of electroorganic processes.

Acknowledgments

One of the authors (B.V.T.) wishes to thank Hooker Chemical Company and another (S.S.) wishes to thank Union Carbide for permission to publish this chapter. Useful discussions with Dr. D. H. Ridgley (Hooker Chemical Company) and Prof. J. M. Honig (Purdue University) during the preparation of this manuscript are gratefully acknowledged.

References

1. K. B. Keating, paper presented at the 78th National AIChE Meeting, Salt Lake City, Utah, August 1974.
2. Ya. M. Kolotyrkin, *Denki Kagaku,* **47,** 390 (1979).
3. Ya. M. Kolotyrkin, V. V. Losev, D. M. Shub, and Yu. E. Roginskaya, *Sov. Electrochem.,* **15,** 245 (1978).
4. M. Ya. Fioshin, *Sov. Electrochem.,* **13,** 1 (1977).
5. F. Goodridge and C. J. H. King, *Technique of Electroorganic Synthesis,* Part I, N. L. Weinberg, Ed., Techniques of Chemistry Series, Vol. 5, A. Weissberger, Ed., Wiley-Interscience, New York, 1974, Ch. II, pp. 7–25.
6. K. Koster and H. Wendt, *Comprehensive Treatise of Electrochemistry,* Vol. 2, J. O'M. Bockris, B. E. Conway, and E. A. Yeager, Eds., Plenum Press, New York, 1981, p. 251.
7. N. L. Weinberg, Ed., *Technique of Electroorganic Synthesis,* Part I, Wiley-Interscience, New York, 1974.
8. N. L. Weinberg, Ed., *Technique of Electroorganic Synthesis,* Part II, Tech-

niques of Chemistry Series, Vol. 5, A. Weissberger, Ed., Wiley-Interscience, New York, 1974.

9. J. W. Kuhn-Von Burgsdorff, *Chem. Ing. Tech.*, **49**, 294 (1977).

10. N. L. Weinberg, in *Technique of Electroorganic Synthesis*, Part II, N. L. Weinberg, Ed., Techniques of Chemistry Series, Vol. 5, A. Weissberger, Ed., Wiley-Interscience, New York, 1974, p. 1.

11. M. R. Rifi, in *Technique of Electroorganic Synthesis*, Part II, N. L. Weinberg, Ed., Techniques of Chemistry Series, Vol. 5, A. Weissberger, Ed., Wiley-Interscience, New York, 1974, p. 83.

12. M. M. Baizer, *Organic Electrochemistry*, M. M. Baizer, Ed., Dekker, New York, 1973, p. 693.

13. N. L. Weinberg, *ACS Audio course, Electroorganic Synthesis*, 1979.

14. R. T. Sanderson, *Introduction to Chemistry*, Wiley, New York, 1954.

15. W. E. Dascent, *J. Chem. Educ.*, **40**, 130 (1963).

16. M. Pourbaix, *Atlas of Electrochemical Equilibria in Aqueous Solutions*, Pergamon Press, New York, 1966.

17. B. V. Tilak, unpublished investigations.

18. A. C. C. Tseung and H. L. Bevan, *J. Electrochem. Chem.*, **45**, 429 (1973).

19. V. L. Bystrov and O. P. Romashin, *Sov. Electrochem.*, **11**, 1141 (1975).

20. E. A. Kalinovskii, R. V. Bondar, and N. N. Meshkova, *Sov. Electrochem.*, **8**, 1430 (1972).

21. V. L. Bystrov, *Sov. Electrochem.*, **11**, 1774 (1975).

22. A. Uzbekov, V. G. Lambrev, L. F. Yazikov, N. N. Rodin, L. M. Zabrodskaya, V. S. Klementeva, and Yu. M. Vlodov, *Sov. Electrochem.*, **14**, 997 (1978).

23. R. V. Bondar and E. A. Kalinovskii, *Sov. Electrochem.*, **14**, 633 (1978).

24. N. L. Weinberg and T. B. Reddy, *J. Appl. Electrochem.*, **3**, 73 (1973).

25. J. O. Besenhard and H. P. Fritz, *J. Electroanal. Chem.*, **53**, 329 (1974).

26. W. J. Settineri and L. D. McKeever, in *Technique of Electroorganic Synthesis*, Part II, N. L. Weinberg, Ed., Wiley-Interscience, New York, p. 397.

27. B. V. Tilak, P. W. T. Lu, J. E. Colman, and S. Srinivasan, in *Comprehensive Treatise of Electrochemistry*, Vol. 2, J. O'M. Bockris, B. E. Conway and E. A. Yeager, Eds., Plenum Press, New York, 1981, p. 1.

28. D. M. Novak, B. V. Tilak, and B. E. Conway, *Modern Aspects of Electrochemistry*, J. O'M. Bockris and B. E. Conway, Eds., Vol. 14, Plenum Press, New York, 1981.

29. C. N. R. Rao, Ed., *Modern Aspects of Solid-State Chemistry*, Plenum Press, New York, 1970.

30. R. S. Roth and S. J. Schneider, Eds., *Solid-State Chemistry*, NBS Special Publication 364, 1972.

31. N. M. Tallan, Ed., *Electrical Conduction in Ceramics and Glass*, Part A and B, Dekker, New York, 1974.

32. C. N. R. Rao, Ed., *Solid-State Chemistry*, Dekker, New York, 1974.

33. J. B. Goodenough, in *Progress in Solid-State Chemistry*, Vol. 5, H. Reiss, Ed., Pergamon Press, New York, 1973.

34. R. W. Vest and J. M. Honig, in *Electrical Conduction in Ceramics and Glass*, Part B, N. M. Tallan, Ed., Dekker, New York, 1974, p. 343.

35. J. M. Honig, in *Defects and Transport in Oxides,* M. S. Seltzer and R. I. Jaffee, Eds., Plenum Press, New York, 1974, p. 315.

36. C. N. R. Rao, and G. V. Subba Rao, *Phys. Stat. Sol. A.,* **1,** 597 (1970).

37. J. P. Carr and N. A. Hampson, *Chem. Rev.,* **72,** 679 (1972).

38. *Handbook of Chemistry and Physics,* Chemical Rubber Publishing Co., Cleveland, Edited by C. D. Hodgman, 36th ed. (1954).

39. V. B. Lazarev and I. S. Shaplygin, *Russ. J. Inorg. Chem.,* **23,** 163 (1978).

40. G. V. Samsonov, *The Oxide Handbook,* Plenum Press, New York, 1973.

41. K. H. Simmrock, *Hydrocarbon Process.,* **Nov. 1978,** 105.

42. T. D. Binns and D. C. G. Gattiker, U.S. Patent 3,635,803 (1972).

43. J. O'M. Bockris and A. K. N. Reddy, *Modern Electrochemistry,* Vol. II, Plenum Press, New York, 1970.

44. R. Parsons, *Trans. Faraday. Soc.,* **54,** 1053 (1958).

45. A. J. Appleby, *Catal. Rev.,* **4,** 221 (1970).

46. A. J. Appleby, *Mod. Aspects Electrochem.,* **9,** 369 (1974).

47. A. A. Balandin, *Problems of Chemical Kinetics: Catalysis and Reactivity,* Academy of Science, Moscow, 1955, p. 462.

48. R. Parsons, *Surf. Sci.,* **18,** 28 (1969).

49. J. P. Hoare, *Encyclopedia of Electrochemistry of Elements,* A. J. Bard, Ed., Vol. 2, Dekker, New York, 1974, Ch. 5, p. 191.

50. B. D. McNicol, *Catalysis,* **2,** 243 (1978).

51. E. Yeager, "Oxygen Electrodes for Energy Conversion and Storage," Annual Report from October 1, 1977–September 30, 1978; DOE Contract #DE-AC02-77ET25502 (1980).

52. J. O'M. Bockris and S. Srinivasan, *Fuel Cells: Their Electrochemistry,* McGraw-Hill, New York, 1969.

53. H. Kita, *J. Electrochem. Soc.,* **113,** 1095 (1966).

54. A. T. Kuhn, *Electrochemistry—The Past Thirty and the Next Thirty Years,* H. Bloom and F. Gutmann, Eds., Plenum Press, New York, 1977.

55. B. V. Tilak, C. G., Rader, and B. E. Conway, *Electrochim. Acta,* **22,** 1167 (1977).

56. V. Jalan K. Cahill, D. Demuth, and J. Giner, Extended Abstracts, St. Louis ECS Meeting, Abstract No. 35, Vol. 80-1, May 11-16, 1980.

57. N. V. Korovin, N. I. Kozlova, and O. N. Saveleva, *Sov. Electrochem.,* **16,** 496 (1980).

58. N. V. Korovin, O. N. Saveleva, and N. I. Kozlova., *Sov. Electrochem.,* **16,** 498 (1980).

59. P. Ruetschi and P. Delahay, *J. Chem. Phys.,* **23,** 1167 (1955).

60. M. H. Miles, E. A. Klaus, B. P. Gunn, J. R. Locker, W. E. Serafin, and S. Srinivasan, *Electrochim Acta,* **23,** 521 (1978).

61. A. J. Appleby and C. J. VanDrunen, *J. Electroanal. Chem.,* **60,** 101 (1975).

62. M. Inai, C. Iwakura, and H. Tamura, *Denki Kagaku,* **48,** 173 (1980).

63. P. Ruetschi and B. D. Cahan, *J. Electrochem. Soc.,* **105,** 369 (1958).

64. A. Damjanovic, A. Dey, and J. O'M. Bockris, *J. Electrochem. Soc.,* **113,** 739 (1966).

65. J. J. McDonald and B. E. Conway, *Proc. Roy. Soc. Ser. A.,* **269,** 419 (1962).

66. M. H. Miles, Y. A. Huang, and S. Srinivasan, *J. Electrochem. Soc.*, **125**, 1931 (1978).
67. T. Hurlen, Y. L. Sandler, and E. A. Pantier, *Electrochim. Acta*, **11**, 1463 (1966).
68. M. Inai, C. Iwakura, and H. Tamura, *Denki Kagaku*, **48**, 229 (1980).
69. G. Singh, M. H. Miles, and S. Srinivasan, in "Electrolysis on Non-Metallic Surfaces," A. D. Franklin, Ed., NBS Special Publication, 455, p. 289 (1976).
70. K. Mund, G. Richter, R. Schulte, and F. v. Sturm, *Z. Electrochem.*, **77**, 839 (1973).
71. F. v. Sturm, *Electrode Materials and Processes for Energy Conversion and Storage*, J. D. E. McIntyre, S. Srinivasan, and F. G. Will, Eds., The Electrochemical Society, Princeton, NJ, Vol. 77-6, 1977, p. 247.
72. K. Mund, G. Richter, and F. v. Sturm, *Proceedings of the Workshop on the Electrocatalysis of Fuel Cell Reactions*, Vol. 79-2, W. E. O'Grady, S. Srinivasan and R. F. Dudley, Eds., The Electrochemical Society, Inc., Princeton, NJ, 1979, p. 47.
73. K. Mund, G. Richter, and F. v. Sturm, *J. Electrochem. Soc.*, **124**, 1 (1977).
74. *Proceedings of the Workshop on the Electrocatalysis of Fuel Cell Reactions*, Vol. 79-2, W. E. O'Grady, S. Srinivasan and R. F. Dudley, Eds., The Electrochemical Society, Inc., Princeton, NJ, 1979.
75. *Electrocatalysis on Non-Metallic Surfaces*, National Bureau of Standards Special Publication 455, 1976.
76. A. Damjanovic and V. Brusic, *Electrochim. Acta*, **12**, 615 (1967).
77. A. Damjanovic and V. Brusic, *Electrochim. Acta*, **12**, 1171 (1967).
78. V. Brusic, Cited in Ref. 52, p. 468.
79. W. M. Vogel and J. T. Lundquist, *J. Electrochem. Soc.*, **117**, 1512 (1970).
80. E. Yeager, EPRI Report No. EPRI-EM-505; Research Project 634-1, June, 1977.
81. P. N. Ross, EPRI Report No. EPRI-EM-1553, Project 1200-5, Sept. 1980.
82. B. V. Tilak, R. S. Yeo, and S. Srinivasan, *Comprehensive Treatise of Electrochemistry*, Vol. 3, 3rd ed., Plenum Press, New York, 1981, p. 39.
83. S. Srinivasan, J. McBreen, and W. E. O'Grady, unpublished report in the "Survey of Status of Electrode Performance in Phosphoric Acid Fuel Cells."
84. J. D. E. McIntyre and W. F. Peck, Jr., Extended Abstracts, St. Louis ECS meeting, Abstract No. 40, Vol. 80-1, May 11–16, 1980.
85. T. Yokoyama and M. Enyo, *Electrochim. Acta*, **15**, 1921 (1970).
86. M. Morita, C. Iwakura, and H. Tamura, *Electrochim. Acta*, **24**, 639 (1979).
87. A. T. Kuhn and C. J. Mortimer, *J. Electrochim. Soc.*, **120**, 231 (1973).
88. V. Srb, R. Mraz, and S. Tichy, *Sci. Paper Inst. Chem. Technol.*, *Prague*, **1972**, 93.
89. T. Arikado, C. Iwakura, and H. Tamura, *Electrochim. Acta*, **23**, 9 (1978).
90. F. Hine and M. Yasuda, *J. Electrochem. Soc.*, **121**, 1289 (1974).
91. L. J. J. Janssen and J. G. Hoogland, *Electrochim. Acta*, **15**, 941 (1970).
92. T. Arikado, C. Iwakura, and H. Tamura, *Electrochim. Acta*, **22**, 229 (1977).
93. F. Chang and H. Wick, *Z. Phys. Chem.*, **A172**, 448 (1975).
94. W. D. Robertson, *J. Electrochem. Soc.*, **100**, 194 (1953).

95. M. Inoue and S. Yoshizawa, *J. Electrochem. Soc. (Jap.)*, **31**, 168 (1963).
96. A. G. Doughty, M. Fleischmann, and D. Pletcher, *J. Electroanal. Chem.*, **51**, 456 (1974).
97. N. Hackerman, E. S. Snavely, Jr., and L. D. Fiel, *Corr. Sci.*, **7**, 39 (1967).
98. B. Beden, C. Lamy, and J. M. Leger, *Electrochim. Acta*, **24**, 1157 (1979).
99. K. V. Rao, S. C. Das, and C. B. Roy, *J. Electrochem. Soc. (India)*, **27**, 135 (1978).
100. K. D. Snell and A. G. Keenan, *Chem. Soc. Rev.*, **8**, 259 (1979).
101. S. Motoo, *Denki Kagaku*, **48**, 328 (1980).
102. W. E. Vander Linden and J. W. Dieker, *Anal. Chem. Acta*, **119**, 1 (1980).
103. J. Weber and L. Kavan, *Chem. Listy*, **74**, 803 (1980).
104. M. Noel, P. N. Anantharaman, and H. V. K. Udupa, *Trans. SAEST*, **15**, 1 (1980).
105. R. R. Adzic, *Isrl. J. Chem.*, **18**, 166 (1979).
106. R. W. Murray, *Acc. Chem. Res.*, **13**, 135 (1980).
107. J. F. Evans, T. Kuwana, M. T. Henne, and G. P. Royer, *J. Electroanal. Chem.*, **80**, 409 (1977).
108. M. F. Dautartas and J. F. Evans, *J. Electroanal. Chem.*, **109**, 301 (1980).
109. J. B. Kerr, L. L. Miller, and M. R. Van de Mark, *J. Am. Chem. Soc.*, **102**, 3383 (1980).
110. A. Merz and A. J. Bard, J. Am. Chem. Soc., **100**, 3222 (1978).
111. N. Oyama and F. C. Anson, *J. Electroanal. Chem.*, **88**, 289 (1978).
112. A. P. Brown and F. C. Anson, *J. Electroanal. Chem.*, **92**, 133 (1978).
113. B. F. Watkins, J. R. Behling, E. Kariv, and L. L. Miller, *J. Am. Chem. Soc.*, **97**, 3549 (1975).
114. B. E. Firth, L. L. Miller, M. Mitani, T. Rogers, J. Lennox, and R. W. Murray, *J. Am. Chem. Soc.*, **98**, 8271 (1976).
115. B. E. Firth and L. L. Miller, *J. Am. Chem. Soc.*, **98**, 8272 (1976).
116. T. Matsue, M. Fujihira and T. Osa, *J. Electrochem. Soc.*, **126**, 500 (1979).
117. P. Grimes, unpublished data, 1981.
118. S. Trasatti, Ed., *Electrodes of Conductive Metallic Oxides*, Part A, Elsevier, New York, 1980.
119. R. W. Zurilla, R. K. Sen, and E. Yeager, *J. Electrochem. Soc.*, **125**, 1103 (1978).
120. P. Bindra and E. Yeager, Abstract No. 28, Extended Abstracts, Boston ECS Meeting, Vol. 79-1, May 6-11, 1979.
121. E. Yeager, EPRI Report, EM 505, June 1979.

Chapter **V**

SCALE-UP OF ELECTROORGANIC PROCESSES: SOME EXAMPLES FOR A COMPARISON OF ELECTROCHEMICAL ·SYNTHESES WITH CONVENTIONAL SYNTHESES

D. Degner

BASF AG,
Main Laboratory,
Ludwigshafen, West Germany

1 INTRODUCTION

Although electrochemical reactions involving organic compounds are among the oldest known reactions, they have not achieved any great commercial significance. After "organic electrochemistry" reached its peak at the turn of the century, and some of the many organic electrosyntheses, such as the oxidation of glucose to gluconic acid [1], had been carried out industrially, the field fell largely into oblivion from 1930 to 1940.

A renaissance of organic electrochemistry started in about 1965. The large-scale application of two organic electrosyntheses (*Monsanto, Asahi*: cathodic hydrodimerization of acrylonitrile to adipodinitrile [2, 3]; *Nalco*: electrosynthesis of tetraethyl lead [4]) made a substantial contribution to this renaissance. Subsequently a number of other organic electrosyntheses were accomplished industrially [3, 5], but a more extensive breakthrough in the use of electroorganic techniques which would have justified the high expectations frequently expressed [6], has not yet been achieved. Rather, the preconditions for the introduction of a new technology—as represented by organic electrochemistry—have even worsened over the years because the growth of the chemical industry has slowed down considerably; however, the preparative potential of these techniques is so attractive that an increasing number of companies are nevertheless occupied with intensive research in the field of electroorganic chemistry.

In the following sections, the problems involved in conversion to the industrial scale and possible solutions to these problems are discussed. Reference is made to a few organic electrosyntheses that have been developed at BASF and in some cases tested on an industrial scale. The industrial advantages and the problems specific to electroorganic reactions are dealt with first; some solutions to these problems are then presented with the aid of examples. From this, some general criteria as preconditions for the successful industrial application of organic electrosyntheses are then derived (see also Chapter IX).

2   INDUSTRIAL ADVANTAGES OF ELECTROORGANIC REACTIONS

Present-day knowledge of numerous electrolyte systems having extremely positive or negative decomposition potentials [7-9] makes it possible to carry out a large number of oxidation or reduction reactions by an electrochemical process, with the result that one can frequently dispense with the use of chemical redox systems. Anodic substitution reactions [10, 11] are of particular industrial value; they can be used to introduce functional groups into olefins or aromatics that are available at low cost.

Industrial electrosyntheses can be carried out without difficulty in the potential range from about -1.8 to $+1.8$ V, relative to a normal hydrogen electrode, using aqueous or at least water-containing electrolytes. Electrosyntheses that require higher potentials can only be applied industrially when they can be carried out in undivided cells because, in this case, mostly nonaqueous, largely aprotic, electrolyte systems having extremely low conductivities must be used. For energy reasons, this requires very small interelectrode distances, which have not hitherto been achieved industrially in the case of a divided cell. Furthermore, membranes or diaphragms that are suitable for industrial purposes are not yet available for these electrolytes.

There are two fundamental routes for carrying out organic electrosyntheses:

1. *Direct electrosynthesis:*

$$O + ne \rightleftharpoons R$$

2. *Indirect electrosynthesis:*

$$X + ne \rightleftharpoons Y$$
$$Y + O \rightleftharpoons X + R$$

The usual choice is route 1, in which the electron transfer takes place directly between electrode and substrate (O/R); in some cases, however, the indirect route 2, in which a catalytic amount of a chemical redox system X/Y is regenerated *in situ* [12], offers advantages in terms of the yield of product and the current efficiency that can be achieved.

In electrochemical reactions the electrode potential and the current density are two adjustable parameters by means of which the selectivity and rate of the reactions can be controlled within wide limits. This constitutes a substantial advantage of electroorganic reactions over conventional chemical oxidation and reduction processes, the course of which can only be influenced by the pressure, temperature, and concentration. Electrochemical reactions can be carried out under mild reaction conditions (atmospheric pressure, temperatures of 10 to 90°C) and these conditions make it possible to use inexpensive materials (for example, those produced from plastics). Expensive pressure equipment, frequently required for conventional processes, is not necessary. The low level of environmental pollution (low level of pollution from outgoing air and effluent) and simple process control constitute further industrial advantages.

3 PROBLEMS INVOLVED IN THE INDUSTRIAL APPLICATION OF ELECTROORGANIC REACTIONS, AND SOLUTIONS TO THESE PROBLEMS

There are a number of problems that have to be solved in the context of an industrial application of electroorganic processes. These problems can be divided into three groups:

1. Carrying out organic electrosyntheses industrially requires special reactors, and only some of the know-how gained from inorganic electrolyses can be used for this purpose.

2. Typical interface problems frequently arise at the electrodes during electroorganic reactions.

3. The necessary use of auxiliary reagents, which serve as electrolytes, gives rise to a number of process engineering problems, in particular during workup of the electrolysis products.

The synthesis section of an electrochemical plant is represented schematically in Fig. 5.1. The section consists essentially of an electrolysis cell, an electrolysis circuit with a pump and a heat exchanger, and also feed and product vessels, including metering devices.

The principal requirement of an industrial cell for electroorganic processes can be seen from the equation for the ohmic voltage drop ΔU in the electrolyte:

$$\Delta U = IR_E = j\frac{d}{\chi}$$

where

I = current
R_E = resistance of the electrolyte
j = current density
χ = specific conductivity of the electrolyte
d = electrode distance.

Because the specific conductivities of organic electrolytes (χ_{OE}) are substantially smaller than those of inorganic electrolytes (χ_{IE}) (χ_{OE}: 0.02 to

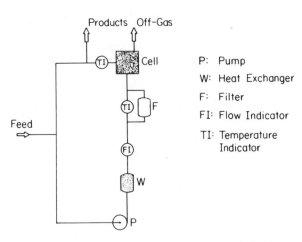

Fig. 5.1.   Electrolysis section of an electrochemical plant.

0.002 ohm^{-1} cm^{-1}; χ_{IE}: 0.1 to 0.4 ohm^{-1} cm^{-1}), it is necessary to develop electrolysis cells with very small interelectrode distances so that the voltage drop in the electrolyte, and hence the energy requirements for the electrolysis, remains within tolerable limits.

In recent years, a number of different cells [13–20] have been proposed for electroorganic reactions, and of these, essentially only three types of cell have hitherto found industrial application. Insofar as the process requires separation of the anode chamber and cathode chamber, a divided cell, which is constructed according to the principle of a filter press, is employed in most cases. Ion-exchange membranes are usually employed to separate the two electrode chambers.

More attractive from the industrial and economic point of view is the use of undivided cells, which virtually consist of only one electrode packet and one vessel. In particular cases, such as, for example, when sacrificial electrodes and small current densities are used, packed bed and fluidized bed electrolysis cells [21–24] are employed.

Figure 5.2 schematically represents the design principle of the plate-and-frame or filter press cell used at BASF.

Generally, the electrodes are connected in series in this cell. Figure 5.3 shows the multiple components of a cell unit.

Problems are numerous, but where a divided cell is required distances between electrode and membrane of less than 2 to 3 mm cannot be achieved industrially. The large number of leakproof interfaces presents a particular problem. The escape of combustible organic compounds can never be completely excluded. Particularly during continuous operation, there is the danger of bridging between two adjacent electrodes, which can lead to short-circuiting and cause a fire. Therefore, the room in which the cell is located must be equipped with special safety devices (for example, an automatic fire

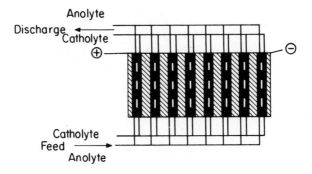

Fig. 5.2.   Plate-and-frame cell (design principle).

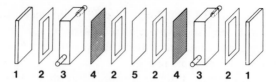

Fig. 5.3. Plate-and-frame cell (components of a cell unit). (*1*) Electrodes, (*2*) gaskets, (*3*) electrolyte frames, (*4*) support grids for membrane, (*5*) membrane.

detector, a smoke alarm or a system for monitoring the air in the room). Figure 5.4 shows a typical leakage occurring during operation that has led to a short-circuit and start of a fire.

The construction and dismantling of such a cell is very time-consuming and in most cases involves replacing the membranes. Because it is not possible, as a rule, to make used membranes leakproof, electroorganic reactions using plate-and-frame cells are only economical in the case of very long troublefree operating times.

If the process permits, the use of an undivided cell is substantially easier. A disc-stack cell [25, 26], which constitutes a further development of the capillary gap cell [13], is currently being employed industrially at BASF. Figure 5.5 shows the design principle of a particularly simple embodiment of this cell, which makes it possible to carry out organic electrosyntheses industrially even when electrolytes with very low conductivities have to be used.

The cell consists of a base plate and a stack of round electrodes with a central bore. The electrode distances are adjusted by means of plastic strips that permit distances reaching 0.3 to 0.5 mm. The electrode stack is in contact with the base plate. The stack is held in position by compressing the cover

Fig. 5.4. Leakage in a plate-and-frame cell.

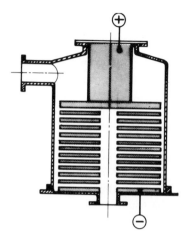

Fig. 5.5. Disc-stack cell.

and base plate, which act as current conductors. The electrolyte inlet is in the bottom plate and the outlet is in the upper part of the outer casing.

This cell does not present problems of leakage. The electrodes can be changed within a few hours and disturbances can therefore be eliminated rapidly. This cell is characterized by ease of operation and low investment and repair costs. Fundamental investigations, in particular into the hydrodynamic conditions in this cell, have been carried out by Dworak and Wendt [27], Ashworth and Jansson [28], Fleischmann et al. [29], and Horn [30].

Because both plate-and-frame cells and disc-stack cells are operated virtually at atmospheric pressure, inexpensive plastics, such as polyethylene and polypropylene, can be used as the construction material for the cell and the cell circuit. In many cathodic processes, it is advisable to employ plastics as the construction material because even traces of metal can completely deactivate the cathodes.

In industrial processes, the electrodes must satisfy the following general requirements:

1. The electrodes must not become deactivated.
2. The electrodes must not become coated.
3. The electrodes must not become corroded.

A critical step in the industrial development of an organic electrosynthesis is the test relating to the service life of the electrodes. Investigation of the long-term behavior of electrodes requires expensive pilot-scale experiments over a period of at least 1000 hr. In these experiments, the electrolysis products must be worked up at least until the electrolyte can be recycled in order to detect any by-products that may be accumulating. Electrode problems, such as cell voltage increase, formation of coatings, and electrode corrosion, which usually are not observed in laboratory experiments, often occur only during long-term experiments. These problems can be solved in many cases by varying the electrolysis conditions and by modifying the electrolyte composition. Even when a large number of experimental data are available, only limited predictions can be made concerning long-term behavior, because the results obtained for the optimization of one process can only rarely be applied to another process.

In addition to the long-term behavior of the electrodes, the discovery of a suitable electrolyte for the industrial application of an organic electrosynthesis is of decisive importance because the solution to this problem often determines the efficiency and hence the fate of the whole process. The electrolyte must satisfy the following general requirements:

1. adequate conductivity.

2. stability under electrolysis conditions.

3. easy workup and recycling of the electrolyte.

In contrast to the large amount of work carried out on the development of new cells, the search for a suitable electrolyte is still receiving too little attention.

4   EXAMPLES OF THE SCALE-UP OF ELECTROORGANIC PROCESSES

The advantage of electroorganic processes and the problems involved are explained in greater detail at this point, with reference to some electrosyntheses developed at BASF and carried out on a semiindustrial or industrial scale. Because there are already a number of publications [31–33] concerning general scale-up problems in electrochemical reactions, the following text only goes into the particulars relating to the scale-up of electroorganic reactions (see also Chapters II and III).

As a rule, the development of an organic electrosynthesis passes through three stages. In the first, the influence of the most important experimental parameters, such as conversion, current density, temperature, electrolyte composition, and electrode material, on the yield of product, and the current efficiency of the electrosynthesis is initially investigated in laboratory experiments. The results of polarographic or voltammetric experiments serve as preliminary information for the preparative experiments. In some cases, this enables the number of necessary optimization experiments to be reduced.

In the second stage, the long-term behavior of the electrodes and the recycling of the electrolytes are then tested in pilot-scale experiments. Only when these experiments have been completed is it possible to state whether the industrial application of the electrosynthesis is possible. The results of the pilot-scale experiments form the basis for an economic analysis in the third stage, which plays a large part in determining whether an experimental facility will be constructed. In this third step, any unsolved process engineering problems are clarified and, if appropriate, the first ton lots are prepared for introduction into the market. The fact that it is necessary to construct pilot plants and experimental plants is often the main reason that electroorganic syntheses cannot gain acceptance in preference to conventional processes, because, in the case of conventional processes, most companies already possess suitable plants, or at least plant sections that can rapidly be converted, for process development and initial production.

The possibilities of using an electroorganic process therefore arise when there is no industrial alternative to this process. In most cases, a cost advantage alone is not sufficient for a decision in favor of electrosynthesis. For example, there was virtually no industrial alternative to the first organic electrosynthesis applied industrially at BASF, namely, the cathodic reduction of phthalic acid to 1,2-dihydrophthalic acid.

Electrosynthesis of 1,2-Dihydrophthalic Acid [34–36]

1,2-Dihydrophthalic acid (DHPA) is required as an intermediate for the synthesis of gasoline additives and special plasticizers and also for the manufacture of polycarboxylic acids. The synthesis described in the literature (reduction of phthalic acid to DHPA with sodium amalgam) cannot be carried out industrially without difficulties and leads to considerable problems of environmental pollution. The electrochemical synthesis was described at the turn of the century by Mettler [37].

However, this synthesis exhibits the disadvantages of low current efficiency of DHPA production and rapid poisoning of the lead cathodes. The use of mercury as the cathode material [38], in which case poisoning is not observed, at least in laboratory experiments, is prohibited for environmental reasons.

In initial laboratory experiments, electrolytes were sought that did not give rise to cathode poisoning and that at the same time permitted higher phthalic acid concentrations in the electrolyte. An increase in the temperature of the electrolyte was desirable to improve the solubility of the phthalic acid;

however, this is only possible to a limited extent because the 1,2-DHPA rearranges into the 1,4-isomer at higher temperatures:

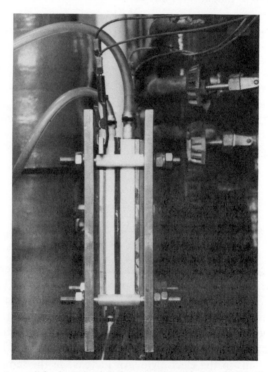

The electrochemical reduction must be carried out in a divided cell because both phthalic acid and 1,2-DHPA are oxidized at the anode. An ion-exchange membrane is employed to separate the electrode chambers. Figure 5.6 shows the laboratory cell that was constructed on the principle of a filter press and was employed for the optimization experiments.

Lead was used as the cathode material. The cathode surface area was 1 dm^2. Because the reduction is only successful in acid solution, sulfuric acid (5%) was employed as the catholyte. When a number of cosolvents were tested, the best results were achieved with tetrahydrofuran or dioxane as the

Fig. 5.6.   Laboratory cell for investigating the cathodic reduction of phthalic acid to DHPA.

Table 5.1.   Influence of Cosolvents on the Yields and the Course of the Reaction in the Cathodic Reduction of Phthalic Acid to Dihydrophthalic Acid.
Cathode: Pb; catholyte: 15% phthalic acid, 55% cosolvent, 25% H_2O and 5% H_2SO_4; current density: 10 A/dm^2; electrolysis using 2.8 F/mole phthalic acid.

Cosolvent	Yield of Product[a] (%)	Current Efficiency (%)	Remarks
Tetrahydrofuran	98	70	
Dioxane	98	70	
Methylglycol	80	57	Workup difficult
Dimethylformamide	75	54	Formation of by-products
Acetonitrile	95	68	Formation of by-products
Ethanol	90	64	Formation of by-products

[a] Relative to phthalic acid converted.

electrolyte additive. Some results of the optimization experiments are summarized in Table 5.1.

When tetrahydrofuran was used as the cosolvent, the phthalic acid concentration could be increased up to 20% by weight. At the same time, cathode poisoning was decisively repressed, as shown by potential/time curves (Fig. 5.7).

In initial experiments of longer duration, sparingly soluble compounds having structures **1** and **2** precipitated in the electrolyte, formed coatings on the electrodes and membranes and led to a progressive increase in the cell voltage. When tetrahydrofuran is replaced by dioxane, this formation of by-products can be almost completely prevented.

1   2

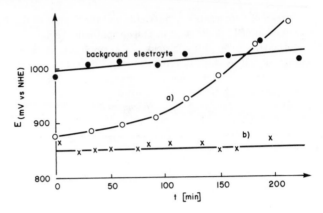

Fig. 5.7. Potential/time curves for the reduction of phthalic acid to 1,2-dihydrophthalic acid. (*a*) Catholyte: 4% phthalic acid and 50% H_2SO_4 in H_2O; T: 85°C (*b*) Catholyte: 20% phthalic acid, 5% H_2SO_4, 55% tetrahydrofuran and 20% water.

After optimization of the electrolyte composition, the influence of the cathode material on the current efficiency was investigated. The results are summarized in Table 5.2.

After successful completion of the laboratory experiments, the synthesis was applied on a pilot scale. The pilot-scale cell consisted of six cell units, as represented in Fig. 5.3. The electrode surface area of one electrode, namely, 0.3 m², corresponded to the size of the planned production electrode. By using special structured electrodes (see Fig. 5.8), the effective surface area could be increased by a factor of 1.8. The results of the pilot-scale experiments are summarized in Table 5.3.

Table 5.2. Current Efficiency in the Cathodic Reduction of Phthalic Acid to 1,2-Dihydrophthalic Acid[a]

Cathode material	Current efficiency (%)
Pb 99.99	68
Pb 95,Sb 5	62
Cd	54
Zn	52
Sn	7

[a]Experimental conditions as described in Table 5.1; cosolvent: dioxane

Fig. 5.8. Lead cathode for the cathodic reduction of phthalic acid to 1,2-dihydrophthalic acid.

Table 5.3. Reaction Conditions and Experimental Results for the Cathodic Reduction of Phthalic Acid to 1,2-Dihydrophthalic Acid in the Pilot Plant

Cathode	Pb
Catholyte	15% phthalic acid, 55% dioxane, 25% water and 5% H_2SO_4
Membrane	cation exchange membrane
Anolyte	5% H_2SO_4 and 95% water
Anode	PbO_2/Pb
Electrolysis using	2.8F/mole of phthalic acid
Temperature	30–40°C
Yield of 1,2-DHPA	94%
Yield of 1,4-DHPA	4%
Current efficiency	70%
Energy consumption	4.3 kWh/kg of 1,2-DHPA

The electrolysis was carried out batchwise and the electrolysis products were worked up continuously. Dioxane was recycled into the electrolysis. To maintain the cathode activity, the current was interrupted every 6 hr and the cell was short-circuited for about 1 min.

During the pilot-scale experiments, tests were carried out, in a laboratory apparatus, on a number of cation-exchange membranes to determine their suitability for use in the DHPA synthesis. Prerequisites for successful use were good electrical conductivity, low permeability to water and dioxane, and satisfactory mechanical stability. Some experimental results are summarized in Table 5.4.

Because increased migration of dioxane into the anode chamber led to rapid destruction of the lead dioxide anodes as a result of pitting corrosion (see Fig. 5.9), the migration of dioxane determined the choice of membrane. On the basis of the laboratory experiments, the membranes from Ionics and Tokuyama were used in the production plant.

After successful completion of the pilot-scale experiments, a plant for the electrochemical reduction of phthalic acid to dihydrophthalic acid, with a capacity of several hundred tons of DHPA per annum, was constructed and brought on stream. Figure 5.10 shows a simplified process diagram of the plant.

The electrolysis itself is carried out batchwise and the workup is carried out continuously. Because the conditions maintained during workup are such that rearrangement of the 1,2-dihydrophthalic acid cannot take place, the DHPA can be further processed immediately without purification.

The experimental results optimized in the pilot plant (compare Table 5.3) were achieved in the production plant on the first attempt. Figure 5.11 shows the electrolysis cells that constitute the core of the plant. The running time of

Table 5.4.   Investigation of Cation-Exchange Membranes for the Electrosynthesis of Dihydrophthalic Acid[a]

Membrane Type	Dioxane Loss (g/kAh)	CO_2 Content in the Anode Off-Gas (vol %)	Migration of Water into the Cathode Chamber (g/kAh)
Asahi Chemicals Aciplex	20	≤ 10	1500
Du Pont Nafion	20–190	> 10	2100
Ionac MC 3470	15	5	1850
Ionics 61 AZG	2	1	2800
Tokuyama CLE-E	< 2	< 1	2000

[a] Experimental conditions as described in Table 5.1; cosolvent, dioxane

Fig. 5.9. Lead dioxide anode condition after operation in anolytes containing dioxane.

the plant was limited by the life of the cation-exchange membranes to approximately 6 to 8 months. Figures 5.12 and 5.13 show sections cut out of a new and a used membrane. The cause of the erosion of the membrane is still unknown. To achieve longer operating times for the plant, it is necessary to improve the life of the membrane.

Electrosynthesis of Acetylenecarboxylic Acids [39, 53]

Acetylenedicarboxylic acid and propiolic acid are structural units for numerous heterocyclic syntheses. Industrially, chromic acid oxidation of the corresponding alkynols is virtually the only method by which these acids can be manufactured; however, the Cr-III waste liquors obtained in this process give rise to environmental pollution. The electrooxidation of alkynols to the corresponding acetylenecarboxylic acids [54] offers an industrially attractive alternative in this context.

Propiolic acid

The electrosynthesis of propiolic acid is only successful in acid solution and aqueous sulfuric acid is therefore preferably employed as the electrolyte:

$$HC \equiv C\text{-}CH_2OH \xrightarrow[+H_2O]{-4e\, -\, 4H^+} HC \equiv C\text{-}COOH$$

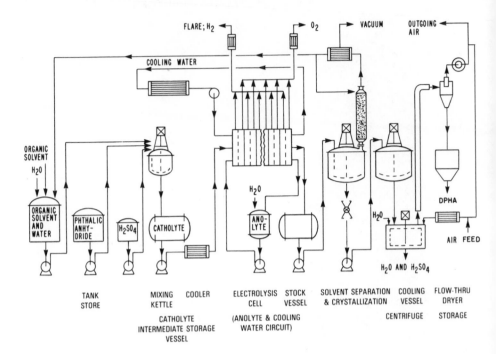

Fig. 5.10. BASF process for the continuous manufacture of dihydrophthalic acid.

Fig. 5.11. Electrolysis cells in the DHPA production plant.

Fig. 5.12. New cation-exchange membrane.

Because both propiolic acid and propargyl alcohol can be cathodically reduced in acid solution, the oxidation must be carried out in a divided cell. The only possible anode materials are lead dioxide and platinum, but, current/potential curves (Figure 5.14) show that the only practical anode material is lead dioxide.

The influence of current density, temperature, and anolyte composition on the yield of product and current efficiency of propiolic acid production was initially investigated in laboratory experiments. Figure 5.15 shows the dependence of the yield of propiolic acid on the temperature at various current densities. Optimum experimental results were achieved under the following conditions:

Anode: PbO_2.

Anolyte: 7% propargyl alcohol; 13% H_2SO_4; 80% H_2O.

Current density: ≤ 12 A/dm^2.

Temperature: $\leq 30°C$.

Electrolysis using $5 F$/mole of propargyl alcohol.

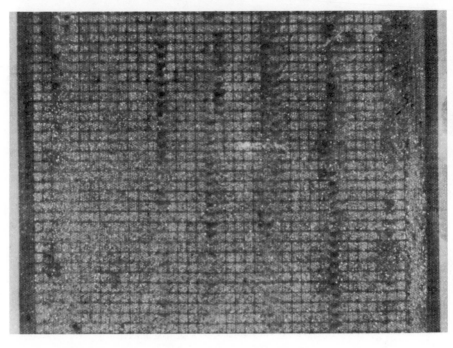

Fig. 5.13. Cation-exchange membrane after an operating time of 6 months.

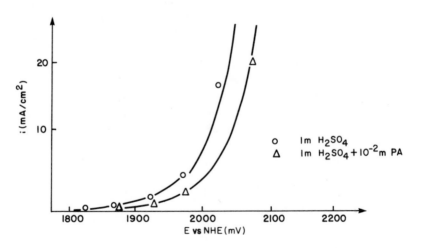

Fig. 5.14. Current/potential curves (oxidation of propargyl alcohol in sulfuric acid solution).

268

The yield of propiolic acid in this case was 76%. Pitting corrosion of the massive PbO_2 anodes was observed in the first laboratory experiments. These anodes were therefore replaced by composite $Ti/Rh/PbO_2$ anodes.

The results of the laboratory experiments were largely confirmed in pilot-scale experiments. Figure 5.16 schematically depicts the process flowsheet for propiolic acid synthesis.

The electrolysis is carried out batchwise and the electrolysis products are worked up continuously. To separate the propiolic acid, the electrolysis product is extracted with high-boiling ethers, such as diethylene glycol dibutyl ether. The aqueous electrolyte, freed from propiolic acid, is recycled into the electrolysis. Propiolic acid is then separated from the extractant by distillation to achieve a purity of about 95%.

Acetylenedicarboxylic Acid

The electrosynthesis of acetylenedicarboxylic acid proceeds substantially as described for the synthesis of propiolic acid:

$$HO-CH_2-C \equiv C-CH_2-OH \xrightarrow[+2H_2O]{-8e - 8H^+} HOOC-C \equiv C-COOH$$

At current densities of about 5 A/dm^2 and a temperature of 10°C, the yield of acetylenedicarboxylic acid is more than 70% and the current efficiency is about 50%. In contrast to the propiolic acid synthesis, no corrosion problems arise with the massive PbO_2 anodes.

The acetylenedicarboxylic acid is also separated from the electrolysis products by extraction. Methyl isobutyl ketone is employed as the extractant. Methyl isobutyl ketone is then distilled off and the acetylenedicarboxylic acid

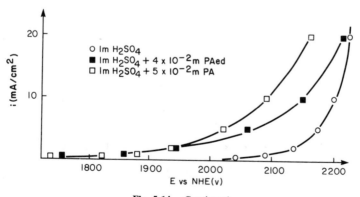

Fig. 5.14. Continued.

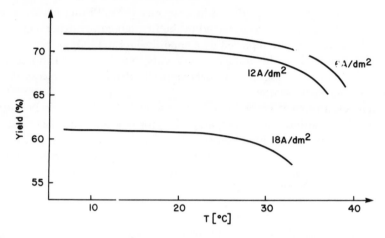

Fig. 5.15. Yields of propiolic acid as a function of temperature and current density.

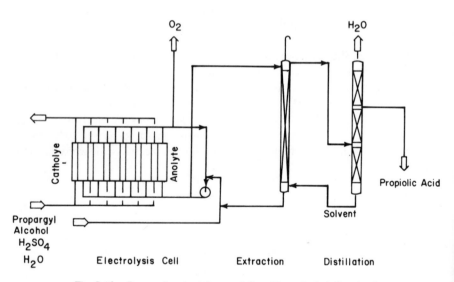

Fig. 5.16. Process flowsheet for propiolic acid synthesis (pilot plant).

obtained is centrifuged and dried. Acetylenedicarboxylic acid thus obtained is about 95% pure. Figure 5.17 shows a simplified process diagram of the acetylenedicarboxylic acid synthesis.

Combined Procedure

Insofar as a divided cell must be employed for an organic electrosynthesis, it is possible, in principle, to use both electrode chambers at the same time

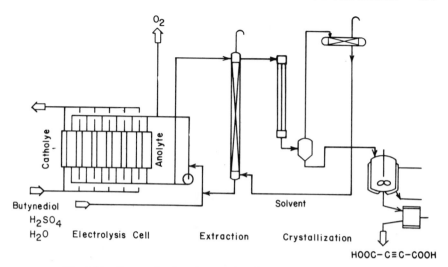

Fig. 5.17. Process flowsheet for acetylenedicarboxylic acid (pilot plant).

one for an anodic process and one for a cathodic process. BASF has succeeded in carrying out the cathodic reduction of phthalic acid to dihydrophthalic acid and the anodic oxidation of acetylene alcohols to the corresponding carboxylic acids simultaneously in one electrolysis cell. In the case of simultaneous production of dihydrophthalic acid and acetylenedicarboxylic acid, the DHPA plant can be used as it stands. The composite $Ti/Rh/PbO_2$ anodes employed for the propiolic acid synthesis cannot be employed in the combined procedure because the current efficiency of the DHPA synthesis drops sharply with increasing operating time. The reason for this behavior is the deposition of small traces of rhodium, which leads to irreversible deactivation of the cathode. Because composite anodes containing noble metals are unsuitable for the combined procedure, a special Ti/PbO_2 anode with a titanium carbide interlayer [39] was developed for the propiolic acid synthesis. Figure 5.18 shows an anode developed for the propiolic acid synthesis.

After completion of the optimization studies, both acetylenedicarboxylic acid and propiolic acid were manufactured in combined production with dihydrophthalic acid.

Electrosynthesis of Sebacic Acid Esters [40–43]

Sebacic acid is required for the synthesis of specific polyamides. Sebacic acid esters are used as special plasticizers and also as synthetic lubricants. Hitherto, sebacic acid has been manufactured from naturally occurring castor oil. However, on an industrial scale, this process is very involved, since

Fig. 5.18. Composite PbO_2/Ti electrode for the production of propiolic acid using the combined procedure.

the manufacture of pure sebacic acid requires several purification operations. Isooctanol is obtained as the coproduct.

An attractive alternative industrial synthesis for sebacic acid or sebacic acid esters is the Kolbe reaction of adipic acid monoalkyl esters:

$$2ROOC-(CH_2)_4-COOH \xrightarrow[-2CO_2]{-2e - 2H^+} ROOC-(CH_2)_8-COOR$$

The first work relating the industrial application of the Kolbe reaction to the manufacture of sebacic acid esters by synthesis was carried out by Offe in Leuna in 1942. Despite a considerable amount of work, the process has not yet gained acceptance.

The anodic oxidation of adipic acid monomethyl ester and adipic acid mono-2-ethylhexyl ester to the corresponding sebacic acid diesters has been investigated at BASF. The synthesis is characterized by the use of an undivided disc-stack cell. In some laboratory experiments carried out to obtain a general picture, the electrolysis parameters and the electrolyte composition were initially optimized. The best results were achieved under the following experimental conditions:

Anode: platinum.

Electrolyte: 36 to 38% adipic acid monomethyl ester; 2 to 4% of the sodium salt of adipic acid monomethyl ester.

Current density: 25 A/dm^2.

Temperature: 40 to 45°.

Electrolysis using 2.7 F/mole of adipic acid monomethyl ester.

Under these experimental conditions, the yield of sebacic acid dimethyl ester was 81% and the current efficiency was about 60%.

Pilot-scale experiments were used to test the long-term behavior of the platinum anodes. In the first experiments, an increase in cell voltage was observed even after short operating times, and the resulting increase in energy costs was no longer economically viable. The formation of coatings is considerably more pronounced in the synthesis of diisooctyl sebacate than in the synthesis of dimethyl sebacate. The increase in cell voltage can indeed be almost completely suppressed by periodic short-circuiting of the electrodes, but this results in an excessive increase in the erosion of platinum from the anodes. In a screening program, electrolyte additives that prevent the formation of coatings were therefore sought. The addition of cosolvents, for example, tetrahydrofuran (about 25% by weight), or of small amounts of formic acid or acetic acid, proved successful in this case, but involved greater expense when working up the electrolysis products. Surprisingly, the electrode problem was then solved by adding small amounts of water (about 0.2 to 2 wt %) to the electrolyte. The use of cosolvents was then no longer necessary. The erosion of platinum was simultaneously reduced from 12 to 4 mg/kg of diisooctyl sebacate.

After successful completion of the pilot-scale experiments, a 100-ton/year sebacic acid diester experimental plant was constructed and brought on stream. The results from the pilot plant were largely achieved. Thus the yield of dimethyl sebacate was about 80%, the current efficiency was 64%, and the energy consumption was 4 to 4.5 kWh/kg of dimethyl sebacate.

Electrosynthesis of 2,5-Dimethoxy-2,5-Dihydrofuran [44]

2,5-Dimethoxy-2,5-dihydrofuran, which is the closed-ring acetal of maleic dialdehyde, can undergo various reactions [45] and can be employed, for example, as a formaldehyde-free cross-linking agent. Catalytic reduction leads to 2,5-dimethoxytetrahydrofuran, which is required for specific heterocyclic syntheses, in particular for the manufacture of tropinones.

The conventional synthesis requires bromine as a reagent and must be carried out at temperatures of about $-5°C$ and in the presence of stoichiometric amounts of auxiliary bases:

$$\text{furan} + Br_2 + 2CH_3OH \longrightarrow \underset{CH_3O \quad O \quad OCH_3}{\overset{H \quad \quad H}{\text{dihydrofuran}}} + 2HBr$$

$$2R_3N + 2HBr \longrightarrow 2R_3NH^+Br^-$$

The principal disadvantages of the chemical synthesis are the high bromine consumption (about 1.6 kg bromine/kg 2,5-dimethoxy-2,5-dihydrofuran), the effluent pollution resulting therefrom, and the formation of brominated by-products. Clauson-Kaas et al [46] have shown that these disadvantages do not arise if the synthesis is carried out electrochemically.

$$\text{furan} \xrightarrow[+2CH_3OH]{-2e, -2H^+} \underset{CH_3O \quad O \quad OCH_3}{\overset{H \quad \quad H}{\text{dihydrofuran}}}$$

In laboratory experiments, tests were first carried out to see whether this synthesis could be applied on an industrial scale and, if so, under what conditions. The laboratory cell used for this purpose is shown in Fig. 5.19.

The laboratory experiments led to two substantial improvements from the point of view of industrial application:

1.   The expensive platinum anodes can be replaced by less expensive graphite electrodes without losses in yield or current efficiency.

2.   The electrolyte system NH_4Br/CH_3OH, which cannot be recycled, can be replaced by NaBr or KBr/MeOH, which can be recycled. Surprisingly, increased storage stability of the electrolyte and the formation of a considerably smaller amount of by-products are observed in this case.

The results optimized in the laboratory are summarized in Table 5.5. The yield of 2,5-dimethoxy-2,5-dihydrofuran and the current efficiency are between 85 and 90%.

After completion of the laboratory experiments, the long-term behavior of the graphite electrodes and the suitability of the NaBr/MeOH electrolyte for recycling were investigated in pilot-scale experiments. The electrolysis was carried out batchwise and the electrolysis products were worked up continously. Methanol and any unreacted furan that were still present were

Fig. 5.19.   Laboratory cell for the electrosynthesis of 2,5-dimethoxy-2,5-dihydrofuran.

Table 5.5   Optimization Results from an Investigation of the Electrosynthesis of 2,5-Dimethoxy-2,5-dihydrofuran

Undivided electrolysis cell
Graphite electrodes
Electrolyte: furan/sodium bromide/methanol
Current density: 10–15 A/dm^2
Temperature: 5–10°C
Electrolysis using 2 F/mole of furan

distilled off under normal pressure, and the sodium bromide that subsequently precipitated was filtered off and recycled into the electrolysis together with the distillate. The crude dimethoxydihydrofuran obtained as the filtrate was further purified by vacuum distillation.

The results obtained in two series of experiments are summarized in Table 5.6.

Table 5.6.   Pilot-Scale Experiments: Electrosynthesis of 2,5-Dimethoxy-
2,5-Dihydrofuran

Running Time (hr)	Yield of Product (%)	Current Efficiency (%)	Purity (%)	By-Products (%)	NaBr Loss [a] (%)
A. Furan conversion 100%; electrolysis using 2 F/mole of furan					
100	90.8	90.8	91	9	10.7
200	89.8	89.8	90	10	12.2
400	89.8	89.8	90	10	14.4
800	90.8	90.8	91	9	14.0

Running Time (hr)	Yield of Product (%)	Current Efficiency (%)	Purity (%)	By-Products (%)	NaBr Consumption [a] (%)
B. Furan conversion 81%; electrolysis using 1.62 F/mole of furan					
100	96.0	96.0	95.5	3.5	8.3
200	95.8	95.8	96.3	3.7	12.3
400	96.1	96.1	96.6	3.4	9.4
800	95.7	95.7	96.5	3.5	7.0
1200	96.0	96.0	96.5	3.5	8.0

[a] NaBr loss based on NaBr employed.

For a furan conversion of 81%, the yield of product and the current efficiency were both 95 to 96%. Recycling the electrolyte did not detract from the results and no electrode problems were encountered in either series of experiments. For a furan conversion of about 80%, the dimethoxydihydrofuran obtained was so pure that it was further processed immediately.

After successful completion of the pilot-scale experiments, a production plant was constructed; the electrosynthesis of 2,5-dimethoxy-2,5-dihydrofuran was the first electrochemical synthesis carried out industrially in an undivided cell at BASF. Figure 5.20 shows the disc-stack cell used for this process.

Electrosynthesis of Naphthyl Acetate [47–49]

α-Naphthol is an important precursor for a number of dyestuffs and for the synthesis of insecticides and is required on a 10,000-ton/year scale for this purpose. At present, essentially two processes are carried out industrially for the synthesis of α-naphthol. In the conventional process, naphthalene is first sulfonated to naphthalenesulfonic acid and the naphthalenesulfonic acid formed is then converted to α-naphthol in an alkaline melt:

The principal disadvantages of this process are moderate yield, low selectivity (the 1-naphthol contains about 10% of 2-naphthol), and high proportion of the generated salt (several tons of waste salt per ton of α-naphthol).

The aerial oxidation process developed and carried out on a large industrial scale by UCC does eliminate these disadvantages, but the process only becomes economical at very high capacities because of the number of stages involved and the numerous recycling operations required.

The electrosynthesis of naphthyl acetate, investigated by Lindstead et al. [50], Eberson and Nyberg [51], and Koehl [52], offers an attractive alternative to the conventional processes. However, this process can only be carried out economically if an electrolyte is discovered that is easy to recycle and if the Kolbe reaction, that takes place as a secondary reaction, can be substantially suppressed. The Kolbe reaction not only causes an increased loss of acetic acid, but also leads to the formation of methylnaphthalenes,

Fig. 5.20.   Industrial disc-stack cell for the electrosynthesis of 2,5-dimethoxy-2,5-dihydrofuran.

which are also acetoxylated and can only be separated from naphthyl acetate by costly distillation operations.

$$2CH_3COOH \xrightarrow{-2e, -2H^+} 2[CH_3] \quad +2CO_2$$

The discovery of distillable conducting salts and the use of plastics containing a graphite filler as the electrode material has made it possible to solve these problems, thus permitting the industrial application of the synthesis. The experimental results optimized in laboratory experiments are summarized in Table 5.7.

After completion of the laboratory experiments, pilot-scale experiments were carried out to investigate the service life of the electrodes and the recycling of the electrolyte. The electrolysis and working up, and also the recycling of the electrolyte, were carried out continuously. The increase in the cell voltage observed after only 100 to 200 hr of operation was so sharp that the process was no longer economically viable. The reason for this increase was the formation of an anode coating of high electrical resistance. By periodically reversing the polarity of the electrodes, the formation of the coating was substantially suppressed: 1000 hr of operation were achieved without observing a prohibitive increase in the cell voltage.

Table 5.7. Optimization Results—Electrosynthesis of Naphthyl Acetate in undivided Electrolysis Cell

Electrodes	80% graphite/20% polypropylene
Electrolyte	naphthalene/acetic acid/trimethylammonium acetate
Current density	4–12 A/dm^2
Temperature	40–80°C
Naphthalene conversion	20–50%
Yield of product	70–85%
Current efficiency	45–60%

After the experiments were completed, the process was tested in a 5-ton/month experimental plant. The results of the laboratory and pilot-scale experiments were substantially confirmed in this plant.

5 POSSIBILITIES FOR THE APPLICATION OF ORGANIC ELECTROSYNTHESES IN THE CHEMICAL INDUSTRY

It is often said that electrochemical reactions in organic chemistry are inferior to conventional chemical syntheses because of unduly high energy and investment costs. This belief, however, is incorrect. An analysis of the investment costs of electroorganic reactions shows that, at capacities up to about 10,000 tons/year, investment for the electrochemical section of the plant is in most cases only 30 to 40% of total investment. The proportion for the cell is only about 2% of total investment when an undivided cell is used. Even when a substantially more expensive divided cell is required for the process, only about 10% of the total investment costs is needed for the cell. The energy costs for the electrolysis are in most cases less than 15% of the manufacturing costs; in rare cases, about 25% of the manufacturing costs are needed for energy.

An essential prerequisite for economic organic electrosynthesis is a simple workup of the electrolysis products to give the desired product, since in most cases more than half the total investment is needed for this operation. Simplifications concerning the solvent/conducting salt auxiliary components therefore lead to a more pronounced reduction in manufacturing costs than do improvements in the electrolysis cell. The most important prerequisites for the industrial application of an organic electrosynthesis are summarized in Table 5.8.

In addition to industrial and economic viewpoints, the interests of the company, the availability of alternative methods, in terms of type and

Table 5.8.   Prerequisites for the Industrial Application of an Organic
Electrosynthesis

a. High yield of product and current efficiency
b. Energy consumption per kilogram of desired product: max. 6–8 kWh/kg
c. Concentration of desired product in the electrolyte: ≥ 8–10%
d. Service life of the electrodes: ≥ 1000 hr
e. Life of the membrane: ≥ 2000 hr
f. Simple workup of the electrolysis products
g. Problemfree recycling of solvent and conducting salt

number, and also the existence of electrochemical plants play important
roles in assessing the prospects of an electroorganic synthesis. These prospects are favored if the synthesis product is used by the company, if the starting material is manufactured by the company, and if electrochemical plants
already exist. A lack of know-how and a lack of electrochemical plants make
the realization of an electrosynthesis more difficult. The synthesis of the
dyestuff precursor *m*-aminobenzenesulfonic acid constitutes an example of
the extent to which the availability of alternative processes determines the application of an electrosynthesis. While smaller companies manufacture the
compound electrochemically, in companies having pressure hydrogenation
plants at their disposal the reduction is carried out catalytically.

$$\underset{O_2N}{\overset{SO_3H}{\bigcirc}} \xrightarrow[-2H_2O]{+6e,\ +6H^+} \underset{H_2N}{\overset{SO_3H}{\bigcirc}}$$

At the present time, it is not easy to assess whether electroorganic syntheses will find a broader application in the chemical industry. Rising raw
material costs and the constraint to change to production processes causing
even less environmental pollution certainly favor a number of electrosyntheses, but the substantially slower growth of the chemical industry, together
with overcapacities in many areas, makes it more difficult to establish a new
technology, such as organic electrochemistry.

References

1. H. S. Isbell, H. L. Frush, and F. J. Bates, *Ind. Eng. Chem.*, **24**, 375 (1932).
2. *Chem. Eng.*, **72**, 23, 238 (1965).
3. *Chem. Eng.*, **82**, 14, 44 (1975).

4.   *Chem. Eng.*, **72**, 23, 249 (1965).
5.   W. J. M. van Tilborg, *Chem. Weekblad*, **1978**, 31.
6.   J. W. Johnson, in *Electrochemistry, The Past Thirty and the Next Thirty Years,* H. Bloom, and F. Gutmann, Eds., Plenum Press, New York and London, 1977, p. 257.
7.   F. Goodridge, and C. J. H. King, in *Technique of Electroorganic Synthesis,* Part I; Norman L. Weinberg, Ed., Wiley, New York, 1974, pp. 28 ff.
8.   H. Lund, and P. Iversen, in *Organic Electrochemistry,* Dekker, New York, 1973, p. 208 ff.
9.   C. K. Mann, and K. K. Barnes, *Electrochemical Reactions in Nonaqueous Systems*, Dekker, New York, 1970.
10. L. Eberson, and H. Schäfer, *Fortschr. Chem. Forsch.,* **21**, 1 (1971).
11. L. Eberson, and K. Nyberg, *Tetrahedron,* **32**, 2185 (1976).
12. R. Clarke, A. Kuhn, and E. Okoh, *Chem. Bri.,* **11**, 59 (1975).
13. F. Beck, and H. Guthke, *Chem. Ing. Tech.,* **41**, 943 (1969).
14. L. Cedheim, L. Eberson, B. Helgée, K. Nyberg, R. Servin, and H. Sternerup, *Acta Chem. Scand. Ser. B.,* **29**, 617 (1975).
15. R. W. Houghton, and A. T. Kuhn, *J. Appl. Electrochem.,* **4**, 173 (1974).
16. P. M. Robertson, F. Schwager, and N. Ibl, *J. Electroanal. Chem.,* **65**, 883 (1975).
17. P. M. Robertson, (Hoffmann-La Roche & Co. AG), DT 2415784, DT 2503819.
18. R. E. W. Jansson, and G. A. Ashworth, *Electrochim, Acta,* **22**, 1301 (1977).
19. J. Ghoroghchian, R. E. W. Jansson, and D. Jones, *J. Appl. Electrochem.,* **7**, 437 (1977).
20. M. Farooque, and T. Z. Fahidy, *AIChE Symp. Ser.* No. 185, **75**, 129 (1979).
21. J. R. Backhurst, M. Fleischmann, F. Goodridge, and R. E. J. Plimley (National Research Development Corp.), DT 1671463.
22. F. Goodridge, *Electrochim. Acta,* **22**, 929 (1977).
23. F. Goodridge, C. J. H. King, and A. R. Wright, *Electrochim., Acta,* **22**, 347 (1977).
24. G. Kreysa, and E. Heitz, *Chem. Ing. Tech.,* **48**, 852 (1976).
25. F. Beck, D. Francke, H. Hannebaum, H. Nohe, and M. Stroezel, (BASF AG), DT 2502167.
26. D. Francke, R. Schwen, and G. Wunsch (BASF AG), DT 2502840.
27. R. Dworak, and H. Wendt, *Ber. Bunsenges. Phys. Chem.,* **80**, 77 (1976).
28. G. A. Ashworth, and R. E. W. Jansson, *Electrochim. Acta,* **22**, 1295 (1977).
29. M. Fleischmann, J. Ghoroghchian, and R. E. W. Jansson, *J. Appl. Electrochem.,* **9**, 437 (1979).
30. R. K. Horn, *AIChE Symp. Ser.* No. 185, **75**, 125 (1979).
31. N. Ibl, *Chem. Ing. Tech.,* **35**, 353 (1963).
32. N. Ibl, E. Adam, *Chem. Ing. Tech.,* **37**, 573 (1965).
33. P. Gallone, *Electrochim. Acta,* **22**, 913 (1977).
34. H. Aschenbrenner, F. Beck, W. Bruegel, H. Nohe, and H. Suter, (BASF AG), DT 1618078.
35. H. Nohe, and H. Suter, (BASF AG), DT 1953259.
36. H. Nohe, and H. Suter, (BASF AG), DT 1953260.

37. C. Mettler, *Ber. Dtsch. Chem. Ges.*, **39**, 2933 (1906).

38. P. C. Condit, *Ind. Eng. Chem.*, **48**, 1252 (1956).

39. W. Habermann, P. Jäger, and H. Nohe, (BASF AG), DT 2344645.

40. F. Beck, J. Haufe, and H. Nohe, (BASF AG), DT 2014985.

41. F. Beck, J. Haufe, and H. Nohe, (BASF AG), DT 2023080.

42. W. Eisele, H. Nohe, and H. Suter, (BASF AG), DT 2039991.

43. H. Nohe, H. Suter, and F. Wenisch, (BASF AG), DT 2248562.

44. D. Degner, H. Hannebaum, and H. Nohe, (BASF AG), DT 2710420.

45. N. Elming, *Ad. Org. Chem.*, **2**, 67 (1960).

46. N. Clauson-Kaas, F. Limborg, and K. Glens, *Acta Chem. Scand.*, **6**, 531 (1952).

47. D. Degner, J. Haufe, and C. Rentzea, (BASF AG), DT 2434845.

48. K. Boehlke, D. Degner, and W. Treptow, (BASF AG), DT 2659148.

49. L. Schuster, and B. Seid, (BASF AG), DT 2706682.

50. R. P. Lindstead, B. R. Shepard, and B. C. L. Weedon, *J. Chem. Soc.*, **1952**, 3624.

51. L. Eberson, and K. Nyberg, *J. Am. Chem. Soc.*, **88**, 1686 (1966).

52. W. J. Koehl, Jr. (Socony Mobil Oil Company), U.S. Patent 3,252,877.

53. P. Jäger, and H. Nohe (BASF AG), DT 2409117.

54. V. Wolf, DT 931409.

Chapter VI

EXPERIENCE IN THE SCALE-UP OF THE MONSANTO ADIPONITRILE PROCESS

D. E. Danly
and
C. R. Campbell

Monsanto Chemical Intermediates Company,
Pensacola, Florida

1 INTRODUCTION

In 1963 the discovery of an electrochemical reductive coupling of acrylonitrile to adiponitrile by Dr. M. M. Baizer was announced [1]. This work was prompted by the very large annual production of adiponitrile as a nylon 66 intermediate, by a U.S. oversupply of acrylonitrile at that time, and by the development by Sohio of a process for the direct ammoxidation of propylene to acrylonitrile offering a significantly improved cost structure for this raw material. Baizer's employer, Monsanto, was both one of the largest producers of acrylonitrile and one of the largest consumers of adiponitrile. While the desirability of converting acrylonitrile to adiponitrile was recognized as early as 1949 [2, 3], the yields were discouragingly low until Knunyants et al. [4] showed that acrylonitrile dissolved in strong mineral acids could be reductively coupled to adiponitrile in 62% yield by treatment with potassium amalgams that were generated electrochemically.

Baizer's development of the electrochemical hydrodimerization (EHD) of acrylonitrile (AN) to adiponitrile (ADN) occurred over the last 11 months of 1960. In this brief time span, he established that the EHD reaction proceeded at very high yields and reasonable current densities and involved the direct electrochemical reduction of AN. He discovered and described the importance of the tetraalkylammonium salts as supporting electrolytes and of the effect of AN concentration and pH on the yield of ADN. A number of cathodes were evaluated, with lead and mercury showing best results, and yields were optimized by the use of divided cells employing separate anolyte and catholyte systems. Baizer utilized Alundum as a separator of the electrolytes, but recognized the potential benefits of a semipermeable ion-exchange membrane, if one could be developed for this application. Platinum wire or carbon rods were utilized as anodes in an Alundum cup containing the strong acid electrolyte. The laboratory cells, described elsewhere [1], were generally contained in 500-ml resin flasks and were operated both batchwise and continuously. The reactions are shown below.

Cathode

$$2CH_2=CHCN + 2H_2O + 2e \rightarrow NC(CH_2)_4 CN + 2OH^-$$

Acrylonitrile   Adiponitrile

Anode

$$H_2O \rightarrow 2H^+ + \tfrac{1}{2}O_2 + 2e$$

Overall

$$2CH_2=CHCN + H_2O \rightarrow NC(CH_2)_4CN + \tfrac{1}{2}O_2$$

Major by-products were formed according to the following reactions:

Electrolytic reactions

$$CH_2=CHCN + 2H_2O + 2e \longrightarrow CH_3CH_2CN + 2OH^-$$
AN   Propionitrile

$$3CH_2=CHCN + 2H_2O + 2e \longrightarrow NCCH(CH_2)_3CN + 2OH^-$$
AN

$$\underset{|}{}(CH_2)_2$$

$$\underset{}{}CN$$

Trimer

Cyanoethylation reactions

$$CH_2=CHCN + H_2O \xrightarrow{\ OH^-\ } HOCH_2CH_2CN$$
AN   Hydroxypropionitrile

$$CH_2=CHCN + HOCH_2CH_2CN \xrightarrow{\ OH^-\ } CN(CH_2)_2O(CH_2)_2CN$$
Biscyanoethylether

The responsibility for the commercialization of the EHD process was transferred to the Nylon Intermediates Development Department at Pensacola, Florida, With an overall charge to develop a commercially attractive continuous process operating at high current density and yield, low power consumption and capital requirements, and utilizing stable materials of construction. To achieve these goals, the following factors required solution and are discussed separately in this chapter:

Cathode/catholyte selection*.

Anode/anolyte selection*.

Membrane development.

Cell design.

Recovery and purification.

Manufacture of supporting electrolyte.

*See Chapter IV for further details

Selection of Supporting Electrolyte

In the initial research, the superiority of the quaternary ammonium salts (QAS) was recognized. The choice of the optimum QAS was based on beliefs that it should be highly conductive, capable of solubilizing 10 to 15% acrylonitrile together with a similar level of products, should not discharge at cathodic voltages below the AN dimerization voltage, should be easily synthesized, and should have favorable solubility characteristics such that ADN and by-products could be conveniently separated from it by solvent extraction with acrylonitrile. Over 30 quaternary ammonium salts were prepared and their physical properties were measured (Table 6.1). From a consideration of their properties and ease of preparation, a limited number of salts were classified as preferred for EHD and evaluated in bench-scale operations (Table 6.2).

While these results show little difference in yields for the first three QAS, tetraethylammonium ethylsulfate was ultimately chosen based on improved conductivity and ease of synthesis. Extended runs comparing QAS of different structures showed that increased size of the quaternary ammonium ion beyond that of the tetramethyl ammonium ion is beneficial in improved yields. This approach was limited, however, because with increased size of the cation it becomes more difficult or impossible to effect a reasonable separation of QAS and cell product by simple extraction.

Cathodes

In his early work, Baizer recognized the importance of hydrogen overvoltage in the selection of a cathode material and described both mercury and lead as giving higher ADN selectivities than other alternatives tested. Mercury was rejected as a cathode candidate in the early development stages because of anticipated design problems. A number of other solid metals with high hydrogen overvoltages were evaluated in short bench-scale runs. As shown in Table 6.3 none was superior to lead. A binary alloy of lead and antimony was tested because of its improved structural properties but gave slightly lower ADN selectivities. Lead/mercury amalgams were equivalent to pure lead. Based on these results, the lead was readily selected as the cathode.

While the selection of lead as the cathode was straightforward, the effect of the composition of the anode on the performance of the cathode was not recognized in the early stages of development. As work progressed and bench-scale runs were lengthened, it became obvious that ADN selectivity decreased with time as shown in Table 6.4. This decline was traced to the deposition of substances on the cathode. Since the fouling material was largely organic, it was reasoned that its accumulation furnished a barrier that retarded the transfer of reactants to the cathode, resulting in increased formation of propronitrile and high boilers. The decreased mass transfer also increased the alkalinity in the layer and contributed to increased rates of formation of cyanoethylation by-product.

Steps were taken to improve filtration and to minimize the concentration of organic impurities, but only a marginal improvement in maintaining ADN selectivity was observed. With decreased organic fouling, it became evident that metals (particularly silver) were being electrodeposited on the cathode. Silver was introduced into the system by corrosion of the silver/lead anode. Pure lead was substituted as an anode and ADN selectivity remained relatively constant. Silver as the sulfate was then added to the anolyte at a rate sufficient to maintain a concentration of 14 ppm. Over a 60-hr period the selectivity of propionitrile increased from 3 to 20%. The addition of silver was then stopped and lead sulfate was added to the catholyte over a 60-hr period. The propionitrile selectivity dropped to 4%.

With these findings, an electrolytic cell with a high cathode surface area was installed to continuously remove metals from the catholyte. The demetallizing cell consisted of a column of 2-mm diameter lead shot serving as a cathode with a platinized titanium or carbon rod anode centrally located in the column and separated from the shot by a plastic mesh sheath. Catholyte was pumped through the cell while a cathode voltage of -1.2 V (SCE) was maintained. At a residence time of 1 min, 70% of the silver was removed when the feed to the demetallizer contained 1 ppm. When the metal concentration was maintained below 2 to 3 ppm a reasonably stable ADN selectivity was observed. The results in Table 6.5 show selectivity values with time. The cathode was pure lead and the anode was the silver/lead alloy.

Anode/Anolyte Selection

In Baizer's original EHD experiments the anolyte consisted of a 35 wt % aqueous solution of tetraethylammonium para-toluene sulfonate (TEAPTS) with a platinum wire serving as the anode. The first efforts to scale up Baizer's pot-cell to a parallel-plate design gave relatively little attention to the anolyte/anode combination. At this point, all that was sought was a system that would allow operation for a sufficient duration to assess the performance of the more-important catholyte/cathode pair. Anolytes of toluene sulfonic acid and sulfuric acid, as well as Baizer's original TEAPTS salt, were successfully employed in short runs with a platinum-plated titanium anode. Graphite and carbon were briefly examined, but both exhibited a severe weight loss after only a few hours at a current density of 0.2 A/cm^2. A pure lead anode scaled badly with toluenesulfonic acid as anolyte, but was somewhat more satisfactory with 1 N sulfuric acid. A workable combination

Table 6.1. Screening of Electrolytes

Quaternary Ammonium Salt	Solubility in QAS, 55°C		QAS Conductivity[a] 55°C(mhos/cm)	Distribution Coefficient[a] in H_2O/in AN	Suitability for EHD
	Pure AN[a] (g/100 g)	50/50 AN/ADN[b] g/100 g			
Tetramethylammonium methylsulfate	7.6[b]	7.9	0.100[b]	—	Rejected—low AN solubility
Tetramethylammonium acetate	<10	—	0.067	180	Rejected—low AN solubility
Tetramethylammonium propionate	8.4[b]	7.9	0.061[b]	—	Rejected—low AN solubility
Tetramethylammonium butyrate	15.2[b]	18.9	0.051[b]	—	Rejected—low AN solubility
Tetramethylammonium glutarate	4	—	0.034	670	Rejected—low AN solubility
Tetramethylammonium adipate	<5	—	0.026	900	Rejected—low AN solubility
Tetramethylammonium benzoate	21.6	—	0.045	110	Rejected—low AN solubility
Tetramethylammonium toluate	>50	—	0.039	95	Tested in cell—unsuitable because of p-toluic acid precipitation on membrane
Tetramethylammonium terephthalate	Insoluble	—	—	—	Rejected—AN substantially insoluble
Tetramethylammonium benzene sulfonate	24	—	0.051	60	Tested—low ADN yield—low AN solubility
Tetramethylammonium toluene sulfonate	>50	—	0.041	30	Extensively tested
Tetramethylammonium ethylbenzene sulfonate	>50	—	0.045	27	Not tested—no benefit over TMATS expected
Tetramethylammonium xylene sulfonate	>50	—	0.041	28	Not tested—no benefit over TMATS expected
Bistetramethylammonium sulfate	Nil[b]	Nil	0.088[b]	—	Rejected—low AN solubility
Tetraethylammonium ethyl sulfate	—	>62	0.042[b]	6.9	Selected for commercialization
Tetraethylammonium formate	—	21.3	0.064[b]	—	Rejected—low AN solubility
Tetraethylammonium benzene sulfonate	—	>60	0.036	12.1	Tested extensively

Salt				
Tetraethylammonium toluene sulfonate	50	—	0.025	5.0	Tested—no apparent advantage over TEABS
Bistetraethylammonium sulfate	10.4^b	10.7	0.061^b	—	Rejected—low AN solubility
Bistraethylammonium phosphate	14.7	14.4	0.046	8.5	Rejected—low AN solubility
Tetraethylammonium sulfamate	—	36.8	0.044	7.4	Borderline solubility for AN; proved to be thermally unstable and rejected
Triethylmethylammonium methyl sulfate	20.3^b	43.0	0.059^b	15.9^c	Not tested—possibly a good salt
Triethylmethylammonium p-toluene sulfonate	$>65^b$	>65	0.032^b	—	Tested
Bistriethylmethylammonium sulfate	8.2^b	—	0.083^d	—	Rejected—low AN solubility
Trimethylethylammonium ethyl sulfate	21.9^b	43.3	0.063^b	19.0^c	Not tested—Performances expected to be below TEAES
Tributylmethylammonium methyl sulfate	55.1^b	64	0.029	0.4	Rejected—low distribution coefficient
Trimethylbenzylammonium sulfate	14	—	0.069	232	Rejected—low AN solubility
Trimethylbenzylammonium formate	17.6^b	26.6	0.086^b	—	Not tested, possibly satisfactory, but high salt cost, low AN solubility
Trimethylbenzylammonium acetate	17.8^b	27.0	0.059^b	—	Rejected—low AN solubility
Trimethylbenzylammonium benzene sulfonate	50	—	0.042	8	Not tested, possibly satisfactory, but high in salt cost
Triethylmethylammonium benzene sulfonate	—	>60	0.041	6.1	Not tested
Bistetrabutylammonium sulfate	8^b	—	0.026	Completely miscible	Rejected unsuitable for product extraction
Trimethylhydroxyethylammonium toluene sulfonate	48	—	0.037	69	Not tested
Trimethylsulfonium methyl sulfate	10.3^b	—	—	—	Rejected—low AN solubility

[a] 60% aqueous salt solution.
[b] 50% aqueous salt solution.
[c] 40% aqueous salt solution.
[d] 35% aqueous salt solution.

Table 6.2. Pilot-Plant Electrolyte Evaluations (0.5 A/cm^2, 50°C, pH 8-10)

Electrolyte		Selectivities (%)	
Cation	Anion	PN + HB[a]	ADN[b]
Me$_4$N	p-Toluenesulfonate	9.0	89.8
MeEt$_3$N	p-Toluenesulfonate	7.1	89.2
Et$_4$N	Benzenesulfonate	7.2	90.9
Et$_4$N	Ethyl sulfate	4.9	93.4

[a]PN = propionitrile, HB = high-boiling AN oligomers.
[b]ADN selectivity corrected for AN recoverable from HOPN and BCE.

Table 6.3. Cathode Screening Tests

Material	ADN Selectivity (%)
Hg	90
Pb	90
Hg/Pb	90
Pb (6% Sb)	88
Cd	85
Zn	84
Ni	82
Carbon steel	77
Ti	75
Graphite	75

Table 6.4. Selectivity Deterioration with Time in Early Bench-scale Runs

Run Time (hr)	Selectivity			
	% ADN	% PN	% HB	% BCE
14[a]	91.4	3.7	2.5	1.8
25	88.5	4.3	4.1	2.4
36	87.7	4.8	4.4	2.5
48	86.8	5.0	4.3	3.1
72	86.3	5.3	4.5	3.1

[a]Insufficient time for full equilibration.

Table 6.5. Selectivity Time Trend with Electrochemically Demetallized Catholyte

Run Time (hr)	Selectivity			
	% ADN	% PN	% HB	% BCE
36	89.3	2.8	5.7	1.7
96	89.2	2.9	5.9	1.5
456	89.0	2.8	6.4	1.3
516	88.1	4.3	5.6	1.2
696	87.7	4.8	5.5	0.6

was found in platinized titanium with 0.5 to 1.0 N H_2SO_4 and this system was used throughout most of the first 3 months of bench-scale testing.

With the accumulation of operating hours on the Pt/Ti anodes it became increasingly clear that the Pt losses far exceeded an economically acceptable level. Typical useful service life using anodes with a 1.25-μm platinum coating was about 40 hr, after which time a rapid rise in cell voltage was experienced. An alloy with 1% Ag as commonly employed in the production of electrolytic Zn from strongly acid sulfate solutions, was found to give very satisfactory results with dilute H_2SO_4 anolyte. At a current density of 0.2 A/cm^2 and 40°C the corrosion rate for the Ag/Pb alloy in 1 N H_2SO_4 was determined to be 0.25 cm/year in agitated, thermostatically controlled baths. The corrosion rate increased with operating temperature to a value of about 0.8 cm/year at 60°C.

The effect of current density on the corrosion rate of the 1% Ag/Pb alloy is shown in Fig. 6.1; for an increase in current density from 0.2 to 1.5 A/cm^2, the corrosion rate rises by a factor of slightly over 3. The 1% silver/lead alloy was used throughout much of the subsequent bench-scale testing, while further optimization of the alloy composition was carried out on the pilot-plant cell.

Rather extensive studies of the stability of lead alloy anodes in sulfuric acid solution were published by Kiryakov and Stender [5] in the USSR in the early 1950s and these served as a starting point for our alloy investigations. The Russian workers had shown improved anode stability by addition of thallium or tin as a third component to the lead/silver alloy, and a quaternary alloy of lead/silver/tin/cobalt was found to be particularly attractive. Some temporary benefit of calcium and selenium as alloying elements was also indicated. Thus a number of possibilities were suggested by this and other published investigations related to electrolytic zinc processing. The major unknown for us at this point was the effect of the migration of organics into the anolyte on the anode life.

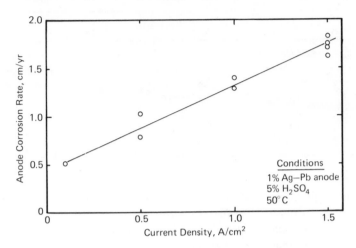

Fig. 6.1. Effect of current density on anode corrosion rate.

Early laboratory testing of anodes involved short runs with the anolyte routinely discarded at the end of each experiment. With the initiation of pilot-plant testing, an increase in anode corrosion was observed with time. Over a 48-hr period the corrosivity of the anolyte, as measured in beaker tests, increased from 0.4 to 8 cm/year, clearly indicating a rapid buildup of detrimental impurities. Isolation of these impurities was carried out by precipitation of sulfate ion with barium carbonate, followed by recovery of the impurities as their soluble barium salts. A variety of analytical techniques was employed to identify the impurities, which ultimately were proven to be nitric and perchloric acid. Typical rates of buildup in pilot plant anolyte are shown in Figure 6.2. The source of the perchlorates was chlorides introduced by way of the catholyte feed streams and these were subsequently controlled by more stringent raw material specifications. The nitrates, however, were evidently formed by oxidation of the acrylonitrile and other nitrogen-containing organics, and therefore were not so readily eliminated. The strong influence of nitrate anion was established in beaker tests in which a 25-fold increase in corrosion rate was observed for a change in nitric acid content of anolyte from 0.05 to 0.6% (Fig. 6.3). In later pilot-plant operations with tetraethylammonium ethyl sulfate in the catholyte, a similar detrimental effect due to the migration of ethyl sulfate ions into the anolyte was noted (Fig. 6.3). The control of the level of these extraneous anions in the anolyte thus became extremely important to the attainment of a satisfactory anode life.

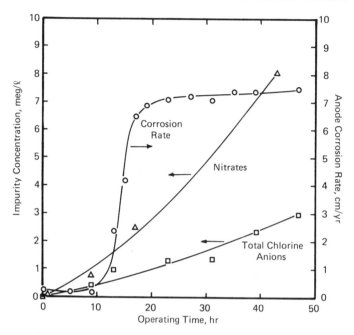

Fig. 6.2. Anode corrosion rate and anolyte impurities as a function of pilot plant operating time.

2 PILOT-PLANT ANODE TESTING

To evaluate various lead alloys under actual cell operating conditions of anolyte composition, temperature, and flow regime, the 76 cm × 91 cm pilot-plant anodes were divided into eight individual panels (roughly 19 cm × 45 cm), so that each could be a different alloy composition. During some of the pilot-plant runs two or three bipolar cells were operated in series, enabling 16 or 24 anode panels to be tested simultaneously. At about 1-month intervals the panels were removed, carefully cleaned of surface scale, and measured at 4 to 8 points to assess the reduction in panel thickness.

With the discovery of the detrimental effect of nitrate, perchlorate, and ethyl sulfate, an expedient means of controlling the impurity levels was adopted by purging anolyte at a rate of about 1 kg/kg of ADN theoretically produced. Simultaneously, laboratory studies were begun to seek more economical means of removing the impurities from the anolyte. Various ion-exchange membranes were being tested in the pilot-plant cell at this s: ˑne time and the diffusion of AN and other organics was a function of the m nˑi-brane properties and stability. Consequently, a fixed anolyte purge rate did

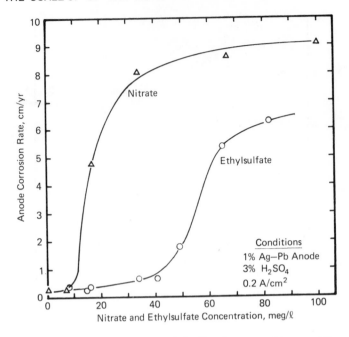

Fig. 6.3. Effects of nitrate and ethyl sulfate ions on anolyte corrosivity.

not ensure a constant level of anolyte impurities. For this reason, comparisons of corrosion rates for different alloy compositions were made from the ratio of the observed corrosion to that for a 1% Ag/Pb standard experiencing the same history. Results for 16 Pb alloys are given in Table 6.6.

Several conclusions were drawn from these studies:

1. The Ag/As/Sn/Co alloy was preferred over the standard 1% Ag/Pb when using TMATS salt, exhibiting about half the corrosion rate.

2. None of the alloys tested with TEAES salt was significantly better than 1% Ag/Pb.

3. Corrosion rates with TEAES salt were 2 to 3 times those with TMATS salt.

At the early stages of the commercial-plant design, the use of TMATS salt was planned and anodes composed of Ag/As/Sn/Co alloy were specified. A later change in plans during plant construction to use TEAES salt in the commercial facility led us back to the original Pb alloy containing simply 1% Ag.

Table 6.6. Corrosion Rates of Pb Alloys Relative to 1% Ag/Pb
(0.5 A/cm^2, 45°C, 5% H$_2$SO$_4$)

Composition (wt %)				Relative Corrosion Rate	
Ag	As	Sn	Co	TMATS[a]	TEAES[b]
1.0	0.2	0.3	0.02	0.45	1.27
1.0	0.1			0.88	0.91
1.0		0.3	0.02	0.98	1.72
1.0				1.00	1.00
1.0	0.2			1.02	1.36
0.5	0.2			1.03	
0.5	0.1			1.23	
7.5		0.3	0.02	1.32	
7.5	0.2	0.3	0.02	1.49	
7.5				1.59	
1.0	0.5			1.61	
7.5	0.2			1.72	
0.1	0.5			1.90	
0.1				4.60	
0.5					1.40
0.5	0.2	0.3	0.02		1.89
Corrosion rate for 1% Ag/Pb (cm/year)				0.43	1.14

[a]TMATS, tetramethylammonium *p*-toluene sulfonate.
[b]TEAES, tetraethylammonium ethylsulfate.

Anolyte Treatment

The removal of nitrate ions and other anionic impurities by extraction of anolyte with a liquid ion-exchange material was investigated in both the laboratory and pilot plant. The liquid ion exchanger was a long-chain aliphatic amine (LA-2, Rohm & Haas Co.) that in an organic solvent, was effective in removing certain acids from the aqueous anolyte in preference to the sulfuric acid. Benzene, xylene, and kerosene were studied as solvents and the extraction efficiencies were 71, 63, and 49%, respectively, in extractions using stoichiometric amounts of the LA-2 solutions. With a tenfold excess of LA-2, the extraction efficiency with a xylene solution was increased from 63 to 88%. Regeneration of the LA-2 solution was accomplished by contacting with an ammonia solution in 10 to 25% stoichiometric excess based on the contained LA-2 nitrate. A demonstration of the effectiveness of this treatment is given by the data in Table 6.7. Pilot-plant anolyte showing a corrosion rate of 6 cm/year was extracted with a 4 : 1 stoichiometric ratio of LA-2 to nitrate and the impurity level improved to lower the corrosivity to 0.5 cm/year.

The LIX treatment of anolyte was tested on an extended bench-scale cell run using LA-2 in kerosene. It was necessary to wash the regenerated LA-2 very thoroughly to prevent significant amounts of ammonium ion from being introduced into the anolyte. Acceptable ammonium ion levels could be achieved by four or five water washings, but this complicated the procedure. The LIX treatment was tested briefly in conjunction with operation of the pilot-plant cell system, but it soon became clear that while the process was workable, considering both capital and operating costs, it would be more economical to control the impurity levels by simply purging part of the anolyte on a continuous basis. Subsequent pilot-plant studies and ultimately the commercial plant operations therefore employed anolyte purging.

3 MEMBRANE DEVELOPMENT

The use of a catholyte containing high concentrations of organic components [AN, ADN, and quaternary ammonium (QA) supporting electrolyte] required separate anolyte and catholyte streams to avoid their anodic oxidation with high consumption of chemicals and the generation of corrosive materials. The classical means of separating electrolytes involves mechanical or semipermeable membranes. In early batch runs, an Alundum cup was used, but the electrical resistance was prohibitively high. Several microporous plastic materials were available, but also exhibited excessive resistance.

Table 6.7. Effectiveness of Liquid Ion-Exchange Treatment of Pilot-Plant Anolyte

Treatment conditions		
H_2SO_4 in anolyte (meq/liter)		500
LA-2 in xylene (meq/liter)		100
Extraction phase ratio, aqueous/organic		5 : 2
Treatment level (meq LA-2/meq NO_3)		4 : 1
	Before Treatment	*After Treatment*
Test results		
Nitrate concentration (mequiv/liter)	9.3	2.1
Anode corrosion rate (cm/year)	6	0.5

Electropositive membranes such as protamine-impregnated collodian and electronegative membranes such as nitro-cellulose were poorly conductive and not very stable under the cell conditions.

Ion-exchange membranes were beginning to find use in electrodialysis applications, particularly water purification, and provided the first encouraging electrolyte barrier. Most of the membranes commercially available at that time were tested, but as shown in Table 6.8 the membranes produced by Ionics, Inc. showed far superior performance.

The Ionics membrane was a homogeneous resin matrix over a reinforcing cloth. The resulting membrane had high dimensional stability, providing a rigid barrier between cell compartments. The exchange resin, a sulfonated polystyrene–divinylbenzene copolymer possessed high ion selectivity and high electroconductivity. Although the initial results with Ionics membranes were encouraging, the membrane lifetime fell far short of commercial acceptability. A joint membrane development program was initiated between the Monsanto workers and the technical department of Ionics, Inc. to provide a suitable membrane. The target was a lifetime of 1000 hr.

4 LABORATORY TEST PROCEDURES

Membrane Life

Service life of a membrane (100 cm^2 exposed area) was first determined in a laboratory cell (Fig. 6.4). Three of these cells were arranged in parallel (termed "Tricell") with a common catholyte system. This was later expanded to six cells to expedite testing. The cells were mounted vertically with upward flow of both anolyte and catholyte. Anolyte pressure was maintained 0.1 to 0.3 kg/cm^2 higher than catholyte pressure to force the membrane firmly against the vertical cathode spacers; this was important to ensure uniform liquid space between membrane and cathode in the flow channels. Adiponitrile was produced under conditions identical to selected plant conditions and the current density was normally 0.4 to 0.5 A/cm^2. Runs were interrupted at intervals to inspect membrane surfaces and to measure mechanical leakage and water transfer per faraday of current. Usually the test was continued until water leakage exceeded 4 ml/(hr)(100 cm^2 membrane) when a hydrostatic head of 1 ft of water was imposed on the anode side of the membrane.

Mechanical Leakage

In the laboratory this test was performed with the 100-cm^2 cell under static conditions, with no current. Both catholyte and anolyte compartments were filled with water to eliminate osmosis. The cell was in the vertical position. A

Table 6.8. Membrane Screening Results

Manufacturer and Designation	Exchange Capacity (mequiv/ml resin)	Water Content of Resin (%)	Burst Strength (psig)	Thickness (mils)	Resistance ohm-cm²		Electrolytic Water Transfer H Form (ml/F)	Type	Comments
					0.5N NaCl	0.5N H₂SO₄			
American Machine & Foundry Co.									
C-60	1.6	35 ± 4	35 ± 6	12	5 ± 2	—	30	Homogeneous	No backing
C-110 (103)	1.2	15 ± 3	55 ± 7	7.5	7 ± 2	1.1	30-60		All of the AMF membranes are physically weak and undergo stretching and distortion in EHD service. Diffusion to anolyte up to 10 times greater than with Ionics membranes
C-313	0.6	12 ± 6	60 ± 7	6.5	5 ± 2	—	80		
Asahi Glass Co., Japan									
CSG-10	—	29	70	10	13	2.0	23	Heterogeneous	These membranes are glass backed Membranes split and broke on bearing surface. Resin binder not stable in catholyte. Membrane wrinkled badly in service
GMG-10	—	32	90	10	6	1.4	21		
CMV-10	—	—	—	—		2.3	—		
Asahi Chemical Co., Japan									
DK-1	2.6	36	—	9	4.2	—	25	Homogeneous	Membrane does not have backing or reinforcing. Broke on bearing surfaces of cell, preventing extended use

Tokuyama Soda, Japan CHR-4	1.8	35	15	8	2.5	0.8	20	Heterogeneous	CHR-4 has PVC backing. CHR-4 OK except for wrinkling over 56 osmotic cycles and adequate for 278 hr EHD service in conjunction with Ionics membrane. Neosepta became badly distended and water transfer doubled in 24 hr service
Neosepta CL 2.5T		35	15	6	2.9	1.2	20		
Permutit of New York 3142	1.4	—	190	6	13	—	—	Heterogeneous	Fabric-supported. Burst after 2 hr EHD service
3235 (Ionac Co.)	2.8	—	165	12	15	2.5	42		Resin became spongy in catholyte (binder not stable)
Permutit of England Permplex C-20	2.0	35	20	(No data available)			80-110	Homogeneous	Burst within 4 hr EHD service
Radiation Applications, Inc. Permian 320 Teflon 1010			(Very thin sheet—Teflon backed)				35		Burst within 2 hr EHD service Two sheets used together for 115 hr. Both were badly wrinkled and blinded. Failed because of pinholes
Ionics AZG 1230-1	2.7	45	—	30	18	—	50-55	Heterogeneous	Ran 255 hr then warped—spalling resin

BENCHSCALE EHD CELL

Polyethylene
Feed Block

Lead
Anode

Neoprene
Gasket

Neoprene
Gasket

Anode
Spacer

Membrane

Cathode
Spacer

Lead
Cathode

Polyethylene
Feed Block

Neoprene
Gasket

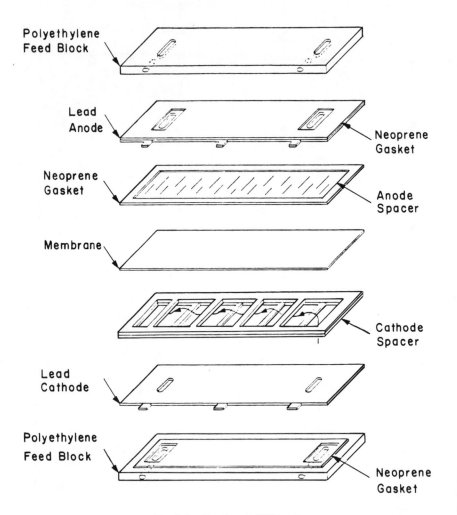

Fig. 6.4. Bench-scale EHD cell.

pressure head of 1 ft of water was imposed on the anode side, and leakage was determined by an increase in volume in the catholyte compartment. The criterion was that a membrane test would be terminated when the static leak was 4 ml/hr or more.

Electrolytic Water Transfer

Water transfer was measured in the standard 100-cm^2 cell with 0.5 M H_2SO_4 in both anolyte and catholyte compartments. The current density was normally 0.05 A/cm^2 and run time was adjusted to allow 100 to 150 ml of volume change through water transfer. Corrections were made for mechanical water leakage and for loss in water through electrolysis. Water transfer was expressed in ml/faraday of current. The preferred range was 40 to 50 ml/F, although a very wide range of porosities was tested—20 to 120 ml/F. The lowest porosity membranes (Ionics DYG Series) were brittle and hard to handle. Higher porosities, over 65 ml/F, allowed too much water and acid transfer and exhibited high AN diffusion.

Membrane Flexibility

This test was performed on the 100-cm^2 cell by filling both compartments with water, placing a positive pressure on the anolyte, and measuring instantaneous volume displacement into catholyte flow channels. Displacement was calculated as percent of liquid volume in the flow channels. The results are most useful for comparing flexibility of membranes in EHD service, as affected by backing and resin formulation or porosity.

Osmotic Shock

The osmotic shock test involved alternate exposure of one side of the membrane first to typical EHD catholyte solution, then to water. The other side was in constant contact with 0.5 M H_2SO_4. This test measures resistance of the membrane surface to damage from cycling between the fully hydrated and partially hydrated state coupled with some degree of cycling between the free acids and the quaternary ammonium salt form. In the early work, this was done in a small cell that held two membranes, and the solutions were changed by hand. The cycles were 30 min long, and generally 20 to 100 cycles were sufficient to cause visible damage from resin spalling. When the membrane testing program was accelerated, this test was automated. Two cells containing four membranes were used. The cycle time was reduced to 10 min and solution changing was done with pumps and a container. As superior membranes were developed, this test became less useful as continually more cycles were required, and it was eventually discontinued.

Membrane Evaluation

Although some of the initial Ionics membranes used glass fabric as a reinforcing material, a number of other backing materials were evaluated, but none was dimensionally stable in cell operations. The only backing materials found to be stable were glass, polyethylene, and polypropylene. In the woven glass-backed membranes, the 9-oz, 21 × 14 thread-count backing was superior to heavier weaves and was particularly effective when used in conjunction with the more porous AZG resin. Heavier weaves and less porous (DYG) resin inevitably led to resin spalling and warping of the membranes. The more porous resin was also more flexible and resistant to loss on flexing.

As membranes showing promise in laboratory testing were moved to pilot-plant testing of full-size 91 × 102 cm membranes, the same general performance rankings were observed, but it was quickly established that a single AZG membrane had inadequate strength and insufficient stiffness. Two AZG glass-backed membranes supplied the necessary rigidity. To conserve on membrane life, one new membrane, adjacent to the anolyte, was used, together with a used membrane on the catholyte side.

Membranes in which the resin was applied to a sandwich of glass fiber between polypropylene felt showed much promise on initial evaluation, but reproducibility of the quality of the felt and, in turn, the performance of the sandwich membrane were unacceptable. The use of two layers of glass backing ultimately proved to be optimum, providing the necessary rigidity for use in full-scale cells and minimizing the effect of resin spalling. When resin flaked away between threads on one side of the membrane, the integrity of the resin on the opposite side was still normally preserved. Three-ply glass membranes did not exhibit any advantage over the two-ply membranes. The performance characteristics of the Ionics glass-backed membranes are shown in Table 6.9.

5   CELL DESIGN

The development of an operable and economical cell was one of the most important and most formidable aspects of the research and development program. The Baizer cell with its mercury pool cathode, platinum wire anode, and Alundum diaphragm clearly did not lend itself to scale-up to commercial size. The desirability of employing solid metal electrodes and an ion-exchange membrane was recognized from the outset and virtually all the cell development activities were concentrated on this approach.

The basic requirements of the commercial cell design were (1) minimum voltage drop, (2) control of mass transfer effects, (3) provision for removal of heat and oxygen, (4) suitable membrane support, (5) ease of maintenance, particularly membrane and anode replacement, (6) amenability to bipolar

Table 6.9. Evaluation of Ionics Glass-backed Membranes

Test No.	Type	Thread Count	Plies	Resin Capacity (mequiv/g)	Water Content (%)	Thickness (mm)	Resistance[a] ($\Omega \cdot cm^2$)	Water Transfer (ml/F)	Results in Cell Tests
1	AZG	21 × 14	1	2.9	50	0.38	12	50	Excessive anolyte leakage at 600 hr
2	AZG	21 × 21	1	2.7	45	0.76	18	52	Warped after 255 hr
3	AZG	30 × 30	1	3.0	49	0.99	27	55	Badly warped after 114 hr
4	DYG	21 × 14	1	2.8	—	—	17	22	Excessive resin loss after 600 hr
5	DYG	21 × 21	1	3.0	34	0.78	20	—	Failed osmotic shock test at 25 cycles
6	DYG	30 × 30	1	2.5	34	1.27	26	—	Failed osmotic shock test at 25 cycles
7	CZG	21 × 14	1	—	—	—	—	35	Developed high resistance at 150 hr
8	AZPGP	21 × 14	3[b]	2.9	45	0.94	18	40	Excellent performance for 2000 hr
9	AZPGP	21 × 14	3[b]	2.5	45	1.30	22	47	Failed from rip at 560 hr
10	AZPGP	21 × 14	3[b]	—	—	1.78	23	49	Failed from rip at 260 hr
11	AZGG	21 × 14	2	2.6	38	1.06	33	32	Ran 2585 hr satisfactorily
12	AZGG	21 × 14	2	2.0	41	1.40	34	42	Ran 2400 hr
13	AZGG	21 × 14	2	2.5	49	1.17	28	48	Ran 1400 hr—still good at run end
14	AZGGG	21 × 14	3	1.8	46	2.20	33	44	Ran 1500 hr—still good at run end

[a]Resistance measured in 0.5 N NaCl.
[b]Runs 8 to 10 used one glass ply between two plies of polypropylene felt.

operation, and (7) economical construction. The team of chemists and engineers undertaking this development program did not include a single person with either formal schooling in electrochemistry or practical experience with electrolytic processes. Their introduction to cell design and the associated voltage drop and heat balance calculations was through several days of consultation with Dr. R. B. MacMullin (R. B. MacMullin Associates, Niagara Falls, N.Y.). This start, together with a thorough review of cell patents, led to an early decision to use a plate-and-frame type cell.

Bench-scale Cells

The first bench-scale cells consisted of multilayer sandwiches of cathode, cathode spacer, membrane, anode spacer, and anode (Fig. 6.4). Three different sizes were operated in the bench-scale studies. The designs were similar, but exposed cathode areas were 30, 100, and 400 cm^2. The 30-cm^2 cathode was operated as a batch unit and was useful for rapid screening of gross effects. Both the 100- and 400-cm^2 cells were operated continuously and were equipped with suitable feed and electrolyte circulation systems (Fig. 6.5).

All components other than the electrodes, gaskets, and membrane were made of polyethylene or polypropylene. Cell parts were clamped between steel backplates and neoprene gasketing was provided around the outside perimeter of the spacers to ensure a leakproof seal for each compartment. The distance from cathode to membrane was established by polyethylene spacers mounted on the electrodes, which formed channels supporting the membrane and established the desired flow patterns through the compartments. Anolyte and catholyte were circulated from external systems, entering and leaving the cell through plenum chambers in polyethylene backing plates for the electrodes. The electrodes were fabricated from 3-mm lead sheets.

Polyethylene spacers, attached to the cathode with polyethylene pegs, provided three 1.8-cm-wide flow tracks running the length of the cathode. The spacers were 0.5 cm wide and their depth was varied to set the desired cathode–membrane spacing. A sheet of expanded polypropylene mesh was inserted in the anolyte compartment to ensure that the membrane was held clear of the anode at all times. Normally the backpressure on the anolyte was adjusted to give a slightly higher pressure in the anolyte compartment than in the catholyte compartment to force the membrane tightly against the cathode spacer strips.

Early studies in the 100- and 400-cm^2 cells revealed the importance of providing turbulence at the cathode surface in order to remove the hydroxyl ions at a rate sufficient to minimize cyanoethylation side reactions. The need for adequate support of the relatively fragile ion-exchange membranes to prevent distortion and rupture also became evident in these small-scale studies.

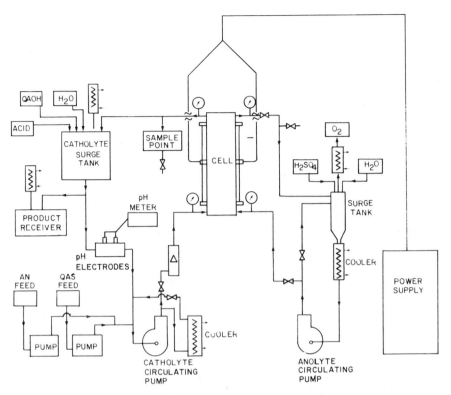

Fig. 6.5. Laboratory EHD bench unit 400-cm^2 cell.

Pilot-Plant Cells

While the small laboratory cells proved extremely helpful in screening membranes, electrodes, and electrolytes, and in preliminary optimization of operating variables, there was little confidence in our ability to scale up from a 400-cm^2 cell to a practical commercial size. It was concluded that a full-scale cell should be designed and tested in a pilot plant so that extension to a plant system would require only replication of cell units rather than further upscaling of dimensions. The ion-exchange membranes available at that time were limited to a maximum width of 91 to 100 cm. While some membranes were produced in continuous rolls, a roughly square configuration fit best with the standard filter press designs and gave the greatest flexibility in use of membranes from different suppliers. Thus the commercial cell design was set to employ a 91 cm $\times$ 102 cm membrane, which provided 0.6 to 0.7

m^2 of available electrode area per cell, after allowing for area lost to gaskets and membrane supports.

The development of the commercial EHD cell was fraught with numerous problems. The need to find electrodes and a membrane with an economically acceptable service life was of obvious importance. However, some of the difficulties in obtaining adequate hydraulic sealing and conducting electricity between electrodes on opposing faces of a bipolar cell leaf were not initially appreciated. The resolution of these problems carried us through three different cell configurations—the Mark I, deNora, and Mark II designs.

Mark I Design

In the Mark I design a 5-cm-thick polyethylene plate served as the supporting element for the electrodes, and cavities were provided in each plate for distribution of electrolyte (Fig. 6.6). Polyethylene frames surrounded the cathode and anode so that the cell assembly would close to its final dimensions with all mating parts in plastic-to-plastic contact. Elimination of electrodes from the gasket area, where cell closing forces would be exerted on them, was critical to establishing precise electrode spacing. Bench-scale cell experience had shown that cold flow of the lead electrodes and permanent deformation of gaskets presented severe problems in maintaining the desired spacing between electrodes.

The Mark I was a single-cell unit so the difficulties of making the anode-to-cathode connection were not faced at this stage. Electrodes were held in place and supplied with current by 12 lead lugs welded into each electrode and pulled into a gasketed recess by cast-in-bronze bolts extending through the plastic plate. Electrical busses were attached to the bolts.

Catholyte hydraulics were never satisfactory. The catholyte was introduced through a central vertical plenum extending down the entire height of the cathode. The cathode was split in the middle, leaving an aperture through which the catholyte flowed normal to the plane of the cathode and then divided into two paths flowing parallel to the cathode toward either outer edge where it was collected in vertical slots. Tests with this cell revealed poor catholyte flow distribution, thought to be due to the jet effect of the 2.5-cm inlet pipe discharging downward into the top of the plenum. This reduced the flow out of the plenum over the top third and gave two nearly stagnant areas. No problems were noted with the anolyte flow, which was vertical from a slot at the bottom of the anode to one at the top.

Among the discarded ideas for series cells prior to the building of the Mark I were many schemes using corner feeds for internal supply of electrolytes. A common drawback to all these approaches was the large leakage current flowing in the electrolyte plenums instead of through the membrane. The Mark I essentially overcame all such problems by feeding electrolytes

MONSANTO MARK I EHD CELL

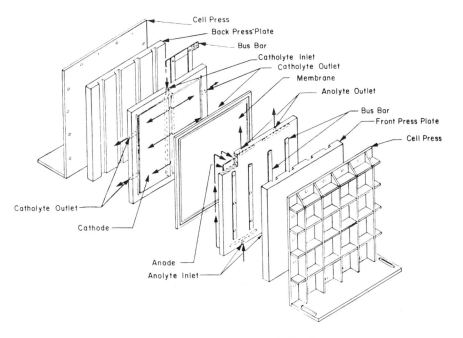

Fig. 6.6.   Monsanto Mark I EHD cell.

through rubber hoses from headers relatively far from the cells themselves. The long thin leakage paths thus formed limited such current losses to less than 1%. This feature was retained in all subsequent designs.

The anode was a toothed member with vertical grooves 1 cm wide × 1 cm deep on 2-cm centers. The grooves were fitted with U-shaped plastic strips whose legs touched the bottoms of the slots. These grooves were to prevent contact of the membrane with the anode during operation. The slots were to aid in discharging oxygen from the anode chamber.

It was determined early in the experiments that operation with a higher pressure in the anode compartment to force the membrane toward the catholyte flow spacers was preferable because of catholyte velocity and voltage drop considerations. This development obviated the necessity for the elaborate anode chamber, and a flat anode with a sheet of large-mesh polypropylene cloth over it was adopted in later designs. The purpose of the polypropylene cloth (National Filter Media Polymax "V") to prevent accidental contact between membrane and anode in the event of pressure reversal.

The cell was designed around the standard 91 cm × 102 cm Ionics membrane. A 1-cm-wide border was provided around the membrane to receive the closing force of the cell. Persistent deflection and leakage in these areas indicated the need for better sealing in future designs. Additionally, the area of the membrane devoted to grip by the gaskets was made too small in a misguided attempt to utilize as much of the membrane as possible.

deNora Cell

Simultaneous with the design of the Mark I cell by the development team at Pensacola, a second design was initiated with the internationally renowned electrochemical engineering firm of Oronzio deNora in Milan, Italy. One of the Monsanto engineers was dispatched to Milan for several weeks to provide guidance in the process requirements, while the deNora expertise in cell design and construction was applied on a crash basis to build a five-cell bipolar prototype unit (Fig. 6.7 and 6.8).

DE NORA EHD CELL

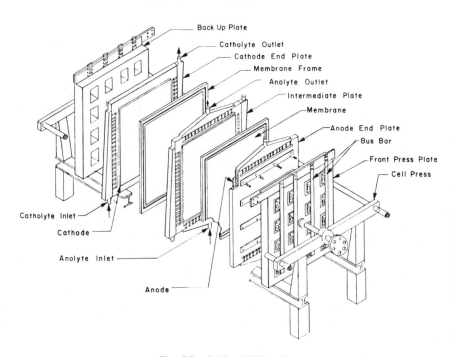

Fig. 6.7.   DeNora EHD cell.

Fig. 6.8. Pictorial view of the deNora cell.

The catholyte flow in the deNora cell was horizontal across the entire face of the cathode, thus reducing the required catholyte feed rate of the Mark I by half, but doubling the pressure drop from one edge of the cathode to the other. Uniform distribution of the catholyte from top to bottom was achieved by providing tapered external distribution chambers that fed through 20 equally spaced holes in the edge of the cell leaf into a U-shaped distribution slot running from top to bottom of the cathode at one edge. Every effort was made to provide horizontal paths of equal pressure drop to achieve a uniform velocity of catholyte at the active surface of the cathode. Catholyte exit passages were mirror images of the inlet except that catholyte fed into the bottom of the inlet plenum and left from the top of the exit plenum to clear any hydrogen generated in the cell.

Anolyte flow was handled from bottom to top of the leaves in a like manner, with external feed chambers and U-shaped slots at top and bottom of the anode. As in the previous cell design, feed of electrolytes to the unit was by rubber hoses from external electrolyte headers.

The cross-flow of electrolytes produced an undesirable quadrantal effect in the differential pressure, forcing the membrane against the catholyte flow spacers. Maintaining a moderate differential pressure in the quadrant containing the inlets of the two streams resulted in excessively high pressures at the quadrant containing anolyte inlet and catholyte outlet.

The deNora cell was originally supplied with firm rubber gaskets 5 cm wide on all four sides of the leaves. At the internal cell operating pressures encountered at high electrolyte velocities, the sealing force required was too great for the light press supplied. Softer gaskets only 3.2 cm wide allowed cell operation at internal pressures up to 2 kg/cm^2, but sealing was tenuous.

A screw closing press with two tension rods on the sides was used to close the deNora cell battery. Leaves rested on the two bottom rails and had a tendency to lean toward one end or the other of the cell during closing. These difficulties indicated the desirability of center-hung leaves as used in a filter press.

Connections between electrodes on intermediate leaves and to the bus bars on the end leaves were made with lead-covered titanium slugs with titanium screws to pull the slug up tightly against the lead electrode. Hydraulic sealing was made by rubber U cups set in recesses around each slug where it penetrated the plastic leaf. Corrosion of the lead-to-lead joints caused frequent hot connections, and sealing of the U cups was poor, particularly when overheating had caused the plastic wall of the recess to melt and run.

Mark II Cell

Continuing difficulties with the deNora cell led to the design of the Mark II (Fig. 6.9). Design goals included parallel flow of anolyte and catholyte, improved sealing of the internal electrical connections, ability to withstand an internal pressure of 2 kg/cm^2, minimum force required for sealing leaf gaskets, generous grip area on the membrane, and center-supported leaves. A P-type gasket set in a recess was employed to seal the leaves and at the same time allow the mating parts to close, plastic-to-plastic, to a fixed dimension.

A border about 5 cm wide around each leaf was reserved to receive the closing force of the cell, resulting in a 97 cm $\times$ 107 cm overall plate size. It was determined that a closing force of 16,000 kg would be adequate to withstand a 2-kg/cm^2 internal pressure and seal the gasketed joints. This force resulted in a compressive pressure around 10 kg/cm^2 on the polypropylene plates and frames, a sufficiently low stress level to resist creep at operating temperatures.

The leaves were maintained in a vertical position during closing of the press by center-suspending them from the side rails of the press. The deNora press was replaced by a more rigid Shriver hydraulic filter press with a 45-ton capacity to facilitate operation and also allow center suspension of the leaves.

Polypropylene was again used for plates and frames because of its higher

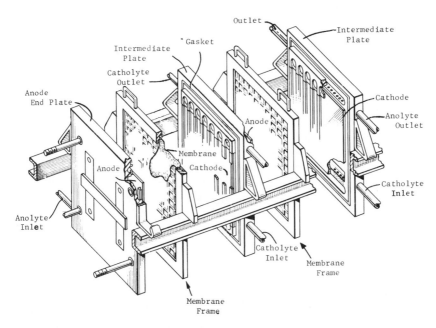

Fig. 6.9. Mark II cell.

creep resistance and stiffness as compared to polyethylene, the only other low-cost plastic with suitable chemical resistance. A plate of Haveg 60 was proposed and tested, but it failed by brittle cracking when exposed to a face-to-face differential pressure of 0.5 kg/cm^2.

One of the prime motivations for designing the Mark II cell was the desire to eliminate external distribution boxes, which had repeated failures at the welds. This was achieved by utilizing the 5-cm thickness of the leaf to contain internal plenums back-to-back located under the top and bottom ends of the cathode. Thus the electrolyte flow was from bottom-to-top in both anolyte and catholyte chambers. Cell feed was through external hoses attached to the edges of the plates by two bolt flanges with O-ring seals. This left a clean rectangular leaf with no projecting plastic parts to be knocked off in handling. The membrane support members evolved through a long course of trial and error. Among the approaches tried and discarded were use of Polymax V polypropylene cloth, Ionics lattice spacers, and harplike arrays of parallel strips supported at the ends in the membrane frame. It eventually became clear that the parallel strips gave the best results as a membrane support. These were located in the catholyte compartment and formed parallel rectangular flow channels from bottom to top of the cathode. (The effects of catholyte flow distribution on yields and cathode fouling are discussed in another section.) Pressure in the anode compartment was about 0.1 kg/cm^2 higher than that of the cathode compartment to force the membrane against the membrane supports on the cathode.

The means of holding the strips in the proper straight, parallel disposition in the catholyte chamber was again the subject of much experimentation. Peening the strips into recesses in the lead was tried and abandoned, as were gluing, stapling, pegging with separate plastic dowels, and sewing them to the cathode, although the last three methods worked well enough to allow evaluation of various strip sizes and spacings. The design chosen for plant use was an injection-molded strip with integral pegs on the back about 6 cm apart. These pegs were inserted through holes in the cathode and upset by heat and pressure on the back side to rivet them to the cathode. A slight countersink was provided at each hole to prevent protrusion of the upset plastic above the surface of the lead.

Considerations in the choice of strip width, thickness, and distance apart were the desire for maximum effective cathode area and thinnest possible channel of flowing catholyte to reduce voltage drop and pumping cost. Working against attainment of these ideals were (1) sagging of the membrane between strips, which reduced the flow channel and caused tearing of the membrane on the strip corners, (2) higher pressure drop with thinner channels, necessitating higher differential pressures in the upper areas of the cathode and aggravating the conditions described above, and (3) blockage of the thinner channels with polymer, resulting in early shutdowns for cleaning. The optimum strip arrangement for the operating conditions being employed at the end of the development were strips 0.3 cm thick $\times$ 0.5 cm wide spaced on 2.4 cm centers.

Materials of Construction

Many different materials of construction were tested in the pilot plant to find those most suitable for plant use. Samples of the material to be tested were either suspended in a moving stream of anolyte or catholyte or made into a spool piece and inserted as part of the line containing the fluid. The samples were removed periodically and examined for changes in weight, volume, and physical characteristics.

Presented in Table 6.10 is a summary of materials of construction tested.

6   OPTIMIZATION STUDIES

Optimization of cell operating conditions virtually commenced with the first bench-scale run and was continued throughout pilot-plant testing and

Table 6.10. Materials of Construction Test Results

Material	Satisfactory	Unsatisfactory
Catholyte		
Nonmetals	Natural rubber	Epoxy-fiberglass
	Neoprene	Polyester-fiberglass
	Polyethylene	Polystyrene
	Polypropylene	Poly(vinyl chloride)
		Polyurethane
Metals	Type 304 stainless	Carbon steel
	Type 316 stainless	
Anolyte (5% H_2SO_4)		
Nonmetals	EPR rubber	Epoxy-fiberglass
	Natural rubber	Polyester-fiberglass
	Neoprene	Delrin
	Polyethylene	
	Polypropylene	
	Saran	
	Poly(vinyl chloride)	
Metals	Carpenter 20	
	Hastelloy C	

into the commercial operations as EVOP experiments. The response of primary interest was generally the ADN synthesis selectivity, since the AN usage was the largest single factor in determining the final ADN manufacturing cost. Statistically designed experiments were used in the initial bench-scale studies, but meaningful results were elusive because of a string of problems encountered in the early runs. These are listed below roughly in order of occurrence.

1. Extreme anode corrosion.
2. Buildup of extraneous metals in recycle streams.
3. Accumulation of polymer in recycled QAS.
4. Nonuniform flow in catholyte flow channels.
5. Silver contamination of the cathode.
6. Short membrane life.
7. ADN selectivity time trends and lack of reproducibility.
8. Impurities in QAS raw materials.
9. Thermal instability of some QAS solutions.

Thus much of the optimization was deferred to pilot-plant testing on the Mark II cell, when many of these problems had been recognized and resolved.

The major operating variables studied were current density, catholyte pH, catholyte flow regime, catholyte temperature, and catholyte composition.

Current Density

The very first 2^3 factorial experiment carried out on the Mark I pilot-plant cell examined current density, catholyte velocity, and AN conversion level. In this study no effect of current density on by-product formation was observed over the range 0.2 to 0.4 A/cm^2.

Preliminary economic studies made at a very early stage indicated the optimal balance of electrolysis capital and operating costs corresponded to current densities in the range of 0.4 to 0.5 A/cm^2, so no further extension was attempted at that time. However, shortly after start-up of the commercial plant there was interest in increasing the capacity of the facilities and additional studies of this variable were made. Pilot-plant runs at current densities ranging from 0.4 to 1.6 A/cm^2 showed virtually no change in the ADN selectivity (Fig. 6.10). The cell was operated for several days at 2.0 A/cm^2, but some overheating of the membrane and the cathode strips was evident. By-product formation was higher than normal, giving an ADN selectivity 3 to 4 percentage points short of standard levels.

One measure of mass transfer at the cathode that emerged during the current density tests was the rate of hydrogen evolution, as indicated by the catholyte offgas rate. At the relative low current densities of 0.3 to 0.5 A/cm^2, a pronounced increase in catholyte offgas was noted with velocities below 1 m/sec. (Fig. 6.11). Thus the existence of a limiting current for the EHD reaction was indicated, but over the range of catholyte velocities normally employed, the limit was well above economical levels.

Catholyte pH

Testing of very low catholyte pH (1 to 2) was inadvertently carried out on several occasions, when serious membrane faults allowed introduction of massive amounts of sulfuric acid solution from the anolyte into the catholyte circulating system. In most of these instances, polymerization of AN took place, turning the catholyte into a fluid with the consistency and appearance of sour cream. Conversely, very high catholyte pH (>11) was shown to lead to excessive production of cyanoethylation by-products (BCE and HOPN). However, over the broad pH range 4 to 10, virtually no effect on cell performance was noted. Thus pH was eliminated at an early stage as an important electrolysis variable.

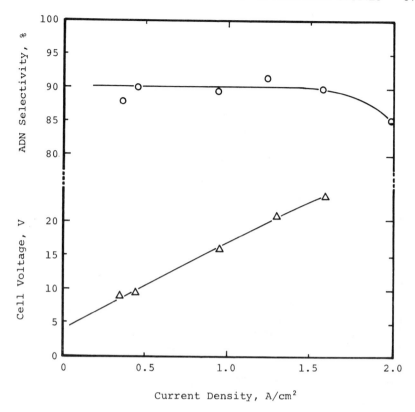

Fig. 6.10. Effects of current density on ADN selectivity.

Catholyte Velocity

In Baizer's original experiments turbulence at the surface of the cathode was provided by a magnetic bar that rode on the mercury surface or rotated at the bottom of the flask with solid metal cathodes. The need for high mass transfer rates at the cathode to minimize hydroxyl ion-catalyzed cyanoethylation reactions was recognized at a very early stage. The means of translating the agitation of the stirring rod into a comparable flow regime in a plate-and-frame cell was, however, not nearly so clear. The first bench-scale experiment with a lead cathode was conducted with an open channel at a linear catholyte velocity of 0.1 m/sec—the cyanoethylation by-products BCE and HOPN represented 60% of the products formed.

With the seriousness of mass transfer in plate-and-frame cells fully

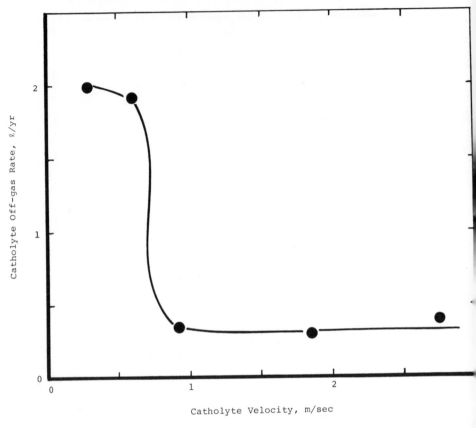

Fig. 6.11.   Effect of catholyte velocity on cathode gas evolution, $0.8A/cm^2$.

demonstrated, the first efforts to enhance the rate of diffusion of the hydroxyl ions away from the cathode involved use of lattice spacers (Fig. 6.4). These were similar in concept to the spacers used in some electrodialysis assemblies, particularly those of Ionics, Inc. As assembled in the catholyte compartment, the pair of stamped polyethylene spacers provided a 3-mm-deep flow track with half-baffles at about 2-cm intervals. At a superficial catholyte flow velocity of 0.6 m/sec the BCE and HOPN produced using this spacer represented less than 2% of the products. While very satisfactory mass transfer was thus achieved by this combination of spacer design and flow rate, the catholyte pressure drop was high even at the start of a run and increased significantly with operating time. Backing off on the catholyte velocity to 0.3 m/sec moderated the pressure drop problem to some extent, but increased the cyanoethylation by-products to 6%.

The selection of a satisfactory catholyte spacer was thus revealed as a critical development challenge to be met. A hydraulic test system was assembled for pressure drop measurements and observation of flow birefringence patterns. A number of different lattice spacers and plastic webbings were compared using a milling-yellow dye solution and polarized light to map the degree of turbulence in different zones.

The lattice-spacer seemed to offer the greatest turbulence at the least cost in pressure drop and a single-pass version was fabricated for the initial pilot-plant studies. These first large-scale runs were of short duration—usually 24 hr or less. The cathode and spacer were cleaned between runs to remove an accumulation of slime, which was identified as a mixture of lead, QAS, and AN polymer. While the lattice spacer gave satisfactory selectivities and pressure drops in these screening experiments, a serious drawback was evident during efforts to make extended runs. In one of the first long trials, which lasted 280 hr, it was necessary to shut down five times for cleaning of the fouled cathode and spacer. Attempts to dislodge the slime by pulsing the anolyte and catholyte flows gave only temporary relief.

Concurrent bench-scale tests were directed at overcoming the problem of fouling of the catholyte spacer, and some encouragement was found in the use of plain parallel ribs, defining fully open flow channels. This approach was tested in the pilot-plant cell using 6-mm-wide polyethylene ribs, providing a membrane-to-cathode gap of 3 mm. With the parallel-rib spacers and catholyte velocities of about 2 m/sec several 240- to 250-hr pilot-plant runs were made.

Following the initial success in the use of open channels at relatively high flow velocities in providing greatly improved run lengths, further definition of the required catholyte velocity was sought. Published investigations relating the limiting current in an electrochemical system to the Reynolds number were studied, but could not readily be translated into prediction of minor losses in the efficiency of the cathodic process at current densities well below the limiting value. The most sensitive index to mass transfer rates at the cathode surface was the yield loss to cyanoethylantion by-products BCE and HOPN. At flow conditions giving rise to extreme polarization at the cathode, increased production of propionitrile and hydrogen was also observed, but the effects of mass transfer on these by-products proved to be of relatively minor economic significance under the flow conditions required to control the cyanoethylation by-products at an acceptable level.

Other factors were also recognized as influencing the BCE and HOPN synthesis, particularly the temperature and the concentration of AN in the catholyte. As might be expected, cyanoethylation of water was dramatically increased at higher operating temperatures and AN levels. Based on data from the first 15 pilot-plant runs employing the parallel-rib spacer the following regression equation was developed:

$$Y = 48.8 + 0.232T + 0.198F + 1.53C$$

where Y = ADN selectivity, adjusted for recovery of AN from cyanoethylation by-products

T = catholyte temperature, °C $(40 < T < 50)$

F = catholyte flow rate, 10^3 kg/hr $(12 < F < 16)$

C = AN concentration in catholyte, wt % $(12 < C < 17)$.

In the series of runs used to develop the above relationship the cathode-to-membrane distance was either 2.4 or 3.2 mm, which did not provide sufficient distinction between Reynolds number and linear velocity to ascertain the better correlating variable. Studies by others in the field of electrodialysis had indicated that power input, as measured by the product of the volumetric flow rate and the pressure drop across the channel, is a useful parameter, particularly when turbulence-promoting spacers are used. Later studies employing gaps from 0.8 to 2.4 mm were made in an attempt to determine the optimal anode/cathode distance. Data from 21 runs (all at the same temperature and AN concentration) were regressed against linear velocity, Reynolds number, and the power input, with the results shown in Fig. 6.12. The highest value of the coefficient determination $(r^2 = 0.77)$ was realized with linear velocity as the correlating variable, while the Reynolds number gave the poorest value $(r^2 = 0.33)$. From an economic viewpoint this result was welcomed, since it suggested that closer cathode-to-membrane spacings could be used not only to reduce the catholyte IR drop, but also to lower the required catholyte pumping rate without affecting mass transfer.

7   DEVELOPMENT OF SUPPORTING PROCESSES

While definition of the electrolysis system was clearly the single most important aspect of the program for commercialization of the EHD process, the conception and proving out of the processing schemes for recovery of ADN from the catholyte and subsequent purification to a quality suitable for catalytic reduction to hexamethylenediamine were also critical to the project's success. Processing steps that were developed on a lab scale and/or in the pilot plant included the following:

1.   Recovery of ADN from catholyte by countercurrent liquid/liquid extraction with AN.

2.   Removal of residual QAS from the AN–ADN extract by countercurrent extraction with water.

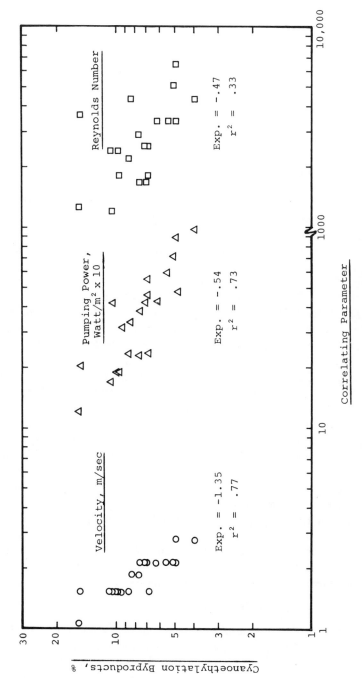

Fig. 6.12. Correlation of cyanoethylation by-products.

3.   Separation of PN from recycle AN by distillation.

4.   Recovery of AN from purged water layer by steam-stripping.

5.   Evaporation of recycle catholyte for water control.

6.   Electrochemical demetallizing of recycle QAS solution.

7.   Multistep distillation of crude ADN to remove low-boiling and high-boiling impurities.

A schematic diagram of the pilot facilities constructed to carry out these studies is shown in Fig. 6.13.

The pilot extraction and distillation facilities served a dual purpose in the development program. They were employed for evaluation of operating conditions and process optimization, and they also fulfilled the important role of closing the loop on a number of prospective recycle streams. The potential pitfalls of buildup of impurities in the various recycle QAS, AN, and water streams was revealed in early bench-scale studies, so the recovery system was married to the electrolysis system from the outset. Continuous operation of the pilot plant, which was conducted on a 24-hr/day, 7-day/week basis, involved electrolysis, ADN extraction, QAS extraction, recycle QAS evaporation, AN stripping, PN distillation, and purging of water through a stream stripper. The ADN refining, the hydrogenation of ADN to HMD, and the HMD refining were carried out only intermittently.

To ensure that the refined ADN did not contain trace levels of impurities that might poison the rather sensitive fixed-bed catalyst used to hydrogenate the ADN to HMD, several metric tons of ADN were processed in pilot facilities through HMD synthesis and refining and then combined with adipic acid to produce nylon 66 salt, which was polymerized in development-scale autoclaves to confirm the suitability of the EHD process in ultimately yielding polymer-grade material. While the quality of the monomers needed for acceptable nylon polymer and yarn was relatively well defined for the monomer processes in use at that time the introduction of a totally new ADN process carried with it the possibility of new impurities in the HMD. The sensitivity of nylon polymer properties to trace levels of certain extraneous organic by-products was well known, so pilot testing completely through to polymer was considered essential.

In addition to development of the above downstream steps, a process for economical manufacture of the makeup quaternary ammonium salt was needed. This constituted a rather major developmental undertaking in itself.

ADN and QAS Recovery by Extraction

In the pilot plant the removal of products from the catholyte of the EHD cell was accomplished by countercurrent extraction with recycle acrylonitrile in a Pyrex column. The column, which was 15 cm in diameter and 3.6 m

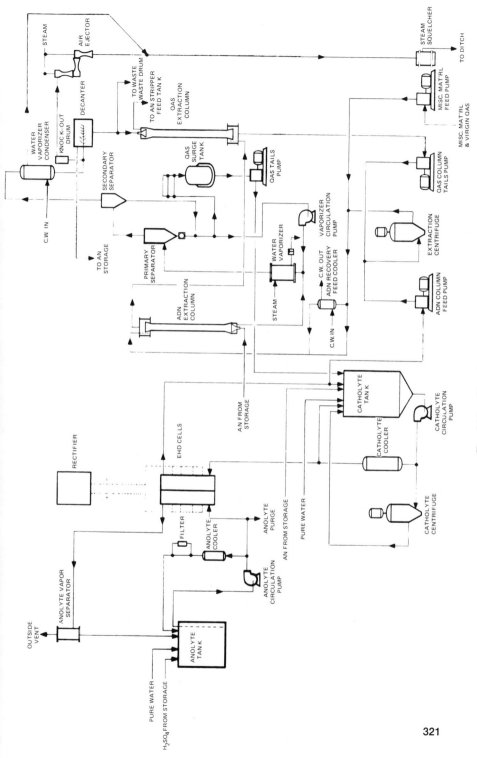

Fig. 6.13. Pilot-plant system

321

high, was normally filled with 1.3-cm ceramic Intalox saddles. The extract from this column, a mixture of AN, ADN, and other reaction products, was fed to the bottom of an identical column, where it was countercurrently contacted with a recycle water stream. High-efficiency removal of ADN from the catholyte was important in minimizing hydrolysis of ADN during the subsequent concentration step. This evaporation of recycle catholyte was necessary to remove water introduced through the membrane from the anolyte by means of electroosmotic water transfer. The effective removal of salt in the QAS extraction step was necessary to minimize the QAS losses from the system and to avoid fouling of the reboiler in the AN stripping column.

Factorial studies were made on both the ADN and QAS extraction columns to determine whether the aqueous or organic phase should be continuous and to test the effect of pulsing the column. In the ADN extraction, it was found that operation with the aqueous phase continuous gave improved ADN recovery. Pulsing the column at 60 cycles/sec with a 9-mm amplitude using a dead-headed piston pump increased the ADN extraction efficiency from 98.1 to 99.5%, but led to emulsion formation. A similar study was made on the QAS extraction column, indicating better performance with the organic phase continuous and with pulsing. The pilot-plant results with respect to the preferred continuous phase were translated directly into the design of the commercial sieve-tray columns. Pulsing, however, was not recommended for the plant system because of the concern over increased emulsion formation.

ADN Refining

The organic phase from QAS extraction was continuously stripped of AN in a 30 cm $\times$ 3 m packed distillation column to produce a crude ADN containing about 90 wt % ADN. This stream was accumulated for ADN refining tests and to prepare sufficient high-purity product for later hydrogenation and polymerization studies. Benchscale distillation of crude EHD ADN, followed by rocking-bomb hydrogenation experiments, revealed that minor levels of either low-boiling or high-boiling impurities could seriously reduce the activity of the ADN hydrogenation catalyst. Recognizing the importance of the ADN quality, it was decided to simulate the entire four-column refining train in the pilot plant (Fig. 6.14). To expedite the study, two existing pilot columns were employed. The design rate for the system was about 15 kg refined ADN/hr.

Two continuous refining train simulations were conducted, each lasting about 500 hr. A number of operating conditions were changed during the two runs to assess the effect on the refined ADN quality and product recovery. Feed composition, column head pressure, reflux ratios, and column base holdup times were varied. Complete material balances were made for all

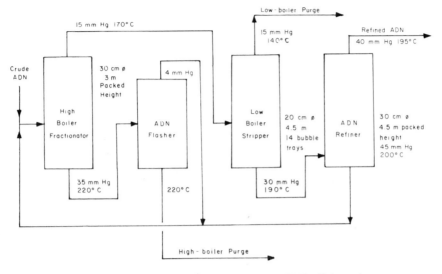

Fig. 6.14. Schematic diagram of pilot-plant ADN refining train.

major impurities, forming the basis for the plant design. Reboiler heat transfer coefficients were also determined as a function of operating time. Corrosion coupons were installed at 11 points in the pilot-plant distillation train to guide selection of materials of construction. Viscosity measurements were made on the overhead and bottoms streams at operating temperatures as an aid to selection of pumps for the commercial system.

One important observation from the pilot-plant simulations was that greater than 90% of the BCE in the crude ADN decomposed in the HB Fractionator and ADN Flasher, drastically reducing the separation requirements of the final ADN refining column. Difficulty encountered in achieving the desired removal of two low-boiling impurities, succinonitrile and methylglutaronitrile, was overcome by feeding high in the column and operating at uneconomical, high overhead purge rates. This problem was avoided in the plant design simply by increasing the number of plates.

Electrochemical Process for Quaternary Ammonium Hydroxides

The quaternary ammonium compounds used as supporting electrolytes in the EHD development program were not available on the scale required for the commercial operations, or even for the pilot-plant studies. The salt used in much of the early development testwork, tetramethylammonium toluene sulfonate (TMATS), was prepared initially by reaction of trimethylamine with the methyl ester of toluene sulfonic acid. However, the limited availability and high cost of the methyl ester led to the search for alternate routes to this quaternary ammonium salt. In addition to the neutral TMATS required for electrolyte makeup, tetramethylammonium hydroxide (TMAOH) was needed for control of the catholyte pH. It was recognized that if an economical source for the TMAOH could be found, the quaternary ammonium salt could readily be produced by neutralization with toluene sulfonic acid. Since tetramethylammonium chloride (TMACl) was commercially available at a relatively low cost ($1 to 1.5/kg), attention was directed toward development of a process for converting the chloride salt to the hydroxide by electrolysis.

The proposed system was similar in many respects to membrane cell caustic-chlorine production, involving electrolysis of TMACl in the anode compartment with liberation of chlorine at the anode. The tetramethylammonium cation in the anolyte migrated through a cation-exchange membrane to combine with the hydroxyl ion formed at the cathode:

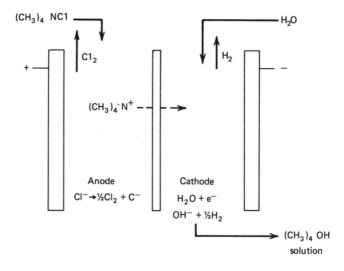

As the EHD development proceeded, it was discovered that the tetraethylammonium ethyl sulfate (TEAES) supporting electrolyte offered certain advantages over TMATS. Therefore, hydroxide cell studies were redirected toward manufacture of tetraethylammonium hydroxide from the corresponding chloride salt.

Preliminary TMACl electrolysis studies were carried out using a parallel-plate bench-scale cell with an electrode area of 170 cm². A graphite anode and a stainless steel cathode were employed. Platinized titanium was tried briefly as an anode, but a short service life was experienced. An Ionics CR61

AZG membrane, with a protective Teflon cloth on each side, was used in most of the variable screening studies. The cell was operated with continuous makeup of TMACl (or TEACl) solution to the anolyte circulation loop and continuous withdrawal of TMAOH (or TEAOH) from the catholyte loop. The normality of the catholyte was controlled by the rate of addition of water and removal of catholyte.

The primary variables studied in the bench-scale cell were the anolyte and catholyte concentrations and the cell current density. Varying the concentration of the tetramethylammonium ions in the anolyte from 1.6 to 3.0 mequiv/liter was found to have virtually no effect on the current efficiency for TMAOH production (Fig. 6.15). In another series of experiments, the current efficiency was shown to be sharply reduced from 40 to 50% to less than 20% by an increase in the catholyte hydroxyl ion content from 0.5 to 1.6 N (Fig. 6.16). Little effect of current density was observed in these runs over the range 0.15 to 0.45 A/cm^2. The drop in current efficiency with increased concentration of the quaternary ammonium hydroxide in the catholyte was expected from published results for sodium hydroxide production in diaphragm cells. In both systems higher hydroxyl ion levels lead to a reduced portion of the current between compartments being carried by the sodium or quaternary ammonium cations, with a consequent loss in current efficiency.

With the preliminary definition of a QAOH process based on relatively few

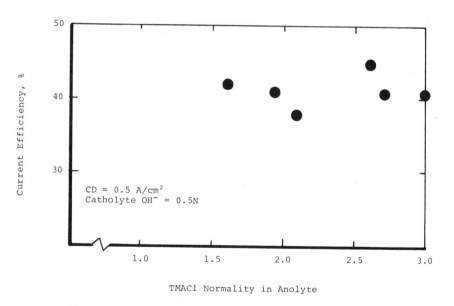

Fig. 6.15.   Effects of anolyte normality of hydroxide current efficiency.

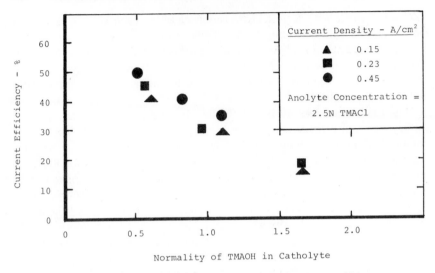

Fig. 6.16.   Effect of catholyte normality on hydroxide current efficiency.

bench-scale tests, the cell was scaled up to a 230-dm^2 unit for further study of components and operating conditions. At that time the pilot-plant cell provided by the Oronzio deNora company for AN electrohydrodimerization had been supplanted by a later Monsanto design (Mark II) and was consequently available for adaptation to electrolytic QAOH production. A semicommercial system was assembled employing up to three bipolar 77-dm^2 cells with a total production capacity of 8 kg TMAOH/hr. Typical cell operating conditions are given in Table 6.11. This cell was employed for production of the TMAOH and TEAOH required in the later EHD process studies and was subsequently moved to the Decatur, Alabama plant for use in the commercial EHD plant.

8   PLANT DESIGN

As is mentioned earlier, the EHD development program was conducted on a very accelerated timetable to ensure the availability of additional ADN capacity in time to meet the rapidly growing nylon 66 demand. To expedite plant design the engineering department was brought in at a very early stage, even before the first successful operation of the pilot-plant cell. A project engineer was assigned to work full-time with research and development to define the proposed commercial facilities and to provide economic guidance in the selection of process alternatives. One of the first such studies was an evaluation of the optimal current density.

Table 6.11. Typical Operation of 2.3 m² QAOH Cell

Cell components	
Anode	National Carbon AGLX-61
Cathode	Type 316 stainless steel
Membrane	Ionics CR61 AZG
Protector cloths	National Filter Media G-200
Cathode spacer	Polypropylene cloth, 4-mesh, 2.4 mm thick
Anode spacer	None
Operating conditions	
Catholyte concentration	0.75–1.0 N OH⁻
Anolyte concentration	2.0 N QA⁺
Operating temperature	40–45°C
Catholyte flow rate	2500 liters/hr
Anolyte flow rate	2500 liters/hr
Cell current	2500 A
Voltage/cell	11–15 V
Operating performance	
QAOH yield	96/99%
QAOH current efficiency	25–40%
Chloride content of product	20–50 ppm

Early laboratory studies had shown that with adequate catholyte circulation the ADN selectivity was virtually independent of current density over a wide operating range. Selection of an optimum design point therefore reduced to a balancing of those costs that varied directly with current density (cell power, cooling water, rectifiers, coolers) and those that varied inversely (cells, electrolyte pumps, piping, operating labor). This first optimization study (Fig. 6.17) required making a number of assumptions, such as the effect of current density on membrane and anode life, which had yet to be determined. It was nevertheless valuable in establishing the range of current densities of commercial interest. With the cost of capital set at 15% after taxes (30% before taxes), the optimum fell between 0.4 and 0.6 A/cm². Ultimately the plant design employed a value of 0.5 A/cm².

The Design Basis for the first commercial plant was something of an evolving document. A preliminary design was proposed after only a few months of pilot-plant test work, but many revisions were made by the time the final approved document was issued, almost exactly 1 year from the date of the first Mark I run. The heart of the process was the electrolysis system, the primary elements of which are defined in Tables 6.12 and 6.13. Initial capacity was 14,500 ton/year. A single anolyte system served the entire electrolysis area,

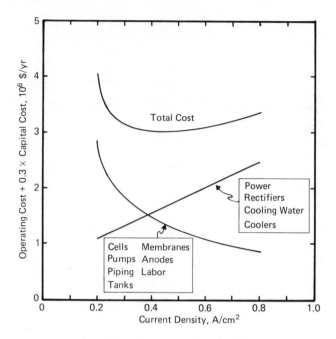

Fig. 6.17. Current density optimization.

Table 6.12. EHD Plant Design Factors

Capacity (ton/year)	14,500
Current density (A/cm^2)	0.45
Cathode area (cm^2/cell)	6350
Cell current (A)	2875
Cell voltage (V)	11.65
Cells/unit	24
Cell units	16
Total cells	384
On-stream time (hr/year)	8000
Catholyte temperature (°C)	50
Catholyte composition (wt %)	
AN	16
ADN + by-products	16
TEAES	40
H$_2$O	28

Table 6.13.   Commercial EHD Cell Components

Anode	1% Ag/99% Pb, 0.64 cm × 86 cm × 91 cm
Anode spacer	National Filter Media, Polymax V polyethylene fabric
Membranes	Two Ionics CR6lAZG, glass-backed, 91 cm × 102 cm
Cathode spacer	Polypropylene strips, pegged to cathode, 0.3 cm thick × 0.5 cm wide on 2.4-cm centers
Cathode	Chemically pure Pb, 0.48 cm × 86 cm × 91 cm
Intermediate plate	Machined polypropylene block, 5 cm × 97 cm × 107 cm

while four parallel catholyte systems were provided to keep the circulating catholyte stream to a reasonable size and avoid excessively large pumps and piping. In addition, the multiplicity of catholyte systems limited the impact of composition upsets to one-fourth of the plant in the event of a serious membrane rupture. Four cell units were served per catholyte system, each unit containing 24 individual cells. Other components were a surge tank circulating pump, cooler, and centrifuge (for solids removal). Similarly, the single anolyte system consisted of a surge tank, circulating pump, and cooler. A schematic diagram of the electrolyte area is shown in Fig. 6.18.

The 16 cell units were located in a building (Fig. 6.19), while the electrolyte surge tanks, circulating pumps, and coolers were installed outdoors. The concrete cell floor was constructed 20 ft above grade to allow room under the floor for electrolyte headers and electrical busses. An area at one end of the building was provided for testing of membranes and the assembly and repair of membrane frames.

In each system catholyte was pumped from the surge tank to the four cell units, which operated with all 96 cells in parallel hydraulically. The catholyte flow rate was controlled to each bank of 24 cells. A flow of 4 liters/sec to each cell was employed in order to give a velocity of 2 m/sec across the cathode surface. The catholyte inlet temperature was maintained at 50°C, with cooling provided by a shell-and-tube heat exchanger located upstream of the cell units. The volumetric flow through the cell units was so high that the conversion of AN per pass was only about 0.2%. Thus conceptually the electrolysis was equivalent to operating a continuous stirred-tank reactor at the terminal conversion condition. Makeup AN and recycle QAS solution were fed to each system at rates sufficient to maintain the desired catholyte composition of 16% AN, 28% water, 40% QAS, and 16% products. A side stream of circu-

EHD PLANT LAYOUT

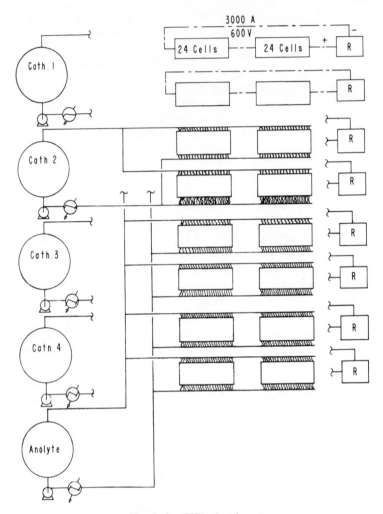

Fig. 6.18.   EHD plant layout.

lating catholyte was withdrawn and sent to the extraction area for recovery of ADN and removal of excess water, which entered the system by means of electroosmotic transfer from the anolyte.

The combined catholyte withdrawn from all four cell systems was cooled to 35°C in a shell-and-tube heat exchanger and centrifuged for removal of solids, which were mostly a finely divided polymer of AN (Fig. 6.20). The

Fig. 6.19. EHD cell room.

centrifuge effluent was then sent to the top of the ADN extraction column, a 25-stage sieve-tray device, where it was countercurrently contacted with an AN stream for 98% removal of ADN. The column bottoms were sent to a vacuum (300 mm Hg) evaporator to concentrate the QAS to a 70 wt % solution for recycle to electrolysis. The ADN extractor overhead, a mixture of AN, ADN, and other reaction by-products, was fed to the bottom of a 35-tray QAS extraction column, where it was countercurrently extracted with water for essentially complete removal of QAS. The extractor make was then sent to a 10-plate distillation column for stripping of AN, PN, and water. Stripper bottoms, crude ADN, was sent to storage for subsequent refining. An organic-phase side stream from the QAS evaporator make was processed through a 50-tray distillation column for removal of PN from the system. The overall water balance was maintained by withdrawal of a portion of the lower layer from the QAS evaporator overhead, stripping it with open steam in a 10-tray column for recovery of dissolved AN and purging the bottoms to waste.

Purification of the crude ADN was accomplished by vacuum distillation in

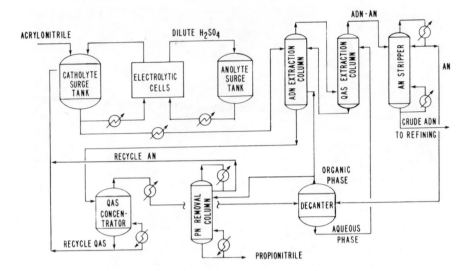

Fig. 6.20.   EHD plant schematic flowsheet electrolysis and recovery areas.

a four-column refining train. To ensure good catalyst life in the conversion of ADN into hexamethylenediamine, a minimum purity of refined ADN of 99.9% was required. To prevent excessive degradation of ADN from overheating during distillation, all four column reboilers were operated using forced circulation and limiting the reboiler outlet temperature to a maximum of 220°C.

The first column in the train contained seven sieve trays and removed high-boiling by-products, mainly the hydrotrimer and hydrotetramer of AN. The long residence time of the bottoms in this column sufficed to thermally decompose nearly all the BCE in the crude ADN, producing equimolar amounts of AN and HOPN. The bottoms was sent to a small trayless column operated at 210°C and 5 mm Hg pressure to recover ADN from the high-boiler stream before it was incinerated.

The overhead from the high-boiler fractionator was fed to a 20-tray column for removal of low-boiling impurities, principally HOPN and water. The bottoms from this column was sent to a 20-tray refining column to eliminate trace levels of BCE, cyanovaleramide, and color bodies formed in the previous fractionating columns. The entire four-column train operated as a continuous system with no intermediate storage being provided.

Initial Operating Experience

The first stages of the start-up of the EHD plant were not too different from other chemical processes; there were the usual problems with instruments, piping, and pumps. After a month or two of operation, however, when these problems had been resolved, several others peculiar to the electrochemical nature of the process became evident.

One of the first problems concerned with cell operation was that of "hot spots." At the relatively high current density employed (0.5 A/cm^2), uniform distribution of electrolyte flow was found to be necessary to prevent overheating and occasional burning of the ion-exchange membranes. In severe cases the temperatures developed were sufficiently high as to cause rupture of the polypropylene membrane frame, which was in several instances followed by ignition of the catholyte. To provide a warning of restricted flow in an individual cell, the voltage of the electrical bus connecting each pair of cell units was continuously monitored. While under normal conditions this voltage was roughly midway between those at the opposing ends of two units (electrically in series), a flow restriction in an individual cell would give rise to a sharp voltage increase in that cell bank, markedly shifting the center-buss voltage. This center-point voltage was interlocked to the electrical supply to the cells so that power to this pair of cell units was automatically shut off when the voltage deviated more than 10 to 20 V from normal. This inexpensive instrumentation proved invaluable as an alarm system for abnormal flow conditions, and the occurrence of cell fires was virtually eliminated.

The life of the ion-exchange membranes in initial operation was only about 100 hr versus a target of > 1000 hr. Most of the premature failures were due to tears at the supporting gasket surface. Theories as to the cause of the problem included (1) excessive differential pressure between anolyte and catholyte compartments, (2) fluctuations in pressure differential across the membrane, (3) high start-up amperage, and (4) trapping of a pocket of water between the pair of membranes during assembly. Tests were conducted in the pilot plant and full-scale plant to check out each of these possibilities, but none appeared to explain the abnormally short membrane life. A mockup of a cross section of the commercial cell was constructed in the development labs and it was observed that upon compression (simulating press closure), the gasket in use at that time was forcing the membrane inward, resulting in creasing and subsequent failure. Once the cause was identified, a simple change in the durometer of the gasket material eliminated the creasing problem.

With the gasket design change the membrane life was extended to 200 to 300 hr, still well short of the desired level. Further increases in life were achieved as a result of a continuing membrane development program conducted by the supplier, Ionics, Inc. A membrane fabricated with two plies of glass cloth proved to be significantly more durable, giving an operating life of over 500 hr in the commercial cells.

With continued recycle of the quaternary ammonium salt solution, the concentration of hydrolyzed polyacrylonitrile and other minor impurities accumulated to a point where they led to deposits on the cathode and reduced the ADN selectivity. Although considerable development effort had been directed at definition and solution of this problem, the extent of the impurities buildup and yield effect were not fully determined until the commercial plant was in operation. Within a few months after plant startup, the need for improved means of purging the offending impurities became apparent. One scheme developed in the plant shortly after startup involved holdup of a two-phase mixture of the recycle QAS, ADN, and AN at about 23 to 35°C for at least an hour, which allowed the hydrolyzed polyacrylonitrile to settle at the liquid/liquid interface. A stream was withdrawn from the interfacial zone and circulated through a filter for removal of this insoluble impurity. The soluble impurities, such as sulfate ions and organic acids, were removed in a slipstream of concentrated recycle QAS by contacting the solution with acrylonitrile to extract most of the tetraethylammonium ethyl sulfate into the organic phase, leaving the sulfate in the aqueous phase. Most of the remaining tetraethylammonium cations were then recovered from the aqueous phase by electrolysis in the QAOH electrolysis system or by means of ion exchange.

Process Improvement

As a consequence of the problems encountered during the first year of plant operation, the AN, membrane, and electrical power usage were all significantly higher than design levels. The need for an aggressive troubleshooting and process improvement program was evident. This was conducted through the joint efforts of the Decatur plant technical group and the Pensacola development engineers and chemists. The results over the first 5 years of plant operation are shown in Table 6.14.

The improvement in power usage from 9.4 kWh/kg during the first year's operation to 7.0 kWh/kg in the fifth year was achieved gradually in a continuing program to reduce the catholyte gap. The 3.2-mm cathode-to-membrane spacing employed in the commercial cells at start-up was lowered first to 2.4 mm and later to 1.6 mm in a progression of tests, first in the pilot plant and then on one or more plant cell units. Virtually all that was required in

Table 6.14. Comparison of EHD Plant Performance with Design

	Design	First Year	Fifth Year
Cell current (A)	2875	3200	4000
Productivity/cell (kg/hr)	5.0	4.8	7.1
Membrane usage (units/10^6 kg)	100	880	350
Power usage (kWh/kg)	6.9	9.4	7.0

the way of cell design changes was the fabrication and installation of thinner polyethylene spacer strips. Since the membrane was held against the cathode strips by the differential pressure between anolyte and catholyte, no change in anolyte compartment design was required as the catholyte compartment was made thinner. The slight increase in anode compartment thickness had only a minor effect on cell voltage. A change in anolyte composition from 5 to 10% H_2SO_4 also contributed to a reduction in the cell voltage drop and power usage. The voltage/current relationships for the commercial cells are shown in Fig. 6.21. The improved cell voltage afforded by the reduced catholyte gap and higher anolyte conductivity was used to increase the cell current and hence the plant productivity. Several units were demonstrated at 4500 A with 37 cells/unit, providing synthesis rates per unit over twice that of the original design. This was accomplished with no additions to the electrolyte circulating systems or rectifiers. Thus the electrolysis capital investment was better than halved relative to the design basis.

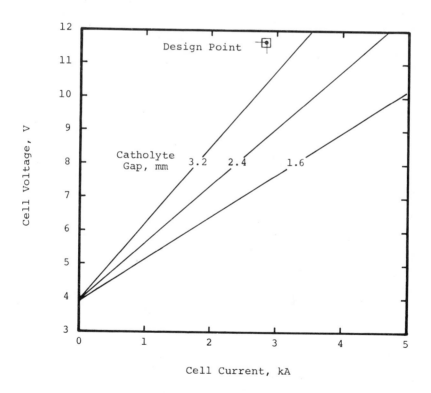

Fig. 6.21. Voltage/current relationship for commercial cell.

Undivided-Cell Process

From a breakdown of the voltage drop components for the divided-cell EHD process (Table 6.15), it is evident that the catholyte and membrane IR drops represent a major portion of the total cell voltage. While process improvement efforts in the Decatur plant were aimed at narrowing the cathode/membrane gap, lab studies were directed toward an undivided-cell process using a more conductive electrolyte. Alkali metal sulfate and phosphate electrolytes had been shown by Tomilov and co-workers [6] to give reasonably high ADN selectivities using a graphite cathode in the presence of low concentrations of tetraethyl ammonium salts.

Monsanto's goals for an undivided-cell process included (1) solid metal electrodes to avoid cell design problems of graphite, (2) highly conductive electrolyte for minimum IR drop, (3) low overvoltage anode to reduce losses of AN and products through oxidation, and (4) stable cell performance, that

Table 6.15. Comparison of Monsanto EHD Processes

	Divided Cell	Undivided Cell
Cathode	Pb	Cd/steel
Anode	Pb (1% Ag)	Steel
Membrane	Ionics CR 61	None
Temperature	50	55
Electrode gap (mm)	7	2
Electrolyte velocity (m/sec)	2	1–2
Electrolyte composition:		
QAS, Cation	Et$_4$N	bisQuat[a]
Anion	Ethyl sulfate	Phosphate
QAS concentration (wt %)	40	0.4
Other salts	None	10% Na$_2$HPO$_4$
		2% Na$_2$B$_4$O$_7$
		0.5% Na$_4$EDTA
Current density (A/cm^2)	0.45	0.20
Voltage drops		
Reaction EMF	2.50	2.50
Overpotential	1.22	0.87
Catholyte IR drop	5.47	0.47
Anolyte IR drop	0.77	—
Membrane IR drop	1.69	—
	11.65	3.84
Power usage (kWh/kg)	6.6	2.3

[a]bisQuat = hexamethylenebis(ethyldibutylammonium) cation.

is, low electrode corrosion and sustained high ADN selectivity/current efficiency. The cathode materials in Table 6.3 were reexamined using an aqueous solution of 10% Na_2HPO_4 and 0.5% ethyltributylammonium phosphate as the supporting electrolyte. Mercury, cadmium, and lead were found to give ADN selectivities of 88 to 89% and ADN current efficiencies of 80 to 85% in short runs. Mercury was eliminated because of cell design complications and cadmium was selected over lead as offering better long-term stability.

Anode materials that were tested included platinum, carbon steel, magnetite, duriron, lead, lead dioxide, and various noble metal oxides on titanium (DSAs). Carbon steel was found to be moderately stable, exhibiting a corrosion rate of about 1 cm/year at 0.2 A/cm^2, 50°C, and pH 8. While this rate of corrosion was acceptable in screening runs, a substantially lower rate was desired for extended operations. A number of additives to the electrolyte were tested in an effort to improve the stability of the carbon steel anode. Those showing some promise included alkali metal borates, carbonates, chromates, polyphosphates, molybdates, and chelating agents, such as EDTA and NTA. A combination of 0.5% Na_4 EDTA and 2% $Na_2B_4O_7$ reduced the anode corrosion rate in bench-scale runs to about 0.5 mm/year. The Na_4 EDTA was found to serve another important role by providing mild corrosion of the cadmium cathode, which kept the working electrode free from anode corrosion products and polymer.

While tetraethylammonium ethyl sulfate was the preferred selectivity-enhancing salt in the divided-cell process, larger homologs gave better results in the sodium phosphate electrolyte (Table 6.16). The best combination of high selectivity and ease of extraction from the organic products was offered by hexamethylene bis(ethyldibutylammonium) phosphate, which could be readily synthesized from hexamethylene diamine (the ADN hydrogenation product).

In addition to the voltage and power saving provided by an undivided cell, the cell construction is greatly simplified. One cell design consists of a stack of carbon steel sheets, plated on one side with several mils of cadmium. From 50 to 200 sheets can be placed in an assembly using 2-mm-thick plastic spacers to set the electrode gap. The bipolar electrode stack is installed in a pressure vessel, suitably baffled to provide electrolyte flow uniformly through the parallel electrode compartments.

Impurities consisting of electrode corrosion products and organic by-products tend to accumulate in the electrolyte of the undivided-cell process, as is also the case in the divided-cell system. With extended operating time these can lead to a deterioration in reaction selectivity and/or current efficiency. The concentration of impurities is controlled by purging 20 to 40 g of aqueous electrolyte per faraday of current to a crystallization unit for recovery and recycle of the Na_2HPO_4 and $Na_2B_4O_7$ and separation and disposal of the detrimental substances. A schematic flowsheet of the synthesis system is shown in Fig. 6.22.

Because of the simplicity of design of the undivided cell, the cost of the cell is less than 10% of that of the plate-and-frame design. This leads to a reduction in optimal current density from 0.45 to 0.20 A/cm^2, for a cell voltage of 3.8 V (Table 6.15). Other economic advantages of the undivided-cell process are (1) no membrane cost, (2) reduced capital cost of rectifiers and electrolyte coolers, and (3) lower maintenance cost.

Since start-up in 1965, the capacity of the Decatur plant has been increased more than fivefold. In addition a 90,000-ton/year EHD facility was brought on line at Teesside, UK in 1978 by Polyamide Intermediates, Ltd., a joint venture between Monsanto and Montedison. The total ADN capacity of these two plants is currently close to 180,000 ton/year. At today's raw material and energy costs the EHD process is believed to be fully competitive with other commercial routes to ADN.

The success of the Monsanto EHD venture, which spans nearly 20 years, can be attributed to a unique combination of favorable raw material (AN) availability, large demand for the premium-priced product (ADN), a breakthrough in the electrochemistry by Baizer, and years of dedicated effort by

SCHEMATIC FLOWSHEET FOR UNDIVIDED CELL EHD PROCESS

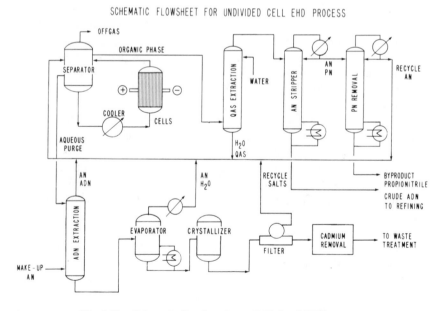

Fig. 6.22. Schematic flowsheet for undivided cell EHD process.

Table 6.16. Quaternary Ammonium Salt Evaluation

10% Na_2HPO_4 Cd cathode 2×10^{-3} mM QAS[a]	ADN Selectivity (%)
Et_4N^+	85
$EtPr_3N^+$	87
$MeBu_3N^+$	88
$EtBu_3N^+$	89
$EtPe_3N^+$	89
$EtHx_3N^+$	87
$Pr_3N(CH_2)_4NPr_3^{2+}$	88
$EtBu_2N(CH_2)_6NBu_2Et^{2+}$	89

[a] $Me = CH_3$, $Et = C_2H_5$, $Pr = C_3H_7$, $Bu = C_4H_9$, $Pe = C_5H_{11}$, $Hx = C_6H_{13}$.

the development and plant technical teams, with the indispensible encouragement and financial support of management.

References

1. M. M. Baizer, *J. Electrochem. Soc.*, **111**, 215 (1964).
2. O. Bayer, *Angew. Chem.*, **61**, 229 (1949).
3. R. M. Leekley, U.S. Patent 2,439,308 (April 6, 1948).
4. I. L. Knunyants et al., *Inzv. Akad. Nauk SSSR, Otd. Khim. Nauk,* **1957** (2), 238–240.
5. G. Z. Kiryakov and V. V. Stender, *J. Appl. Chem. USSR,* **25**, 25–40 (1952).
6. A. P. Tomilov, S. L. Varshavskii, and I. L. Knunyants, British Patent 1,089,707 (Nov. 8, 1967).

Chapter **VII**

THE PHILLIPS ELECTROCHEMICAL
FLUORINATION PROCESS

W. V. Childs

Phillips Petroleum Company,
Bartlesville, Oklahoma

1 INTRODUCTION

Caution

Many of the reactants used and the products made in this process are
known to be extremely hazardous. These include HF, which causes severe

burns, fluoroolefins, which are toxic at the part per billion level, and monofluoroacetates. In addition, most of the products, intermediates, and byproducts have not been tested and must be presumed to be highly hazardous. In a number of cases, symptoms begin several hours after exposure.

The dramatic differences in properties between fluorinated organic compounds and their hydrocarbon analogs have attracted considerable attention and resulted in a wide range of remarkable applications. These applications range widely and include:

1. *Consumer Products.* Refrigeration fluids, aerosol propellants, blowing agents, nonstick coatings for cookware and other items, antistain fabric treatments, leather treating, paper coatings, nonlubricated bearings, and a broad spectrum of hidden applications in the automotive and electronic fields.

2. *Pharmaceuticals and Biologicals.* Herbicides, pesticides, anticancer agents, anesthetics, artificial blood formulations, X-ray contrast agents, and weight control agents.

3. *Technological Applications.* High-performance plastics and elastomers, lubricants and fluids for extreme conditions, catalysts, semiconductor processing, special fire-fighting materials, load bearing pads for bridges and buildings, coatings for a broad range of environmental and corrosion resistance situations, and electrochemical membranes.

The chemistry and technology involved in the introduction of fluorine into organic molecules is basic to most of the discipline of organic fluorine chemistry, as few naturally occurring organic molecules contain fluorine. Generally the techniques used to prepare fluorine-containing molecules in high yield are indirect and produce low-valued coproducts. The direct processes, such as electrochemical fluorination and reactions involving elemental fluorine are typically inefficient, or fraught with extreme complications or hazards. This chapter describes a simple and efficient way to introduce fluorine directly into a variety of organic molecules to produce useful and interesting products.

The initial description of and conceptual basis for the Phillips Electrochemical Fluorination (ECF) Process was given by Fox et al. [1]. It was later discussed by MacMullin upon the occasion of his receipt of the Electrochemical Engineering and Technology Award given by the Electrochemical Society [2]. The Phillips ECF Process is conceptually distinct from the Simons process practiced by 3M and by others [3], and from the process described by Ashley and Radimer [4, 5].

The Simons process works best with organic compounds that have an appreciable solubility in anhydrous hydrogen fluoride and form conducting solutions. These solutions can be electrolyzed in a cell with interleaved nickel anodes and steel cathodes. Hydrogen is evolved at the cathode and highly fluorinated products are formed at the anode. Current densities are low, 0.1 to 0.2 kA/m^2, as are current efficiencies and yields of specific products. Reference 3 provides an excellent entry to the literature of this process.

The Ashley-Radimer patents teach the use of a porous or a foraminiferous anode, preferably made of carbon and rotating, and the use of a molten $KF \cdot 2HF$ electrolyte containing LiF. The feed, often diluted with an inert gas, is supplied to the anode so that it bubbles or is sparged into the electrolyte. Fluorination in the Ashley-Radimer cell apparently takes place in the bubble near the point of breakout.

In contrast to the Simons process, the Phillips ECF Process works best with substrates that are not appreciably soluble in the molten $KF \cdot 2HF$ electrolyte. The reasons for this are given in considerable detail here. A number of compounds that are soluble in cold HF are not soluble in $KF \cdot 2HF$ at 100°C, so there is some overlap in application.

In contrast to the Ashley-Radimer process, breakout of the substrate into the electrolyte is an anathema in the Phillips ECF Process. Fluorination takes place by the reaction of substrate with elemental fluorine within the confines of the porous carbon anode. This mode of operation and its application and advantages are described below.

The Phillips ECF Process has been used to fluorinate a wide spectrum of compounds, including carbon monoxide, light hydrocarbons, chlorinated hydrocarbons, esters, ethers, and acyl fluorides. The result is generally the very efficient, nearly statistical, introduction of fluorine into the substrate.

2   PROCESS DEVELOPMENT

Of interest relative to the Phillips ECF Process is the work Ralph Adams did with the analytical carbon paste anode [6] and A. J. Rudge's work on the preparation of elemental fluorine [7]. At Phillips a process for the electrochemical chlorination of organic compounds was developed using a porous carbon or graphite anode [8]. In one embodiment dodecane was fed to a porous thimble through a hollow current collector. The dodecane thoroughly wetted the carbon, passed through it, and rose to the top of the HCl/water electrolyte. Electrolysis generated chlorine and resulted in chlorination of the dodecane.

Attempts to translate this process into a process for electrochemical fluorination ultimately led to a system using a cell somewhat similar to the Rudge/ICI process for preparing elemental fluorine [7]. The electrolyte is nominally $KF \cdot 2HF$. The melting point/composition diagram for mixtures of KF and HF is shown in Fig. 7.1 and the vapor pressure/composition curve is shown in Fig. 7.2. These data were taken from Cady [9] and clearly show that an electrolyte composition near KF·2HF possesses a favorable combination of melting point and HF vapor pressure. We found, as did Rudge, that graphite is not satisfactory for this use, but ungraphitized carbon performs very well in KF·2HF. At high levels of HF, say KF·4HF, carbon is also unsatisfactory, as it decomposes when current is passed, sometimes with violence.

The early cells were simple, as seen in Fig. 7.3, which is almost exactly like the cell used in the chlorination work, and in operation it was not clearly distinct from the Radimer-Ashley concept. The anode was machined from

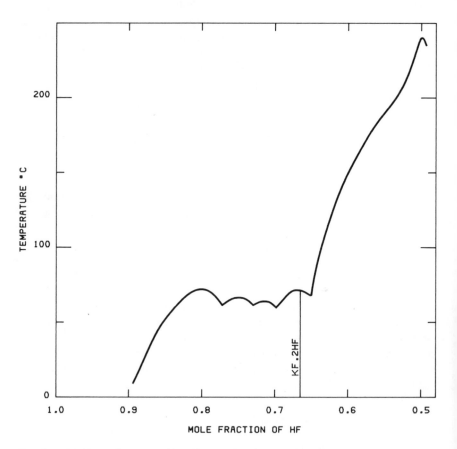

Fig. 7.1. Melting point/composition diagram for the potassium fluoride/hydrogen fluoride system.

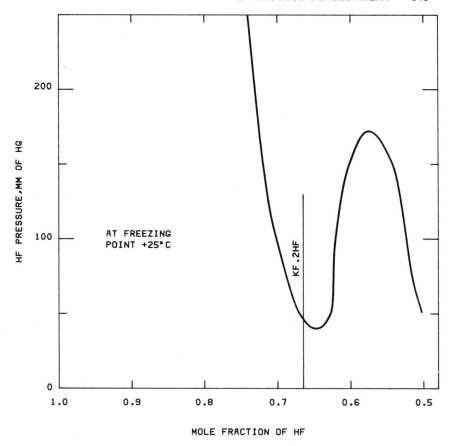

Fig. 7.2.   Pressure of hydrogen fluoride over potassium fluoride/hydrogen fluoride system.

Stackpole 139 porous carbon threaded onto a brass current collector that also served as a feed tube. Other, more permeable, carbons are presently preferred (*see below*). The current collector was wrapped as shown with Teflon tape to minimize corrosion. This wrapping extended over some of the anode to force the gas to bubble out into the electrolyte as taught by Radimer and Ashley. The cathode was nickel mesh mounted on the thermocouple well as shown. The anode and cathode assemblies were mounted in a neoprene stopper with provision for a product line, thermocouple, and HF addition. The electrolyte was nominally KF · 2HF to which a trace of LiF was added. HF was added manually, along with some nitrogen to prevent suck-back, to maintain a constant level. The entire assembly was mounted in a heated silicone oil bath that could be cooled with water as needed.

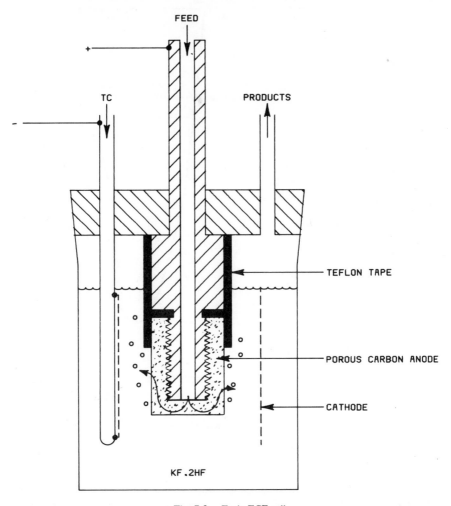

Fig. 7.3. Early ECF cell.

The gaseous feed, usually ethylene, was passed in and simply bubbled out through the pores while current, 3 to 6 A, was passed. The products were caught in a cold trap, or gaseous samples were analyzed by gas chromatography. These thimble cells operated fine for an hour or so and then they simply stopped passing current. They became "polarized." In this usage polarization has a specialized meaning [7] that applies only to fluorine cells (*see below*). Until satisfactory means were developed for dealing with this polarization, there was considerable frustration. Ultimately, a high-voltage

(60 to 80 V) treatment was developed to "depolarize" the anode. Less successful treatments included standing at open circuit, current reversal, surface abrasion, and wrapping with nickel wire. The high-voltage treatment is described in detail below.

The fluorination of ethylene was initially done at low per pass conversion. The major recovered product boiled at about 30°C and was a real puzzle. The expected product was 1,2-difluoroethane and the CRC *Handbook of Chemistry and Physics* [10a] gave 11°C for its boiling point. There was also a question about the stability of 1,2-difluoroethane to the workup, which included a wash with aqueous potassium carbonate. The product was shown to be 1,2-difluoroethane and later editions of the CRC handbook [10b] give 30.7°C for its boiling point.

Efforts to scale up this sparger style anode led to a variety of unsuccessful designs. It is difficult to get bubbles to come out of an unwetted sparger uniformly without going to very high flow rates. The realization that we wanted to fluorinate instead of sparge led to the Phillips ECF Process. Rudge [7, 11], Cady [12], and many others have pointed out that molten KF · 2HF does not wet a porous carbon anode. Rudge developed this into a process for making elemental fluorine in which the fluorine is generated on a porous carbon anode surface, passes into the bowels of the anode, passes up within the anode, and is recovered in the vapor space above the electrolyte. This nonwetting characteristic is sketched in Fig. 7.4. (Carbon means the nongraphi-

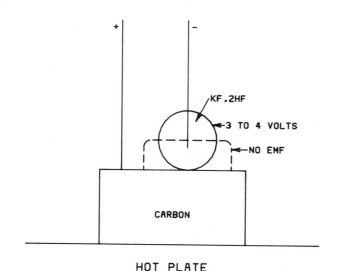

Fig. 7.4.   Molten KF·2HF does not wet a carbon anode.

tized material. Graphite delaminates in this system.) Molten $KF \cdot 2HF$ wets carbon much as mercury does dirty glass. When a cathode is supplied and the carbon is made 3 to 4 V anodic, the melt tucks together and forms a drop with a high contact angle, much as mercury does on clean glass. Rudge reported a contact angle of 150° for the $KF \cdot 2HF$/carbon anode system [11].

Because of this, when a piece of porous carbon with suitable characteristics was made an anode in $KF \cdot 2HF$, the pores did not fill with electrolyte. The electrolyte simply pressed up against the anode rather like mercury does on clean fritted glass. When this was realized the anode shown in Fig. 7.5a was assembled and shown to work reasonably well. This configuration rapidly evolved into the anode of Fig. 7.5b and eventually into the 25-A cell shown in Fig. 7.6. Later versions on this scale worked well at 40 A, but the power supply on hand at the time was a 25-A, 140-V model.

Obviously if you put enough pressure on mercury it will enter the pores of the fritted glass. Likewise, if you put enough pressure on the electrolyte it will enter the pores of the carbon. It turns out that with the best carbons from mechanical and polarization frequency considerations it takes about 12 cm of electrolyte to develop enough pressure to enter the large pores. The result is that some electrolyte enters the feed cavity, is stripped of HF, solidifies, and plugs the cavity. This is no problem at a depth of 10 cm, but below 12 cm it is. Again we went through a variety of designs and fixes, including the addition of HF with the feed to maintain the electrolyte plug in a liquid state. The solution was truly elegant in its simplicity. It was the concept of a gas cap [13] feed system. Figure 7.7 is a sketch of a laboratory cell incorporating an anode with a gas cap feed system. Note the side entry of the Teflon FEP feed tube. This feature eliminates any low spots where electrolyte may accumulate, dry out, and plug the line. The sketch is approximately to scale. The anode is 35 cm long and 3.5 cm in diameter. The current collector is a copper rod screwed or pressed in 10 cm below the electrolyte level. Versions of this cell with one and with four 1000-A anodes have been successfully operated for extended periods.

3   PROCESS DESCRIPTION

The fact that $KF \cdot 2HF$ does not wet a carbon anode and the separate postulate that elemental fluorine is the fluorinating agent can be developed to explain ECF cell behavior in detail. This should be kept in mind throughout the balance of this chapter.

Figure 7.7 is a sketch of a recent laboratory ECF cell. The cell is jacketed (not shown) for temperature control with circulating silicone fluid. It was fabricated from mild steel with Teflon lids. The anode reactor is the heart of the system. It was machined from commercially available ungraphitized

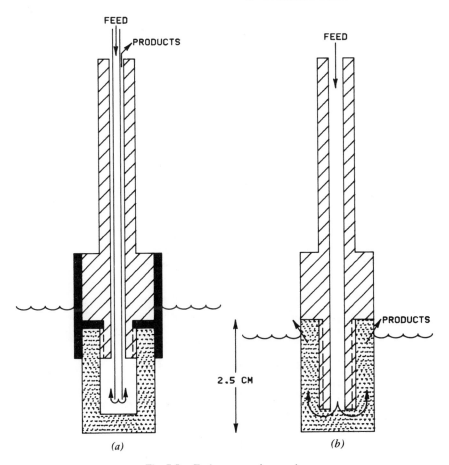

Fig. 7.5.   Early nonsparging anodes.

porous carbon stock [14]. The feed tube is Teflon FEP tubing flared slightly on the lower end to hold it in place. Makeup HF is added on demand through the Teflon FEP tube shown; a slow flow of nitrogen is necessary to prevent electrolyte suck-back. Mild steel level probes were used for HF level control and for high- and low-level shutdown. These probes were cathodes in circuits that energized DC relays when contact was made with the electrolyte. These relays in turn energized solenoid valves and thence air-operated soft Kel-F seat valves to control the HF addition and shutdown. The entire HF system was heated to 35 to 38°C to maintain pressure. A variety of sensors and control devices were used to permit unattended operation. These included high and low electrolyte level, high and low temperature, high cell pressure, low

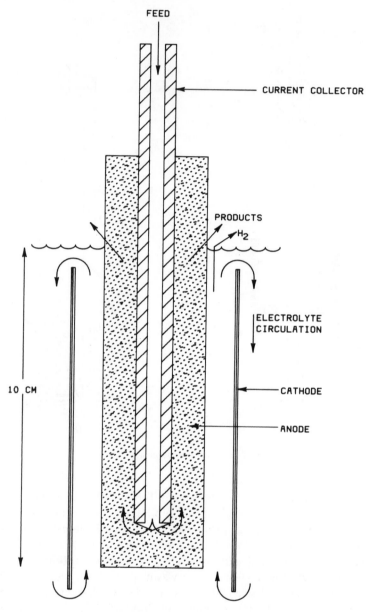

Fig. 7.6. Twenty-five-ampere nonsparging anode.

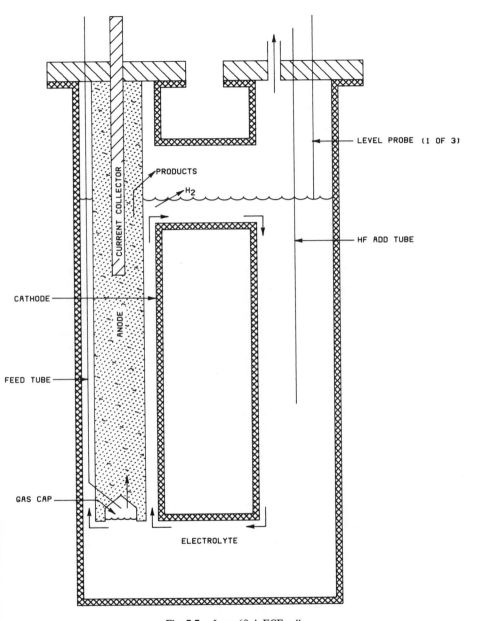

Fig. 7.7. Loop 60-A ECF cell.

feed pressure, low control air pressure, temperature control system, and water pressure. These were set to open under error conditions.

Figure 7.8 is a schematic diagram of the circuitry used to operate the laboratory ECF cell unattended. This third generation design worked reliably. The main feature of this design is the use of on/off sensors in parallel with slow-blow fuses to shut down the apparatus and indicate the cause of the shutdown. There are two series of sensors passing 0.8 A through the 150 Ω, 200 W resistors. When a sensor opens this current goes through its associated fuse, blows it, and causes a shutdown. There are two levels of shutdown. Relay K1 does a partial shutdown, leaving the temperature control equipment functioning to make restart easier. Relay K2 does a complete shutdown.

The function of most of the circuitry is obvious, but S6 and its associations need some explanation. This is a one-shot auto-restart circuit. Oklahoma is noted for its thunderstorms and momentary power failures. With switch S6 in position A and relay K7 latched on, the auto-restart is armed. When a power failure occurs K7 falls out, and when the power comes back on the motor starts to drive S6 at 1 rpm. When S6B is energized, K1 and K2 are energized using K8A and B. K8 is energized for 30 sec and then turned off. If the peripherals have come on satisfactorily, K1 and K2 stay on and the power supply relay, K3, is energized when S6I is energized. When S6J is energized K7 is energized and the motor stops. The auto-restart system is reset by advancing switch S6 to position A by holding down S5 until the pilot light connected to S6A comes on.

For the main power supply two Sorensen DCR40-125 A units were used. These are 40-V, 125-A units with SCR controllers and can be run under either constant current or constant voltage with automatic crossover. For routine operation one unit is used at constant current and a 15 to 20 V crossover limit. For depolarization the units were used in series. This connection required the use of auxiliary power diodes as specified in the instruction manual.

Polarization/Depolarization

The word polarization has a variety of distinct, albeit related, meanings in the sciences. In the present context it is used in a restricted sense that dates back to the early years of fluorine generation with carbon anodes [7, 11]. The ECF cell is operated at constant current and the normal terminal voltage varies between 7 and 9 V. On occasion, the voltage will increase rapidly and the cell is said to be polarized. Generally, before this rapid increase the cell spends some time in a condition of incipient polarization, and a recording of the voltage shows very erratic behavior with numerous spikes and excursions before the cell is firmly polarized. The power supply is usually set to limit or

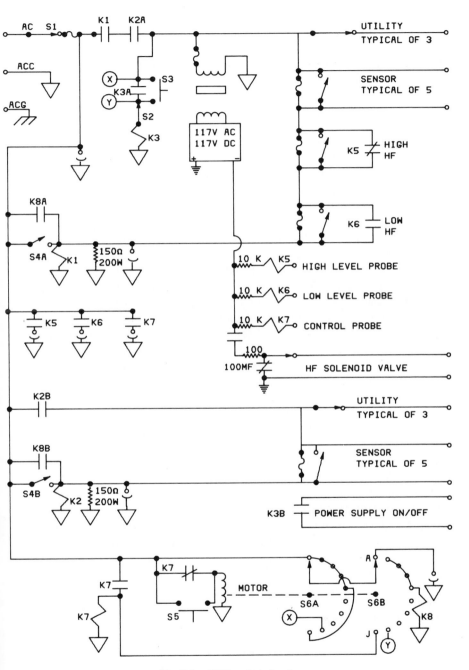

Fig. 7.8. ECF control circuitry.

crossover to voltage control at 15 to 20 V. When a cell is polarized only a small fraction of the normal current can be passed at twice the normal voltage.

The cell can be depolarized by permitting the voltage to rise at constant current. Normally it rises to 60 to 80 V. If the cell is permitted to operate at this high voltage for 1 to 3 min and then turned off and back on, it will operate normally and it is said to be depolarized.

At some later time the cell will polarize again. The initial polarization may occur soon after a new anode is started. With low permeability porous carbons this initial polarization occurs within a few hours. With more permeable carbons, for example, PC45, this may take a day or so. And with very permeable carbons, several days may go by before the first polarization. After the initial polarization/depolarization cycle, subsequent polarizations are less frequent. The polarization frequency is higher with high current densities, more polar feeds, less permeable carbons, and low HF concentrations. The addition of a small amount of lithium fluoride appears to decrease the frequency of polarization. Other slightly soluble fluorides appear to function similarly. With ethane feed, 2.00 kA/m^2, PC45, 41 wt % HF, 0.1 wt % LiF, and 95°C, the polarization frequency was measured in weeks.

Some authors have identified this polarization with the anode effect frequency observed in molten salt electrolysis systems. Generally the anode effect is observed to occur at a characteristic current density and can be brought on almost at will. When an ECF anode has polarized and been depolarized, it is difficult to cause another polarization.

Process Rationalization

The porous carbon anode-reactor is the heart of the system. A variety of porous carbons have been used and work reasonably well in this system. The best combination of physical properties, polarization behavior, and availability is found in commercially available filter carbons [15]. These include Union Carbide's PC25, PC45, and PC60 and Great Lakes' B303. Figure 7.9 is a set of scanning electron photomicrographs of a fractured PC45 carbon sample. Figure 7.9a was taken at 50× and shows the major structure and macroporosity. The other photomicrographs show the microporosity inherent in the porous carbon. This carbon is made from selected calcined petroleum coke that has been ground and classified. The particle size distribution is selected to control the porosity, pore size distribution, and permeability. The classified carbon flour is then mixed with pitch, molded, and carbonized. The rough molded product is then machined to size. These filter carbons have about 50% void volume with fairly narrow pore size distributions. Figure 7.10 is a mercury porosimetry result showing the pore size distribution in PC45 carbon. Note that most of the void volume is included in pores with diameters between 40 and 50 μm.

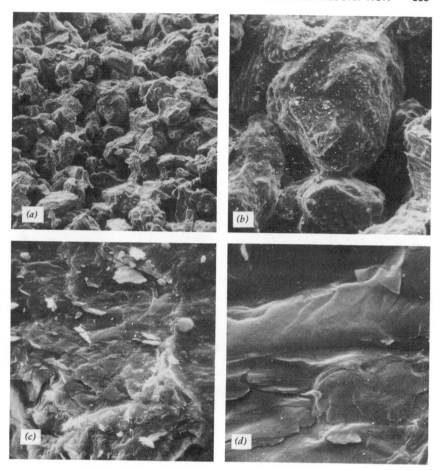

Fig. 7.9. Scanning electron micrographs of PC45 carbon.

The electrolyte does not wet a carbon anode, but it can be forced into the pores by a pressure differential. This penetration by the electrolyte can be controlled by the pore size distribution and, to a lesser extent, by the electrolyte composition and temperature. Larger pores, more HF, and higher temperature increase penetration. Penetration is also affected by anode history. If a fresh piece of PC45 carbon is simply immersed 30 cm into molten KF · 2HF, the pores below about 10 cm of immersion rapidly flood. If the carbon had been made anodic initially, only a fraction of the carbon floods, and if the potential is disconnected from this anode flooding is slow. This is shown in Figure 7.11. This electrolyte is found throughout the anode and it appears to exchange rapidly with the bulk electrolyte. Figure 7.12 is a sketch

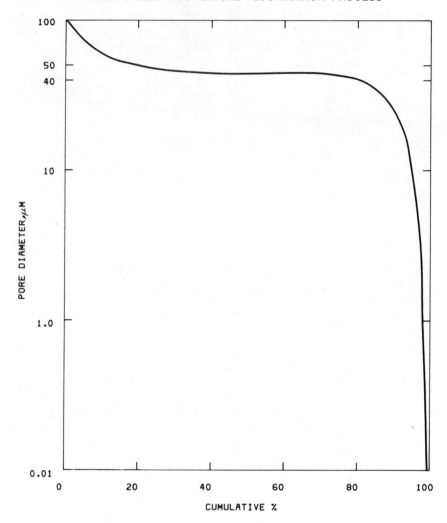

Fig. 7.10.   Pore size distribution in PC45.

of the model just described. Small pores are essentially empty, larger pores contain some electrolyte, and still larger pores provide a pathway for electrolyte to move around within the anode.

The Phillips process works best with reasonably volatile feeds that are relatively insoluble in the electrolyte and produce volatile products. The Simons or 3M process works best with feeds that are soluble in the electrolyte (HF). The Phillips process has been used to produce small amounts of perfluorooctanoic acid and some higher perfluoroalkanes.

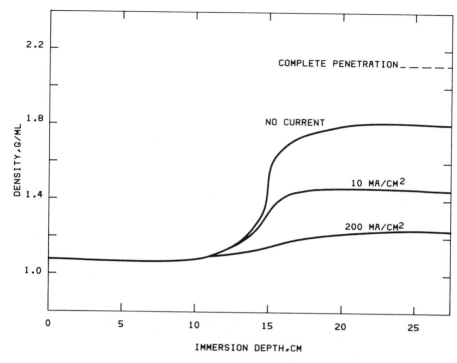

Fig. 7.11. Electrolyte penetration into PC45 carbon.

The reactions on the anode surface and within the anode are of vital interest. Presumably, the electrolytic generation of elemental fluorine initiates a reaction sequence as shown in Figure 7.13. In addition to the main radical processes shown, radical disproportionation to produce an olefin and a saturated molecule is observed. For example, with ethane feed the products include ethylene and vinyl fluoride and dimer.

The porous carbon anode-reactor is unique to the Phillips ECF Process. One reactant, the feed, is introduced into the bottom of the reactor and flows up the reactor. It passes up through the partially flooded reticulated network of pores until it escapes from the anode above the electrolyte. It does not bubble out into the electrolyte.

The other reactant, fluorine, is generated along the lateral anode surfaces, enters the pore network, and reacts with the feed inside the anode. The porous carbon and contained electrolyte appear to moderate the reaction very efficiently. The heat of reaction, about 400 kJ per gram-equivalent of hydrogen replaced with fluorine, is rejected to the carbon and thence to the bulk electrolyte and temperature control system.

Figure 7.12 is an idealized sketch of the situation in the anode near the

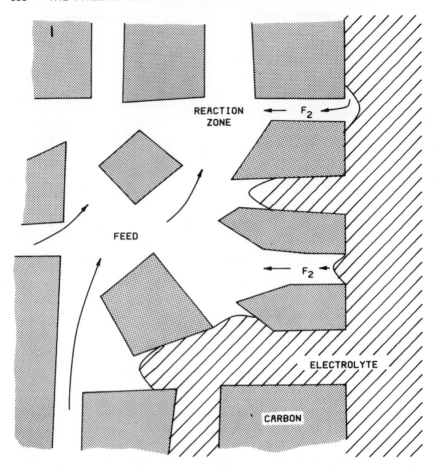

Fig. 7.12.   Electrolyte vapor/carbon normal relationship.

surface. Generally, fluorine generated on the anode surface enters the pore network near where it was generated. If the fluorine cannot enter the anode-reactor as fast as it is generated it forms lens-shaped bubbles on the anode surface. These bubbles may react with hydrogen below the electrolyte level, with the principal effect being a decrease in the current efficiency; or they may react above the electrolyte level in unmoderated, small-scale explosions. Enough of these little explosions can destroy anodes, feed tubes, and cells and degrade products.

This description of the Phillips ECF Process implies several operating constraints.

Main radical processes

$$CH_3-CH_3 + F_2 \rightarrow CH_3-CH_2\cdot + F\cdot + HF$$

$$CH_3-CH_2\cdot + F\cdot \rightarrow CH_3-CH_2F$$

Disproportionation of radicals

$$2 \ CH_3-CH_2\cdot \rightarrow CH_2=CH_2 + CH_3-CH_3$$

$$2 \ CH_2Cl-CHCl\cdot \rightarrow CH=CHCl + CH_2Cl-CHCl_2$$

Production of dimers

$$CH_3-CH_2\cdot + CH_2=CH_2 \rightarrow CH_3-CH_2-CH_2-CH_2\cdot$$

$$CH_3-CH_2-CH_2-CH_2\cdot + F\cdot \rightarrow CH_3-CH_2-CH_2-CH_2F$$

Fig. 7.13. Radical processes. Note that 1-fluorobutane butane, was detected as a product from ethane but butane was not.

1. The rate that feed can enter the anode is limited. An excessive feed rate results in gross bubbling of feed around the bottom of the anode. This is not serious if the object is to convert a hydrocarbon into a perfluorocarbon. Proper selection of the porous carbon usually permits adequate feed rates. The preferred carbons are 40 to 50% porous with a narrow pore size distribution and an average pore size of 10 to 60 μm. When carbonyl fluoride is produced from carbon monoxide, the ratio of feed to current is high and the ability of the anode to take feed is a limit. Typical laboratory anodes, 30-cm immersion and 3.5-cm diameter, operate at 90 A, 3.00 kA/m^2 and accept about 50 l/h of vaporized feed.

2. The feed flow up the porous anode-reactor is limited. This is similar to the first constraint. It is serious only in the case of plugging of the pore network. Conceptually, the anode could be fed and products removed at several locations.

3. The entrance of fluorine into the anode may be limited. This constraint is useful in rationalizing a number of phenomena, including anode polarization and depolarization, anode burning at the electrolyte level, and "frying." This constraint sets a practical limit on usable current density.

This model gives a satisfactory explanation of polarization. When an anode is approaching polarization, the passage of fluorine into the anode is restricted by the plugging of pores as shown in Fig. 7.14. A gas film or lens-

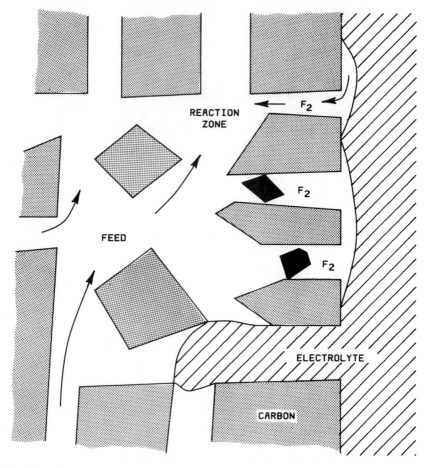

Fig. 7.14. Electrolyte vapor/carbon relationship near polarization. Note plugs and increased contact angle.

shaped bubble builds up on the surface and blocks it off to increase the electrical resistance of the cell. This causes the cell to operate at an increased and erratic voltage. The voltage is erratic because of the growth and collapse of these bubbles. If the bubble escapes it causes an audible frying noise and flashes of light when it breaks into the vapor space and reacts violently with hydrogen or cell products. Initially, this involves a small part of the anode, but with time the entire anode will be covered with a film, the cell becomes polarized, and the current cannot be maintained at twice the normal voltage.

The cell is depolarized by allowing the voltage to increase. At 2.00 kA/m^2

it increases from 7 to 10 V to 60 to 80 V. This creates and stabilizes a gas film over the entire anode by a Leidenfrost effect.* The film is supported by heat from arcs across the anode/electrolyte gaps. These arcs are easily seen and can be detected in the megahertz range. The high power of these arcs, about 150 kW/m^2, strips off the outer layer of the anode and unplugs the pores to permit freer passage of fluorine into the anode.

The exact nature of the plugs is speculative. Considerable machining debris is present on a new anode. With time, accumulation of reaction debris and corrosion products could cause plugging.

The rate at which fluorine can enter the porous carbon anode-reactor thus appears to place a limit on the effective current density. If the fluorine cannot get into the anode to make useful products it does little good to make more fluorine by running the anode at a higher current density. Continued operation at these higher currents, where frying and explosions are evident, can result in the destruction of an anode. Damage in such a situation is particularly severe about the electrolyte level.

4.   A slow reaction rate of the feed with elemental fluorine has been proposed to explain the observed limited current density with some feeds, such as acetyl fluoride (~ 1.50 kA/m^2 compared to > 4.0 kA/m^2 with ethane). A more likely explanation is that the wetting characteristics of the carbon/electrolyte vapor systems are different with acetyl fluoride vapor and this limits the flow of fluorine into the interior of the anode-reactor.

5.   Heat removal from the system is a significant constraint. Heat removal from the anode must be sufficient to keep the carbon cool enough to minimize product degradation and carbon burning. The temperature differential between the electrolyte and the cooling system is limited because the electrolyte freezes on the cooled walls when the coolant is below about 60°C. It is impractical to design a system that can maintain the temperature during depolarization; the temperature is simply allowed to rise during this operation.

Efforts to fluorinate high-boiling feedstocks such as n-heptane are subject to constraint number 2. With time, the anode pores fill with high boiling, nonvolatile materials, principally lightly fluorinated dimers. (Ethane produces about 4% dimers, and heptane is thought to produce a like amount.) When the pores are filled, feed bypasses the reactor and some of the fluorine is not used up until it reacts noisily with the bypassed feed and hydrogen in the vapor space. The result is burning of the anode at the electrolyte level and complete anode failure.

---

*A phenomenon describing the maintainence of a vapor film between a solid and liquid by heat transfer into the surface of the liquid being vaporized.

This behavior suggests that operation at higher per pass conversion would be advantageous. It is well known that partially fluorinated hydrocarbons often boil at much higher temperatures than the corresponding hydrocarbon or perfluorocarbon; for example, ethane, $-88°C$; hexafluoroethane, $-78°C$; 1,2-difluoroethane, $+31°C$. The high per pass conversion would minimize the concentration and the accumulation of high-boiling intermediates. It had been assumed that current-to-feed ratios approaching 100% conversion, that is 2F per hydrogen equivalent, should be avoided. Poor mixing in the anode would result in the accumulation of pockets of elemental fluorine with the same expected undesirable results.

Operating procedures limited per pass conversion to about 50% of the feed hydrogen content. This limit was supported by experience and by a variety of incidents such as small explosions when feed rates dropped. Experiments designed to define a clear limit as the fluorine/feed ratio gave some surprising results.

A simple experiment was performed. A laboratory ECF cell was set to running smoothly with ethane at 50% per pass hydrogen conversion and 2.00 kA/m^2 current density. A geophone was mounted on the current collector and monitored remotely with an oscilloscope and loudspeaker. The feed rate was dropped to give 120% per pass conversion of hydrogen. Nothing serious happened; in fact, it was difficult to tell that anything unusual was occurring. The voltage rose slightly and the copper current collector warmed, but there were no explosions or fires. Examination of the anode showed no evident damage. Subsequently, the conversion was varied smoothly from very low to very high with no problems; the cell even operated smoothly with no feed.

Apparently, the fluorine reacted smoothly with hydrogen from the cathode within the pores of the anode above the electrolyte. This warmed the current collector and caused corrosion at the copper/carbon joint. Feeding additional hydrogen to the top 2 cm of the anode through a hollow current collector prevented this corrosion. The Teflon feed tubes were occasionally penetrated at the electrolyte level. This was eliminated by running the feed tube down through a hollow current collector and on down through the anode. Such anodes were run for several hundred hours at currents corresponding to 150 to 200% per pass conversion with ethane feed. The only significant byproduct was about 4% perfluorobutane with a barely detectable amount of perfluoromethane.

New insights have been gained that permit a detailed explanation of the Phillips ECF process. These insights are based on the postulate that the fluorinating agent is elemental fluorine. This fluorine is generated at the anode surface and readily enters the porous carbon anode, which is not wetted by the electrolyte. The fluorine may move some distance within the anode-reactor before it reacts with the feed. This reaction is considerably more efficient than most direct fluorination reactions because the anode-reactor efficiently sinks the high energy from the reaction and rejects it to the bulk electrolyte.

These new insights led to the development of a mode of operation that permits complete one-pass fluorination. This mode of operation with excess fluorine has value in at least three cases:

1. Production of a desired perfluoro compound on a small scale. This simplifies recovery and purification, as the penultimate product and the feed are frequently difficult to separate from the ultimate product.

2. Fluorination of materials that yield high boiling intermediates. This mode of operation keeps the concentration and accumulation of these products low.

3. Fluorination of thermally sensitive materials. Isopropyl trifluoroacetate and the partially fluorinated esters tend to cleave to olefins and trifluoroacetic acid in the cell and in fractionation for recycle. This technology permits complete one-pass conversion to the perfluoroester, which is easily converted to hexafluoroacetone, and trifluoroacetyl fluoride [17].

Practical Considerations

Hydrogen Fluoride Handling

Hydrogen fluoride is dangerous. A seemingly trivial contact with it may not appear to be serious, especially if it is washed off rapidly, but can lead to an excruciatingly painful blister if prompt medical treatment is not obtained. Before any work is done with HF, the manufacturer's literature should be carefully perused and adequate medical resources should be established.

Figure 7.15 shows schematically the HF handling system used with the unattended operation facility. The portion enclosed by dashed lines was heated to 37°C. A thermostat was used and the heater was sized to prevent excursions above 40°C. Piping and tubing were of mild steel. The valves had stainless steel bellows seals with Kel-F seats except for the throttle valve, which had a V-stem. The backup and control valves were selected to prevent HF leakage into the control air system in case of seal failure and to fail in a closed position upon air loss.

The electrolyte level probes used to control the addition of HF to the cell were $3/16$-in. mild steel rods with blunt ends. They were the cathodes in an electrochemical cell with the case serving as the anode. The power supply delivered about 120 V DC through the coil of a sensitive relay and a limiting resistor that permitted about 5 mA of current to pass. The case of the cell was grounded and the connections to the probes were insulated to prevent accidental shocks.

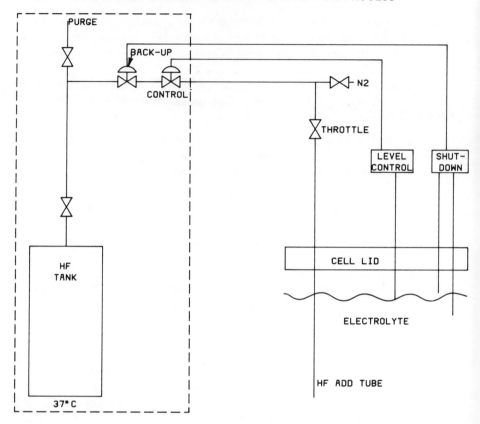

Fig. 7.15.   HF control system.

Electrolyte Preparation

Preparation of the KF·2HF electrolyte exposes the operator to considerable hazard. Adequate protective equipment should be used and each operation should be carefully thought through. Manufacturer's literature should be consulted for details about hazards and needed protection.

The electrolyte is best prepared from technical grade potassium bifluoride, KF·HF. This is available from some supply houses (Matheson Coleman & Bell PX 1350) and in bulk from various manufacturers. Preparation from anhydrous potassium fluoride, KF, is not recommended, as addition of HF to KF results in an extremely vigorous reaction that flings reactants around.

Preparations should be made for electrolyte analysis before the electrolyte preparation begins. Needed are standard acid—1 *m* HCl; standard base—

1 mequiv/g NaOH—and this should be checked occasionally; phenolphthalein solution; several clear Teflon FEP bottles with caps; a small Teflon dipper; and a 50-ml buret. The electrolyte is analyzed by adding excess base to a weighed sample and back-titrating with acid. A 5-g sample is removed using a heated dipper and weighed to about 10 mg in a capped Teflon bottle. This contains about 105 mequiv of HF. About 10% excess of standard base is added; and the mixture is titrated to the phenolphthalein end point with standard acid. Single analyses are done during preliminary adjustments and triplicate determinations are made to check the final composition. Comparison of this titration procedure with neutron activation analyses for total fluorine gave excellent checks. Many, if not most, instances of unstable cell operations can be traced to changes in the electrolyte composition. Compositions of 41 to 42 wt % HF usually give satisfactory operation; 41.7 wt % is the stated goal when fresh electrolyte is prepared. The biggest source of drift in the electrolyte composition is the removal of electrolyte with an anode that is being replaced. Generally this loss is replaced with $KF \cdot HF$ or $KF \cdot 2HF$. After a cell has been in operation for an extended period material from cell corrosion accumulates as potassium fluoroferrate sludge and suspended solid. This gives a precipitate when the base is added, but it apparently does not interfere with the analysis. Occasionally some of this sludge must be removed to maintain circulation around the loop. This can be done by allowing it to settle and drawing it off into a Teflon bottle.

The electrolyte is prepared by weighing out sufficient $KF \cdot HF$ to fill the cell to somewhat above the operating level with $KF \cdot 2HF$ when combined with HF. The density of the electrolyte is about 1.90 g/ml at 100°C [16]. The cell is loosely filled with a portion of the $KF \cdot HF$ and HF is added from the HF system (see preceding section). A mild steel tube is used to introduce HF near the bottom of the cell. A Teflon FEP bellows tube is used to connect the tube to the HF system. A nitrogen bleed into the HF system is essential to prevent suck back. After a substantial liquid volume is established the system should be frequently and carefully stirred. The balance of the $KF \cdot HF$ is added in installments as the volume works down.

After the electrolyte composition is adjusted to the desired value, lithium fluoride is added to give about 0.5 wt % LiF. Less than 0.1 wt % LiF is soluble, but polarization/depolarization procedures work better at this level.

The cell is then assembled and the electrolyte level is adjusted to the control level by drawing out excess electrolyte. This is conveniently done with a 1-liter Teflon FEP bottle equipped with a rubber stopper and a length of ¼-in. Teflon FEP tubing. The excess electrolyte is poured into a plastic ice cube tray and allowed to freeze. These cubes are used for makeup when anodes are removed. They are hygroscopic and should be protected from moisture. They also have a significant HF pressure and unless care is taken, this will corrode nearby items. Storage behind two layers of polyethylene is recommended.

Cell Start-up

After the anode is installed and the electrolyte level is adjusted the cell should be started within a few minutes to minimize electrolyte invasion. Typically ethane feed is started at 20 to 30 liters/hr. Before the current is turned on the ethane can be heard "bubbling" around the anode, but when the current is turned on to about 10 A this noise stops. Clearly the anode has been rendered nonwetting by the action of the current. The current is increased in steps to 60 A over about an hour. The cell is allowed to run-in for a day or so to dry the electrolyte and get through the first polarization/depolarization cycle. After the run-in period the electrolyte composition is checked and adjusted if needed.

Ethane is the routine feed of choice for several reasons. It runs well, the product is gaseous and can be analyzed easily, and it is readily available. Another good feed is 1,1-difluoroethane. It gives a simpler product mix, but it is considerably more expensive.

Product Collection

Figure 7.16 is a sketch (not to scale) showing the essential details of a heated induction assembly for collecting ECF products at liquid nitrogen temperatures. It is designed to screw loosely into a stainless steel cylinder, which is then immersed in liquid nitrogen. This assembly is needed to prevent the plugging of the induction tube with frozen products. The heater is made from Nichrome wire. Below the liquid nitrogen level it has a resistance of 2.6 Ω/m and above this level it is doubled and twisted and has a resistance of 1.3 Ω/m. The wire is heated by passing 2 to 4 A of current from a controlled current power supply isolated from the mains and grounded. The Teflon T insulates the ends of the wire from each other. The top of the wire is passed out through a hole in the stainless tubing as shown and then silver soldered down in a loop. The lower end of this stainless tubing is reduced slightly to make a snug fit with a flared piece of Teflon tubing that extends down into the collection cylinder as shown. Another piece of stainless tubing with a gas vent and threaded bushing is attached to the bottom of the Teflon T. This tubing guides the Teflon tubing and provides a means to complete the electrical circuit. The heating wire is silver soldered to the bottom of this tubing. If it is kept clean and dry between uses this assembly will last for a long time. Corrosion of the silver solder joint is a problem if moisture gets to it.

Some of the products are extremely volatile, so considerable care should be exercised not to overpressure the equipment. It is a good idea to collect the product with liquid nitrogen, remove the induction assembly, attach a valve

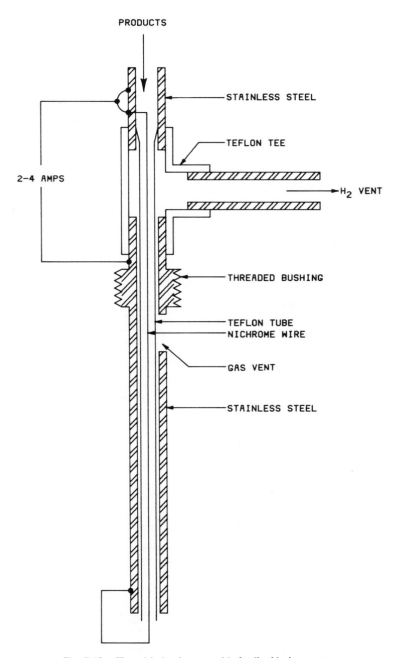

PRODUCTS

STAINLESS STEEL

TEFLON TEE

2-4 AMPS

H₂ VENT

THREADED BUSHING

TEFLON TUBE
NICHROME WIRE

GAS VENT

STAINLESS STEEL

Fig. 7.16. Heated induction assembly for liquid nitrogen trap.

367

and pressure gauge assembly, and allow the products to warm slowly in a hood behind a shield. If signs of overpressure are noted, either vent the apparatus or cool it down again. *Caution.* These products are almost certainly toxic.

Figure 7.17 is a sketch of the temperature control system. It provides tempered cooling, > 60°C, during cell operation to maintain a cell temperature of 95 to 100°C. On standby, the circulating silicone fluid is heated to 90°C to prevent electrolyte freeze-up. Tempered cooling is necessary to prevent electrolyte from freezing on the cell wall and blocking the electrolysis current path.

4 PROCESS APPLICATIONS

Results from a limited selection of applications are used to demonstrate the breadth of application of the ECF process and to provide insight into the operation of the process. Typically the current efficiency to products was near 100%.

Methane, Ethane, Ethylene, and 1,1-Difluoroethane

Most of the laboratory work was done with this set of feedstocks. They are readily available and the products can be collected as gases and analyzed with reasonable ease and accuracy.

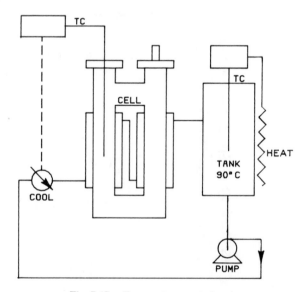

Fig. 7.17. Temperature control system.

Methane can be fluorinated easily and efficiently to produce the expected fluoromethanes and a trace of fluoroethanes. The production of fluoroethanes could not be accounted for by the amount of ethane in the methane used.

Ethane was used extensively. It fluorinated smoothly, gave interesting products, and the results were instructive about the mechanisms of the ECF process. The "mainstream" products were produced in roughly statistical yields. The expected yields were calculated from second-order kinetics using statistically weighted rate constants. The highly fluorinated products were produced in higher than expected yields. Operation at lower current densities and operation with helium dilution of the feed pushed the product composition toward the expected composition. Products not on the main stream included vinyl fluoride, ethylene, 1-fluorobutane and its derivatives, and a trace of fluoromethanes. The major route to fluorobutanes appears to be radical disproportionation to produce olefins. An ethyl radical then adds to the olefin followed by a fluorine radical cap. This is shown in Fig. 7.13.

Radical disproportionation also explains the production of vinyl fluoride. Presumably other olefins besides vinyl fluoride and ethylene were produced but not detected. The ultimate yield at complete conversion was 96% C_2F_6, 4% C_4F_{10}, and a trace of CF_4.

Ethylene fluorinated fairly well. The polarization frequency was about twice that observed with ethane. At very low conversion the product mix was dominated by 1,4-difluorobutane and its derivatives. At these low conversions well over half the product was fluorobutanes, 15% vinyl fluoride, 5% ethyl fluoride, 1% 1,1-difluoroethane; the balance was 1,2-difluoroethane and its derivatives. Passage of ethylene through the system without current gave no products. At conversions higher than 10% the products resulting from the addition of fluorine to ethylene were dominant, but 10 to 15% of the products were still butanes.

1,1-Difluoroethane was a nearly ideal feed from an operational standpoint and from the information it yielded. It fluorinated smoothly, the polarization frequency was low, and the expected products were observed. The only significant by-product was about 2% fluorobutanes. In an interesting study the ratio of 1,1,2-trifluoroethane to 1,1,1-trifluoroethane was measured at various conversions. Statistically three times as much of the 1,1,2-isomer is expected. At the usual conversions, 25 to 50% of the hydrogen replaced, the amount of 1,1,2-isomer is only two times the amount of 1,1,1-isomer, but at very low conversions this factor approaches 3. This behavior is explained by the high-boiling 1,1,2-isomer (b.p. $+5°C$) being adsorbed more tightly than the 1,1,1-isomer (b.p. $-48°C$) and being exposed to more fluorination as it moves more slowly through the anode.

Propane, Butane, and Isobutane

These fluorinated smoothly and several hundred kilograms of the per-fluoro derivatives were prepared. When isobutane was fluorinated at low conversions the yield of 2-fluoro-2-methylpropane was about 10% higher than statistically expected relative to 1-fluoro-2-methylpropane. This yield ratio can be extrapolated to the expected 1:9 at zero conversion.

Distillation of the propane, butane, or isobutane products from the ECF cell gives a minimum boiling azeotrope of the hydrocarbon and perfluorocarbon. This fraction is the lightest of the major fractions and is taken over and the balance is recycled. The azeotropes can be broken by toluene extraction or by chilling to about $-70°C$ and drawing off the heavy perfluorocarbon phase. This phase is then redistilled to produce a high-purity product. The product contains a trace of highly toxic fluoroolefins, which can be removed by treatment with alcoholic KOH.

1,2-Dichloroethane

Considerable work was done with 1,2-dichloroethane (EDC) feed. It is readily available, it fluorinates smoothly, and nearly all of the products have commercial value. Table 7.1 is a list of the products that were identified in the cell effluent. A large number of products were identified and some others were doubtless missed. About 90% of the products would ultimately go to the desired CF_2Cl-CF_2Cl (114) and 98% would be included if CF_3-CF_2Cl (115), and $CF_2Cl-CFCl_2$ (113) were included as desired products. In addition to its utility as a refrigerant, CF_2Cl-CF_2Cl has potential as a chemical intermediate. It can be hydrodehalogenated to tetrafluoroethylene and cleaved with chlorine to produce CF_2Cl_2, which is widely used as a refrigerant and propellant.

The intermediate chlorofluoroethanes can be recovered and are of value as chemical intermediates. This is an unusual situation. The perfluoro compound has an established volume market and if one of the partially fluori-nated materials had value, the by-product would not have to be sold at a dis-tressed product value or even disposed of at a penalty.

The multiplicity of by-products can be explained as the results of inter-molecular shifts of chlorine by various radical species by a radical dispropor-tionation mechanism.

Tetrafluoroethylene from Ethylene

The dimerization of ethylene with tetrafluoroethylene to produce the cyclo-butane C354,

$$CH_2{=}CH_2 + CF_2{=}CF_2 \longrightarrow \overline{CH_2-CH_2-CF_2-CF_2}$$

Table 7.1. Products from 1,2-Dichloroethane

Fluorocarbon Number	Structural Formula	Amount Present[a]	Ultimate Product If Recycled
14	CF_4	Tr	14
32	CH_2F_2	Tr	14
13	CF_3Cl	Tr	13
12	CF_2Cl_2	Tr	12
116	CF_3-CF_3	Tr	116
125	CF_3-CHF_2	Tr	116
134a	CF_3-CH_2F	Tr	116
134	CHF_2-CHF_2	Tr	116
143	CHF_2-CH_2F	Tr	116
1141	$CH_2=CHF$	Tr	116
115	CF_3-CF_2Cl	Tr	115
1131a	$CH_2=CFCl$	Tr	115
1140	$CH_2=CHCl$	Tr	115
124a	CHF_2-CF_2Cl	Tr	115
142a	$CH_2F-CHFCl$	Minor	115
133a	CF_3-CH_2Cl	Tr	115
133b	CH_2F-CF_2Cl	Tr	115
133	$CHF_2-CHFCl$	Tr	115
150	CH_2Cl-CH_2Cl	Feed	114
114	CF_2Cl-CF_2Cl	Minor	114
123a	$CF_2Cl-CHFCl$	Minor	114
132	$CHFCl-CHFCl$	Minor	114
132b	CF_2Cl-CH_2Cl	Minor	114
141	$CHFCl-CH_2Cl$	Minor	114
132c	$CFCl_2-CH_2F$	Tr	114a
132a	CHF_2-CHCl_2	Tr	114a
113	$CF_2Cl-CFCl_2$	Minor	113
122	$CHCl_2-CF_2Cl$	Minor	113
122a	$CHFCl-CFCl_2$	Minor	113
131	$CHCl_2-CHFCl$	Minor	113
131a	$CH_2Cl-CFCl_2$	Minor	113
140	$CH_2Cl-CHCl_2$	Minor	113

[a]The classification of "amount present" was done by the analyst and must be viewed with the knowledge that routine analysis showed that the five components that ultimately go to F-114 make up 90% of the product.

offers a route to tetrafluoroethylene. Fluorination of C354 produces octafluorocyclobutane (C318), which can be pyrolyzed to two molecules of tetrafluoroethylene. These steps were demonstrated on a blocked out basis.

1,1,2-Trichlorotrifluoroethane from 1,1,1-Trichloroethane

One of the fluorochemicals with a respectable market is trifluoroacetic acid. It is manufactured by the isomerization of $CF_2Cl\text{-}CFCl_2$ to $CF_3\text{-}CCl_3$ followed by hydrolysis. It seemed obvious to fluorinate $CCl_3\text{-}CH_3$, which is readily available, reasonably volatile, and readily fluorinated to produce three products. However, the products obtained were not the expected three products. The intermediate radical undergoes a nearly quantitative intramolecular chlorine shift and thus the products were identified as derivatives of $CHCl_2\text{-}CH_2Cl$.

Esters

Shreeve (quoted in Ref. 19) used the reaction

$$
\begin{array}{ccccc}
\text{CF}_3 & \text{O} & & \text{CF}_3 & \text{O} \\
| & \| & & | & \| \\
\text{F-C-O-C-CF}_3 & & \overset{\text{M}^+\text{F}^-}{\rightleftharpoons} & \text{C=O} + \text{F-C-CF}_3 \\
| & & & | & \\
\text{CF}_3 & & & \text{CF}_3 &
\end{array}
$$

and analogous reactions to prepare a variety of perfluoroesters. At low temperatures this equilibrium lies to the left and can be used for preparative purposes. For example, trifluoroacetyl fluoride was dimerized over cesium fluoride at $-180°C$ and the product was identified as perfluoro(ethyl acetate), which could be pumped off and stored at room temperature. Various ionic fluorides catalyzed the reverse reaction at or near room temperature.

Isopropyl trifluoroacetate, isopropyl acetate, ethyl trifluoroacetate, and ethyl acetate were fluorinated with no problems at per pass hydrogen conversions approaching 80%. As expected all these produced a number of products, most of which were fluorinated esters, including the perfluoroester. The perfluoroesters appear to be reasonably stable in stainless steel in contact with HF at room temperature.

At $80°C$ the half-life of heptafluoroisopropyl trifluoroacetate was over 3 hr in contact with HF; the half-life was about 30 min in contact with $KF \cdot 2HF$ at $80°C$. Cleavage was rapid over KF, NaF, and BaF_2.

Cleavage of heptafluoroisopropyl trifluoroacetate offers a route to hexafluoroacetone from isopropyl alcohol. Thirty grams of high-purity hexafluoroacetone as prepared in this manner.

The fluorination of ethyl acetate and ethyl trifluoroacetate offers a route to trifluoroacetyl fluoride and thus trifluoroacetic acid. This route was used, but a simpler process is described in the next section.

Perfluorocarboxylic Acids by Way of Acyl Fluorides

Perfluorocarboxylic acids and their derivatives have a variety of uses, including:

Catalysts for organic reactions.

Derivatizing and protecting agents.

High performance surfactants.

Leveling agents in paints and waxes.

Antisoil agents for fabrics.

High-performance polymers.

Inert fluids.

X-Ray contrast agents.

Artificial blood formulations.

Agricultural chemicals.

These acids are prepared commercially by 3M using the Simons process, and by Halocarbon Corporation from 1,1,2-trichlorotrifluoroethane by isomerization and hydrolysis. The Phillips process is based on the ECF cell and a surprising equilibrium.

The metathetical reaction of acetic acid and trifluoroacetyl fluoride was not surprising, but the value of the equilibrium constant was a surprise.

$$CF_3COF + CH_3COOH \overset{K \geq 100}{\rightleftharpoons} CF_3COOH + CH_3COF$$

The large equilibrium constant is the basis for the process.

The initial charge of acetyl fluoride was made from the reaction of acetic anhydride with HF:

$$(CH_3CO)_2O + HF \rightleftharpoons CH_3COF + CH_3COOH$$

The equilibrium lies far enough to the right to permit distillation in glass equipment with little etching. This reaction is general for the production of acyl fluorides, but with the higher analogs distillation in glass results in appreciable etching.

Acetyl fluoride fluorinated reasonably well. Per pass hydrogen conversions of 50% were easily obtained with current densities of 1 to 3 kA/m^2. More permeable carbons permitted higher current densities before the onset of noise and destructive reactions. About 95% of the reacted feed retained the acyl fluoride structure; that is, CH_2FCOF, CHF_2COF, and CF_3COF. The remainder produced CF_4, COF_2, and C_2F_6. At 80°C the current efficiency to products was 90%, and at 110°C it was 80%.

Similar results were obtained with propionyl fluoride and butyryl fluoride, but detailed measurements were not made because of the complexity of the product mixture.

Trifluoroacetyl fluoride was recovered by cryogenic distillation and was reacted with acetic acid according to the metathesis step shown above. The recovered trifluoroacetyl fluoride contained some HF so a little acetic anhydride was added to convert it to acetyl fluoride. A slight (about 10%) excess of acetic acid was used. When excess acetic anhydride was used, tar was formed along with 1,1,1-trifluoroacetone. This ketone can be made in a better, but still modest, yield from acetic anhydride and the anhydride of trifluoroacetic acid [18].

The product from the metathesis step was worked up by taking the acetyl fluoride overhead in a nonequilibrium distillation. (In an equilibrium distillation the very volatile CF_3COF would be the overhead product.) Some CF_3COF was recovered. High-purity CF_3COOH was easily recovered. The metathesis, fluorination, and nonequilibrium distillation worked well with both acetic and butyric acids, and several hundred kilograms of the perfluoro acids were prepared and sold. Figure 7.18 summarizes the process.

Perfluorohexanes from Hexafluoropropene

Under electrochemical fluorination conditions ethylene produced considerable dimer (see above). More dimer was produced under low-conversion and low-current-density conditions. This nuisance was noted and was turned into an excellent synthesis for perfluorohexanes (and perfluoropropane). Hexafluoropropene is sold commercially and it can be handled with reasonable precautions. Hexafluoropropene is not inclined to polymerize explosively as does tetrafluoroethylene.

Hexafluoropropene was fluorinated in the loop cell (Fig. 7.7). The feed rate was 2 moles/hr. The current level was 53.6 A or 2 F/hr. The run time was 54.5 hr. The product was collected in a 5-gal/cylinder chilled in a Dry Ice/Freon 11 bath.

The product efficiency to perfluoropropane and perfluorohexanes was 98%. About 1% of the product was monohydroperfluorohexanes and 1% was perfluorononanes. The current efficiency was about 86%. Sixty-three percent of the charged hexafluoropropene was converted to perfluoropropane and perfluorohexanes.

The perfluorohexanes were shown by F^{19} NMR to have the following compositions:

Compound	Mole %
$CF_3-CF_2-CF_2-CF_2CF_2-CF_3$	10
$CF_3-CF(CF_3)-CF_2-CF_2-CF_3$	40
$CF_3-CF(CF_3)-CF(CF_3)CF_3$	50

Lower current densities and lower conversions gave higher yields of perfluorohexanes relative to perfluoropropane.

5 PILOT-PLANT AND COMMERCIAL-SCALE OPERATIONS

The 1000- and 4000-A pilot-plant designs were based on a conceptual plant design study done for Phillips by R. B. MacMullin Associates [19]. These studies resulted in a commercial cell design with 36 anodes in parallel and a nominal rating of 36,000 A. These are connected in series to permit the use of higher voltages. The cell body was a callandria or tube and shell heat exchanger with anodes occupying about half of the tubes and the other tubes serving as downcomers for electrolyte circulation driven by the gas lift from the cathode [20]. The design case anodes were 20 cm in diameter and 60 cm long. This size was based on existing fabrication facilities for porous carbon cylinders.

A major problem in the operation of anodes of this size is the voltage loss from top to bottom because of the electrical resistance of the porous carbon matrix. The current density at the top of a long anode is significantly higher than that at the bottom. This results in higher power costs and lower production than would be the case for uniform current distribution. Improved designs provided better current paths [13, 21]. Figure 7.19 is based on these references. The principal feature is the use of a dense carbon core to protect the copper current collector. Without the dense carbon core, electrolyte will eventually penetrate to the current collector, and cause corrosion and the corrosion products will split the anode.

The cell shown in Figure 7.7 was actually designed by scaling down the pilot-plant and commercial concepts [22]. The basic design has the anode in one leg of a gas lift heat exchanger with another leg to serve as a return path for the electrolyte. In the larger cells the cross-sectional area for the electrolyte return was matched to that between the anode and cathode. This return area was generally distributed fairly evenly around the anodes. Figure 7.20 is a section of a single-anode pilot-plant cell. A photograph of this cell is shown in Fig. 7.21. The velocity of the electrolyte in such a cell operating at 1000 A was measured at 30 cm/sec in a downcomer (one of four) and 45 cm/sec in the anode/cathode gap [19].

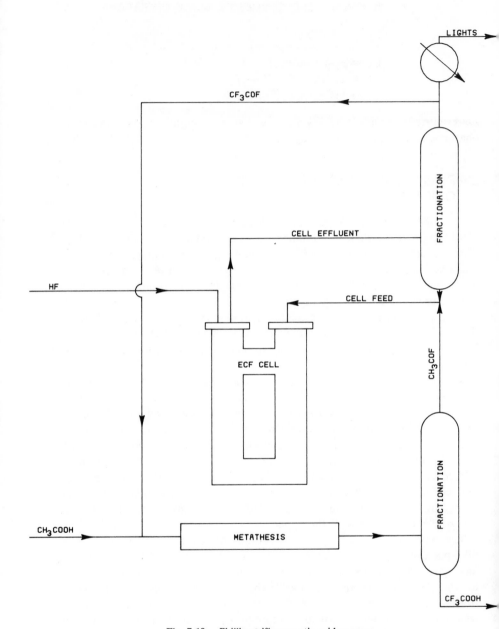

Fig. 7.18. Phillips trifluoroacetic acid process.

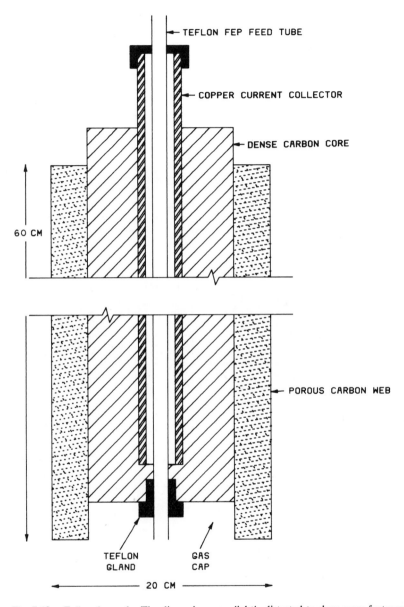

Fig. 7.19. Full-scale anode. The dimensions are slightly distorted to show some features.

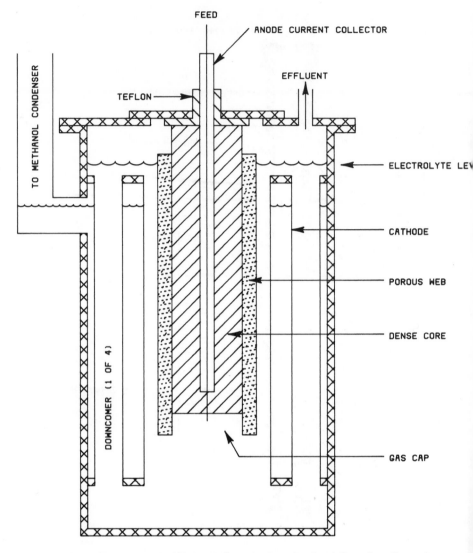

Fig. 7.20. Sketch of 1000-A cell. Some details omitted and dimensions distorted.

Fig. 7.21. 1000-A cell and controls.

The lab cell was cooled with circulating silicone fluid. This was satisfactory despite the poor heat transfer properties of the fluid because of the relatively large heat exchange area. The pilot-plant cells had relatively less heat exchange area and were cooled with boiling, pressurized methanol, which was condensed in a conventional tube and shell heat exchanger with cooling tower water. Hydrogen accumulation on the methanol side caused pressure buildup that had to be monitored and vented occasionally. This pressure apparently was due to hydrogen migration through the steel from the cathode. Some blistering of the cathode was noted from hydrogen embrittlement.

Downstream processing on the pilot-plant scale was a significant problem, particularly in closed loop recycle situations. The cell effluent ranged from high-boiling partially fluorinated products to the perfluoro products to CF_4 and hydrogen. The procedures used varied with the feed and the desired product. If the desired product was partially fluorinated, it was generally necessary to operate on a blocked out basis because of the complexity of the purification process.

Generally the desired product was the perfluoro compound and it, or an azeotrope of it, as the most volatile condensable; thus the cell and initial purification stages were operated with closed loop recycle. The cell effluent was cooled with an atmospheric pressure knockout to remove the highest boiling compounds. The liquid from this stage was pumped directly to the cell feed tank; sometimes a de-oiler was used. The vapor from the atmospheric knockout was then compressed in two stages with interstage cooling, and the compressed product was subjected to cryogenic distillation under pressure. The hydrogen and lights that escaped through the cryogenic condenser passed through a liquid nitrogen cooled trap, an aqueous potassium hydroxide scrubbing tower, and anything that was left was flared. Figure 7.22 is a sketch of a representative ECF closed loop operation.

The overhead product from this distillation column was then subjected to further processing and purification depending on its nature and end use. Despite its relatively high boiling point, hydrogen fluoride showed up nearly everywhere in the processing system. Provision must be made to remove it at some stage in nearly every process.

The pilot plant was used to prepare substantial quantities of a variety of products in yields ranging from excellent to fair. These included the following:

Perfluoropropane

Perfluorobutane

Perfluoroisobutane

Octafluorocyclobutane

1,1,2,-Tetrafluorocyclobutane

Tetrafluoroethylene

Trifluoroacetyl fluoride

Trifluoroacetic acid

Heptafluorobutyryl fluoride

Heptafluorobutyric acid

Heptafluorobutanol

Perfluorodimethyl ether

Bisdifluoromethyl ether

Various chlorofluoroethanes

Hexafluoroacetone

Perfluorohexanes

Obviously, the preparation and purification of some of these products was not straightforward.

6 SUMMARY

The Phillips Electrochemical Fluorination Process is a simple and efficient means of introducing various amounts of fluorine directly into a variety of

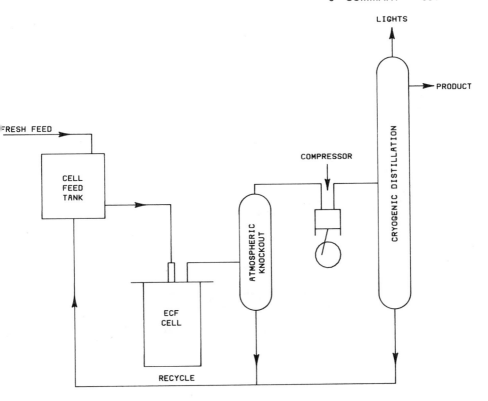

Fig. 7.22. Sketch of representative ECF closed-loop operation.

organic molecules. The process has been used to prepare partially or completely fluorinated alkanes, chloroalkanes, cyclic alkanes, acids, ethers, esters, and others. Details are given on the development and use of the process with laboratory-scale apparatus. Pilot-plant design and operation are discussed, but the complexity of the operation and variety of applications make it necessary to do so at an abstract level.

The process has been satisfactorily rationalized in considerable detail from the observation that the electrolyte does not wet the porous carbon anode and the postulate that fluorination takes place when the feed reacts with elemental fluorine within the confines of the porous carbon anode.

Acknowledgments

Special thanks are due to B. H. Ashe (deceased), H. M. Fox, P. S. Hudson, R. B. MacMullin, and F. N. Ruehlen, who have been deeply involved in and made significant contributions to the development of this process.

Thanks are also due to Phillips Petroleum Company for the support of this work and for permission to publish it.

Appendix: Issued U.S. Patents Related to ECF Process

3,461,049, W. V. Childs, Electrochemical Production of Oxygen Difluoride.

3,461,050, W. V. Childs, Production of Carbonyl Fluoride.

3,511,760, H. M. Fox and F. N. Ruehlen, Electrochemical Fluorination of Organic Compounds.

3,511,761, W. V. Childs and F. N. Ruehlen, Electrochemical Fluorination of Organic Compounds.

3,511,762, W. V. Childs, Electrochemical Conversion.

3,551,307, E. H. Gray, Temperature Control in Electrochemical Conversion Processes.

3,558,449, W. V. Childs, Process for Electrochemical Conversion.

3,558,450, B. H. Ashe, Jr., and W. V. Childs, Process for Electrochemical Conversion.

3,582,482, L. W. Pollock, Temperature Control in Electrochemical Fluorination Processes.

3,616,315, W. V. Childs, Inducing and/or Eliminating Polarization in One Cell of a Series of Cells in Operation.

3,616,336, W. V. Childs and F. N. Ruehlen, Method of Conditioning Anodes.

3,616,353, W. V. Childs and F. N. Ruehlen, Maintaining Electrolytic Cell in Standby Condition.

3,617,453, W. V. Childs, Temperature Control in Electrochemical Conversion Process.

3,620,941, F. N. Ruehlen, Electrochemical Fluorination of 1,2-Dichloroethane and 1,1,2-Trichloroethane.

3,630,880, M. B. Howard, Current Collector and Electrode Assembly.

3,645,863, W. S. Stewart and J. W. Vanderveen, Electrochemical Fluorination.

3,650,917, F. N. Ruehlen, Recovery of Products from Electrochemical Fluorination.

3,655,525, W. V. Childs, Sludge Removal from Electrochemical Cell.

3,655,535, F. N. Ruehlen and H. M. Fox, Multi-Porosity Electrode for Electrochemical Conversion.

3,655,548, W. V. Childs, Treated Porous Anode for Electrochemical Fluorination.

3,657,100, W. V. Childs, Current-Carrying Sparger for Introducing Feed to Porous Electrode.

3,657,101, H. M. Fox, F. N. Ruehlen, and K. A. Williams, Sparger for Introducing Feed Adjacent to Bottom of Porous Electrode.

3,658,685, W. V. Childs and F. N. Ruehlen, Combination Electrode.

3,659,001, K. L. Mills, Electrolytic Cell.

3,660,254, R. O. Dunn, Recovery of Products from Electrochemical Fluorination.

3,660,255, H. M. Fox, F. N. Ruehlen, and K. A. Williams, Process for Electrochemical Conversion.

3,662,009, W. M. Hutchinson, Preparation of Unsaturated Fluorocompounds.

3,663,380, K. L. Mills, Electrodes for Electrolytic Conversion.

3,676,324, K. L. Mills, Composite Carbon Electrode Having Improved Electrical Conductivity.

3,686,082, F. N. Ruehlen, Process for Recovery and Separation of Perhalogenated Fluorocarbons.

3,691,051, K. A. Williams, Porous Electrode Having Cavity with Impervious Dome.

3,692,660, R. B. MacMullin, H. M. Fox, F. N. Ruehlen, and W. V. Childs, Electrolytic Cell.

3,707,457, L. W. Pollock, Apparatus for Controlling the Temperature of the Electrolyte in an Electrolysis Cell.

3,708,416, F. N. Ruehlen and H. M. Fox, Multiporosity Electrode for Electrochemical Conversion.

3,709,800,	H. M. Fox, Process for Preparing Perfluorocarbon Compounds.
3,711,396,	B. H. Ashe, Jr., and H. M. Fox, Porous Electrode Having Open Feed Cavity.
3,720,597,	R. H. Ashe, Jr., and W. V. Childs, Multiporosity Electrode for Electrochemical Conversion.
3,721,619,	F. N. Ruehlen, Electrolytic Cell.
3,728,233,	B. H. Ashe, Jr., and H. M. Fox, Porous Electrode Having Open Feed Cavity.
3,730,858,	K. A. Williams, Porous Electrode Having Cavity with Impervious Dome.
3,772,201,	K. L. Mills, Electrode for Electrolytic Conversion Cells Including Passage Means in the Electrode for Electrolyte Flow Through the Electrode.
3,798,144,	W. V. Childs, Method for Separating Liquid.
3,806,432,	K. L. Mills, Method for Electrolytically Fluorinating Organic Compounds.
3,840,445,	R. A. Paul and M. B. Howard, Two-Stage Electrochemical Octafluoropropane Production
3,853,737,	W. V. Childs, Shallow-Bed Electrochemical Cell.
3,882,001,	K. L. Mills, Method for Electrochemically Forming Fluorocarbon Compounds.
3,896,022,	W. V. Childs, Apparatus for Separating Liquids.
3,900,372,	W. V. Childs, B. H. Ashe, Jr., and P. S. Hudson, Recycle of Acyl Fluoride and Electrochemical Fluorination of Esters.
3,981,783,	W. V. Childs and H. C. Walters, Electrochemical Fluorination Process Utilizing Excess Current and Hydrogen Addition.
3,983,015,	W. V. Childs, Electrochemical Fluorination Using Excess Current.
3,990,988,	W. V. Childs, Constant Boiling Admixture.
3,990,989,	G. Bjornson, Azeotropic Composition Containing Fluorotrichloromethane and 1,1,1,3,3,3-Hexafluoroisopropyl Trifluoroacetate.
4,003,807,	W. V. Childs and F. N. Ruehlen, Electrochemical Fluorination of Ketones within the Pores of an Anode.
4,022,824,	W. V. Childs, Perfluorocarboxylic Acids from Carboxylic Acids and Perfluorocarboxylic Acid Fluorides.
4,026,930,	G. Bjornson, Azeotropic Distillation of 1,1,1,3,3,3-Hexafluoroisopropyl Trifluoroacetate with Fluorotrichloromethane.
4,029,552,	G. B. Fozzard, Process for Obtaining High Purity Perfluoro-n-Heptane.
4,035,250,	H. C. Walters and W. V. Childs, Production of Perfluoro-n-Heptane.
4,038,310,	G. Bjornson and H. C. Walters, Acid Anhydride to React with Impurities in the Production of Perfluorocarboxylic Acids.
4,059,633,	W. V. Childs, Recovery of Hexafluoroacetone from a Hexafluoroacetone–HF Complex.
4,062,795,	W. M. Hutchinson, Azeotropes of 1,2-Dichloro-1,1,2-Trifluoroethane.

References

1. H. M. Fox, F. N. Ruehlen, and W. V. Childs, *J. Electrochem. Soc.*, **118**, 1246 (1971); see also U.S. Patents 3,511,760, 3,511,761, 3,511,762, 3,728,233, and 3,983,015.

2. R. B. MacMullin, *J. Electrochem. Soc.*, **123**, 359c (1976); see also R. B. MacMullin, K. L. Mills, and F. N. Ruehlen, *J. Electrochem. Soc.*, **118**, 1582 (1971).

3. As described by A. J. Rudge, in *Industrial Electrochemistry*, A. T. Kuhn, Ed., Elsevier, New York, 1971, Chap. 2; J. H. Simons, Electrochemical Process of Making Fluorine-Containing Carbon-Compounds, U.S. Patent 2,519,983 (1950).

4. K. J. Radimer, Process for the Production of Fluorine-Containing Compounds, U. S. Patent 2,841,544 (1958).

5. P. E. Ashley and K. J. Radimer, Fluorination Process, U.S. Patent 3,298,940 (1967).

6. R. N. Adams, *Electrochemistry at Solid Electrodes*, Dekker, New York, 1969 p. 280.

7. A. J. Rudge, *The Manufacture and Use of Fluorine and Its Compounds*, Oxford University Press, 1962; also see *Chem. Ind. (Lond.)* **1949**, 247 and Ref. 11.

8. F. N. Ruehlen, G. B. Wills, and H. M. Fox, *J. Electrochem. Soc.*, **3**, 1107 (1964).

9. G. H. Cady, *J. Am. Chem. Soc.*, **56**, 1431 (1934).

10.(a) C. D. Hodgman, Ed., *Handbook of Chemistry and Physics*, 38th ed, Chemical Rubber Publishing Co., Cleveland, 1956, p. 928.

  (b) R. C. Weast, Ed., *Handbook of Chemistry and Physics*, 50th ed., Chemical Rubber Co., Cleveland, 1969, p. C-286.

11. A. J. Rudge, in *Industrial Electrochemistry*, A. T. Kuhn, Ed., Elsevier, New York, 1971, Chap. 1.

12. G. H. Cady, D. A. Rodgers, and C. A. Carlson, *Ind. Eng. Chem.*, **34**, 443 (1942).

13. B. H. Ashe, Jr., and H. M. Fox, Porous Electrode Having Open Feed Cavity, U.S. Patents 3,711,396 (1973) and 3,728,233 (1973).

14. For example, Union Carbide PC45 porous carbon, Union Carbide Corp., Carbon Products Division, 270 Park Avenue, New York, NY 10017; and grade B303 porous carbon, Great Lakes Carbon Corporation, 299 Park Avenue, New York, NY 10017.

15. A. Davies and A. J. Rudge, French Patent 1,218,692 (1962); see also Ref. 11.

16. W. C. Schumb, R. C. Young, and K. J. Radimer, *Ind. Eng. Chem.*, **39**, 244 (1947).

17. R. A. DeMarco, D. A. Couch, and J. M. Shreeve, *J. Org. Chem.*, **37**, 3332 (1972).

18. W. V. Childs, U.S. Patent application filed 1977.

19. R. B. MacMullin, R. B. MacMullin Associates, in a series of reports to Phillips Petroleum Co., 1967–1969.

20. R. B. MacMullin, H. M. Fox, F. N. Ruehlen, and W. V. Childs, Electrolytic Cell, U.S. Patent 3,692,660 (1972).

21. F. N. Ruehlen, H. M. Fox, Multi-Porosity Electrode for Electrochemical Conversion, U.S. Patent 3,655,535 (1972).

22. R. B. MacMullin, *Electrochem. Technol.*, **1**, 5 (1963).

Chapter **VIII**

USE OF ROTATING ELECTRODES FOR
SMALL—SCALE ELECTROORGANIC PROCESSES

H. V. K. Udupa and K. S. Udupa

Central Electrochemical Research Institute,
Karaikudi, India

1   INTRODUCTION

The use of rotating electrodes in electroanalytical applications, as well as in the electrowinning of metals, is well known [1]. Applications of the use of rotating electrodes, especially for organic preparations, have been meager, but systematic investigations have been carried out in the last three decades initially in the Department of Chemistry, The Ohio State University, Ohio, and later continued in the Central Electrochemical Research Institute, Karaikudi, India. This chapter describes the use of rotating electrodes in the development of electroorganic processes for the preparation of a variety of organic compounds. The reactions are described with examples (of laboratory-scale experiments) arranged according to the functional group. The design of higher amperage cells employing rotating electrodes is described with examples toward the end of the chapter.

In the majority of electroorganic processes, the electrode potential depends on the current density, nature of the electrode surface, composition of the electrolyte, pH, temperature, and so forth. In general, the slowest step in such reactions is that of the diffusion of the reactants *to*, or that of the reactions product *from*, the electrode. If the product of reaction becomes adsorbed or forms a film on the electrode, as in the case of benzyl alcohol during the reduction of benzoic acid, the dielectric constant of the layer close to the electrode surface is considerably reduced, as is also the active surface area of the electrode, which determines the kinetics of the reaction. By the rotation of the electrode, the product often can be dislodged and moved away from the electrode. As the electrolysis progresses at an appreciable rate, the concentration of organic molecules near the electrode leads to a decrease of its concentration, which is normally made good by diffusion or transfer of more material from the bulk of the solution. If the replenishment of organic molecules by diffusion is lower than the number of molecules used up at the electrode, any of the competing electrode processes, such as oxygen evolution at an anode and hydrogen evolution at a cathode, are facilitated, thus leading to a lowering of overall current efficiency for the formation of the product.

It is generally accepted that the layer or film next to the surface of the electrode is important, since its composition and properties are different from those of the bulk. The rate of diffusion of reactant species is proportional to the concentration gradient across which it diffuses. According to Fick's diffusion equation, the rate of diffusion of ions toward the electrode is proportional to $C_s - C_e/\delta$ per square centimeter of an electrode surface, where C_s is the concentration of the species in the bulk of the solution, C_e is that at the surface of the electrode, and δ is the thickness of the diffusion layer. Thus

$$\frac{dc}{dx} = D_0 \frac{C_s - C_e}{\delta}$$

where D_0 is the proportionality constant called the diffusion constant of the substance. The current density, i_d, resulting from the electrolysis of a diffusing substance is

$$i_d = nFSD_0 \frac{C_s - C_e}{\delta}$$

where S is the surface area of the electrode. Any factor that decreases the value of the thickness of the diffusion layer permits usage of a higher current density. Rotation of the electrode brings down the diffusion layer thickness considerably, permitting higher mass transfer rates. In such reactions, which are controlled by the rate of diffusion of the reactants, the rate of reaction can be increased by the use of rotating electrodes. It was found by Wilson and Udupa [2] that the yield of p-aminophenol increased with speed of rotation up to 2500 to 3000 rpm and any further increase in rotation rate of the electrode had no effect on the yield of p-aminophenol.

In this context the view expressed by Prof. Charles Tobias [3] in his Acheson Medal address entitled "New Directions in Electrochemical Engineering" is of importance and is reproduced below:

"Reduction of boundary layer thickness by flow is attractive only up to a point. The power requirement decreases sharply for constant R_θ as we increase the inter-electrode gap. However it is interesting to note that if we fix the boundary layer thickness, the power required increases with electrode separation. Clearly, increasing flow rate is not the direction in which we need to proceed; rather, interference with the boundary layer locally by either *moving the electrode* (*rotating cylinder,* oscillation, fluidised bed) or by gas evolution, wiping, slurry flow are promising avenues. Still another method, the through-flow electrode, can provide high transport rates, although at the expense of non-uniform accessibility."

In industrial practice, flow conditions are generally maintained in the turbulent zone. With rotating electrodes, the laminar-turbulent transition occurs at comparatively low values of the Reynolds number, in the range of 50 to 200 depending on the disposition and the distance between the two electrodes, as well as on the surface roughness. Thus use of rotating electrodes leads to establishment of turbulent flow conditions at very low speeds. In practice, a peripheral velocity of 2 to 2.5 m/sec is maintained while scaling-up these cells. Cells employing rotating electrodes can be used in all cases where the depolarizer is present as an emulsion, solid suspension, or homogenous solution (see Chapter II for additional details).

2   GENERAL EXPERIMENTAL METHODS

The general conditions for experiments employing rotating electrodes are similar to those that are commonly used with stationary electrodes and include the material used for electrodes, electrolyte media, separator, and reference electrode. The only change is the arrangement of the working electrode and the method by which the direct current is fed to the rotating electrode so that it acts as one of the terminal electrodes during electrolysis. In principle, electrolysis is carried out by the application of controlled current or by controlling the potential of the working electrode at a predetermined value. The latter requires the use of a reference electrode and a luggin capillary placed very close to the working electrode, which is rotating.

Figure 8.1 shows the general set up of the cell used for the reduction of organic compounds with rotating electrodes. It is usual practice to rotate only the working electrode, the counterelectrode being kept stationary. The cells employing rotating electrodes have been useful for batch-type operations.

The simplest type of an electrolytic cell is made up of a glass beaker as shown in Figs. 8.1 and 8.2. Figure 8.1 shows a divided cell in which the rotating working electrode and the counterelectrode are separated. Two types of separators are used in electrolytic cells. The first type includes porous separa-

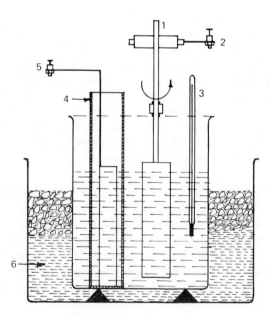

Fig. 8.1. Electrolyte cell. (*1*) Rotating cathode, (*2*) carbon brush contact to cathode, (*3*) thermometer, (*4*) porous diaphragm, (*5*) anode connection, (*6*) ice and salt mixture.

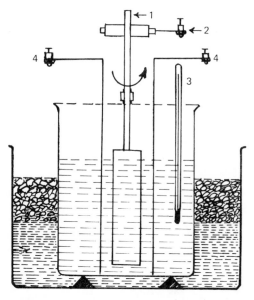

Fig. 8.2. Electrolytic cell. (*1*) Rotating cathode, (*2*) carbon brush contact to cathode, (*3*) thermometer, (*4*) anode connection.

tors, such as those made of ceramic, sintered glass, plastic, and asbestos which act as a physical barrier between the two sides and at the same time provide a conducting path by way of electrolyte present in their pores. The second type makes use of semipermeable ion-exchange membranes. Whenever plastic, asbestos, or ion-exchange membranes are used, they are reinforced to provide mechanical ingrity, occasionally, by a cage or frame made out of the metal to be used as the counterelectrode. Such separator-cum-anodes have been reported by Balakrishnan et al. [24].

Figure 8.2 shows an undivided cell in which the electrodes are arranged very close together to minimize the voltage drop through the electrolyte. Here the products of the reaction or the reactant neither directly react at the counterelectrode nor take part in the reaction by manipulation of the ratio of the area of two electrodes. The cell voltage in these types of cells is considerably less than that in cells employing separators. In addition to the above two types of cells, the H-shaped cell with a sintered glass disc, usually of porosity between 5 to 10 μ, between the two parts of the cells, has also been used for small-scale preparative work. These H-type cells can be fabricated with standard ground joints and can be used both for polarization studies and preparative electrolysis.

The rotating electrode may be either a cylindrical (Fig. 8.3) or a disc electrode (Fig. 8.4), consisting of a number of discs suitably interspaced. In the case of the cylindrical electrode, the current distribution is uniform throughout the electrode surface, but in the case of disc-type electrode, there is variation of current density on the disc as well as on the spacer between two discs. Because of the variation of current density, the electrode potential at these points also varies, the difference being dependent on the width of the disc (difference in the diameters of disc and the spacer), the distance between the two discs (height of the spacer), and the distance between the working electrode and the counterelectrode. But a disc electrode is lighter in weight, has

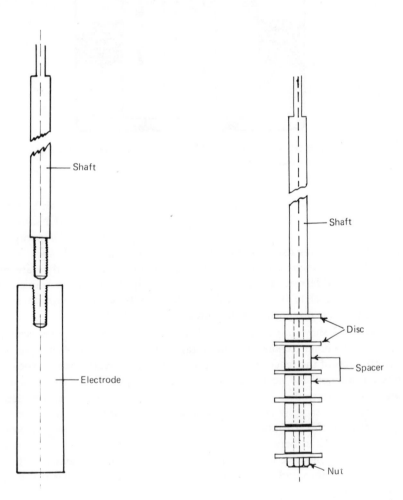

Fig. 8.3.   Cylindrical-type electrode.   **Fig. 8.4.**   Disc-type electrode.

more area than the cylindrical electrode of similar size, and gives a better agitation, especially when used in an emulsion or solid suspension.

For laboratory-scale operations a cylindrical electrode is fabricated as follows: A rod 2.5 cm in diameter by 10 cm in length is cast out of pure metal or cut from a rod of the particular electrode material. At one end of this a bore of 0.5-cm diameter is drilled having threads to a depth of 1 cm. A copper rod 1.0 cm in diameter by 30 cm in length is provided with matching threads up to a length of 1 cm after the diameter of one end of the rod is reduced to 0.5 cm. The two portions are tightened together well. The directions of the threads must be such that they are in clockwise position so that they do not become loosened during rotation of the electrode. The other end of the copper rod is attached to the lower end of the shaft of a fractional horsepower motor and can be rotated through a glass guide.

A rotating disc electrode is made by assembling the discs and spacers on a rod. The diameter of the rod where the discs and spacers are assembled is reduced and the end is provided with threads so that the discs and spacers are held in position without wobbling. A typical electrode used for laboratory experiments is shown in Fig. 8.4. A rod 1.0 cm in diameter by 30 cm in length is used and the diameter is reduced to 0.5 cm for a length of 8 cm and is provided with threads for a length of 1.0 cm. Discs having 2.5-cm outer diameters, and 0.5-cm inner diameters and 0.158 cm thick are assembled alternatively with cylindrical spacers 1.25 cm long and with 1.25-cm outside diameters and 0.5-cm inner diameters. The whole assembly is held secure by tightening a nut on the threaded end of the rod. The number of discs and spacers could be varied depending on the area required, as well as the depth to which it can be immersed in the electrolyte. For a typical electrode having six discs and five spacers, as shown in Fig. 8.4, the area of the electrode works out to 65 cm^2. Since the electrode is assembled on the 1.0-cm diameter rod, this rod should also be of the same material as the disc and spacers, thus restricting the construction of disc-type electrodes to only few electrode materials. However, this can be overcome by electrodepositing the required metal onto a suitable substrate such as copper, which has good electrical conductivity.

Electrical contact to the rotating electrode is provided through a mercury cup or a carbon brush. A mercury cup arrangement is shown in Fig. 8.5. It is made of nickel-plated copper. The mercury cup also acts as a guide to ensure smooth rotation. To remove friction, between the stationary mercury cup and the shaft of the electrode, the mercury cup is provided with ball bearings at the bottom portion. In the case of the carbon brush arrangement for electrical contacts, the carbon brushes, which are normally used in electric motors, are pressed onto the shaft of the electrode by a set of springs. Direct current is fed through the lead attached to the carbon brush.

The use of rotating electrodes in various reactions involving reduction and

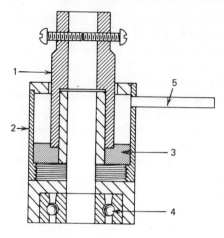

Fig. 8.5. Mercury cup arrangement. (*1*) Contact part, (*2*) body of mercury cup, (*3*) mercury, (*4*) ball bearing, (*5*) electrical contact to mercury cup.

oxidation of organic compounds is described in Sections 3 and 4, comparing the current efficiency and the yields obtained with rotating and stationary electrodes. Key design considerations involved in scaling-up of the laboratory cells to high amperage cells are outlined in Section 5.

3   ELECTROCHEMICAL REDUCTION OF VARIOUS FUNCTIONAL GROUPS

Unsaturated Acids

Maleic and Fumaric Acids

$$\begin{matrix} CHCO_2H \\ \| \\ CHCO_2H \end{matrix} + 2H^+ + 2e \longrightarrow \begin{matrix} CH_2CO_2H \\ | \\ CH_2CO_2H \end{matrix}$$

The electrolytic reduction of maleic acid reported by earlier workers employed a divided cell and sulfuric acid medium, and of the various electrodes studied, mercury cathodes gave the best yield. However, little attention has been paid in the past to how to avoid the use of a separate anode chamber. The need for a separate anode chamber is a serious drawback in cell design for larger-scale adoption. Therefore, the electrolytic reduction of maleic acid and fumaric acid was studied with a rotating lead cathode in a modified cell [4, 5] with two cylindrical lead-lined copper rods, covered with asbestos fiber used as anodes to avoid a separate anode chamber. The current efficiency data obtained for the reduction of maleic and fumaric acids at rotating, as well as at stationary, lead cathodes (see Table 8.1) shows that the difference in current efficiency between rotating and stationary cathodes is significant at high current densities. It is interesting to note that while maleic acid (having the cis form) is easily reduced at a stationary cathode and at high current densities, fumaric acid (having the trans form) is less easily reduced. The difference in current efficiencies between a rotating and stationary electrode is much more significant with fumaric acid than with maleic acid.

The efficiency of reduction remains high even in the absence of a separator, since the reactants or their reduction products are not destroyed at the anode by oxidation, as the pores of the asbestos rope wound on the anode become filled with sulfuric acid whose concentration is higher than the bulk concentration. In the presence of this high acidity, the dissociation of free organic acid is minimal. This prevents the organic anion from discharging at the anode thus enabling the maintenance of high faradaic efficiency. This may also reduce the anode corrosion due to the organic anion attack, although this cannot be completely prevented.

From the results seen in Table 8.1, it is clear that with maleic acid, the efficiency is high even under stationary conditions and current densities, as high as 20 A/dm^2. A cell employing a stationary cathode was adopted during the scaling up of the process.

Laboratory-Scale Study

Twenty-one grams of maleic acid in 5% H_2SO_4 (230 ml) was reduced at a rotating lead-lined copper electrode (area 22 cm^2) in an undivided cell for 4 hr at a current of 2.4 A and at temperatures between 70 and 75°C. On cooling the electrolyte, crystals of succinic acid separated out, and the mixture

Table 8.1. Reduction of Fumaric and Maleic Acid
Conditions: 230 ml of 5% H_2SO_4 containing 21 g of either fumaric or maleic acid; temp., 70–75°C; cathode, lead; anode, lead-lined copper covered with asbestos fiber.

		Current Efficiency (%)	
Current Density (A/dm^2)	Cathode Motion	With Maleic Acid	With Fumaric Acid
10	Stationary	98.0	80.9
10	Rotation	98.5	98.5
15	Stationary	94.8	28.5
15	Rotation	99.0	69.5
20	Stationary	95.2	18.1
20	Rotation	99.0	51.4

was filtered. Further quantity of the product was recovered by evaporating the mother liquor, and in all 21 g of succinic acid separated corresponding to an assay yield of 95%.

Given below are the results of some of the studies on the reduction of unsaturated acids, such as crotonic acid, cinnamic acid, and sorbic acid. These reactions have been studied on a laboratory scale employing rotating cathodes and they show interesting possibilities for synthesis of some of the compounds.

Crotonic Acid

During the reduction of crotonic acid at high overvoltage cathodes, crotyl alcohol and n-butyl alcohol were reported as the main reduction products by Dineen et al. [6]. They showed that the reduction proceeds directly from the crotonic acid by way of crotonaldehyde. However, when rotating amalgamated cathodes were employed, the reduction products were reported to be butyric acid and dimethyl adipic acid in addition to small quantities of resinous materials that are soluble in benzene [7].

The formation of different products during the reduction of crotonic acid at an amalgamated lead cathode (see Table 8.2) may be rationalized as follows:

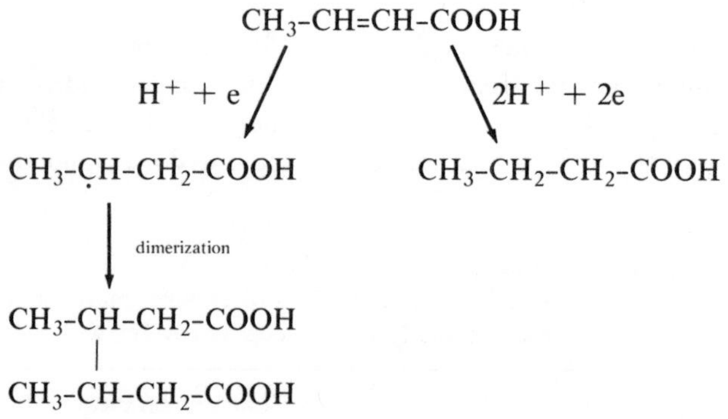

$$CH_3-CH=CH-COOH$$

$$H^+ + e \diagup \qquad \diagdown 2H^+ + 2e$$

$$CH_3-\overset{.}{C}H-CH_2-COOH \qquad CH_3-CH_2-CH_2-COOH \qquad (a)$$

$$\downarrow \text{ dimerization}$$

$$CH_3-CH-CH_2-COOH$$
$$|$$
$$CH_3-CH-CH_2-COOH$$

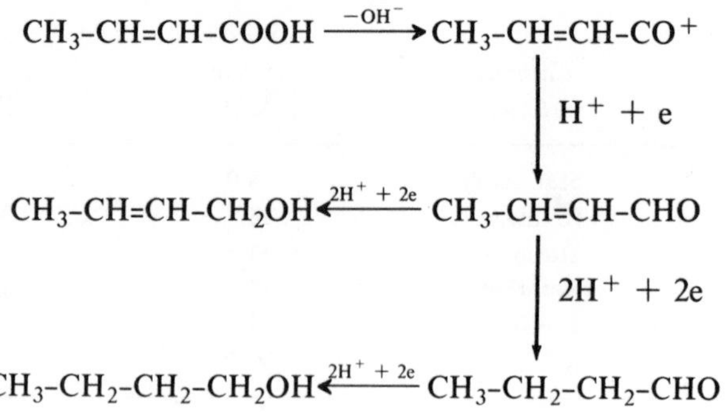

$$CH_3-CH=CH-COOH \xrightarrow{-OH^-} CH_3-CH=CH-CO^+$$

$$\downarrow H^+ + e$$

$$CH_3-CH=CH-CH_2OH \xleftarrow{2H^+ + 2e} CH_3-CH=CH-CHO \qquad (b)$$

$$\downarrow 2H^+ + 2e$$

$$CH_3-CH_2-CH_2-CH_2OH \xleftarrow{2H^+ + 2e} CH_3-CH_2-CH_2-CHO$$

Table 8.2.   Reduction of Crotonic Acid
Catholyte, 500 cc of 10% H_2SO_4; crotonic acid, 40 g. Temp, 16–18°C; current, 12 A
(current density, 4.9 A/dm^2); anolyte, 500 cc 10% H_2SO_4; anode, lead.

Motion of Cathode	Alcohols (g)	Butyric Acid (g)	β,γ-Dimethyl Adipic Acid (g)	Residue (g)
Rotation	5.6	19.0	3.2	12.5
Stationary	1.0	4.6	1.2	5.2

Reduction at amalgamated copper and tin under the above experimental conditions was minor and at an amalgamated cadmium cathode there was no dimer formation at all.

Cinnamic Acid

The electrolytic reduction of cinnamic acid was studied at amalgamated cathodes of lead and copper [7]. The products of the reduction mainly include β,γ-diphenyl adipic acid and β-phenyl propionic acid. In addition to this, a benzene-soluble material was also formed. Although amalgamated lead did not show much difference in yield between stationary and rotating cathodes, amalgamated copper showed a considerable effect on the yield of the products. At an amalgamated lead cathode, the yields of β,γ-diphenyl adipic acid and β-phenyl propionic acid were 43 and 15%, respectively. However, at a rotating amalgamated copper cathode the yield of the former was 55% with very little β-phenyl propionic acid.

$$C_6H_5-CH{=}CH-COOH \xrightarrow[\text{lead cathode}]{\text{amalgamated}} \begin{cases} C_6H_5-CH_2-CH_2-COOH \ (15\%) \\ \\ \begin{array}{c} C_6H_5-CH-CH_2-COOH \\ | \\ C_6H_5-CH-CH_2-COOH \end{array} \ (43\%) \end{cases}$$

$$C_6H_5-CH{=}CH-COOH \xrightarrow[\text{copper cathode}]{\text{amalgamated}} \begin{array}{c} C_6H_5-CH-CH_2-COOH \\ | \\ C_6H_5-CH-CH_2-COOH \end{array} \ (55\%)$$

In addition, there was a considerable difference in the yields of diphenyl adipic acid at rotating and stationary amalgamated copper cathodes as in

Table 8.3. The results showed that the yield of diphenyl adipic acid was not dependent on the current density employed. The yields obtained at amalgamated copper cathodes were comparable to those reported [8] at a mercury cathode.

Sorbic Acid

The study of the reduction of unsaturated acids at cathodes has been extended to sorbic acid. It has been reported [7] that the product of reduction included the dihydrosorbic acid and unsaturated dimeric product. The relative yields of these products depended on the medium, as well as on the cathode material as noted in Tables 8.4 and 8.5.

The dimeric product yield was much greater during the reduction of sorbic acid than during the reduction of sorbate anion. Amalgamation of the copper cathode increased the yield of the dimeric product.

Carbonyl Compounds

Reduction of Glucose

The reduction of glucose to sorbitol and mannitol at rotating cathodes of lead and amalgamated lead are exemplified [9]. The electrochemical reduc-

Table 8.3. Reduction of Cinnamic Acid

Motion of Cathode	Current Density (A/dm^2)	Total Current Passed $(A \cdot hr)$	Yield of Adipic Acid (%)
Rotation	2.45	33	52.3
Stationary	2.45	60	23.3

Table 8.4. Reduction of Sorbic Acid in Alkaline Medium
Conditions: Catholyte, 500 cc of 2 N KOH; anolyte, 500 cc of 2 N KOH; sorbic acid, 50 g.

Cathode	Motion of Cathode	Yield of Dimeric Product (%)	Yield of Dihydrosorbic Acid (%)
Lead	Rotation	3.02	86.2
	Stationary	5.08	83.7
Amalgamated lead	Rotation	20.8	64.4
	Stationary	13.6	77.8
Amalgamated copper	Rotation	34.7	60.0
	Stationary	19.8	68.5

Table 8.5. Reduction of Sorbic Acid in Acid Medium
Catholyte, 600 cc of alcoholic H_2SO_4; anolyte, 500 cc of aq. H_2SO_4 (10%).

Cathode	Motion of Cathode	Yield of Dimeric Product (%)	Yield of Dihydrosorbic Acid (%)
Lead	Rotation	73.3	16.8
	Stationary	73.5	14.7
Amalgamated lead	Rotation	84.5	8.7
	Stationary	86.0	4.3
Amalgamated copper	Rotation	86.4	4.9
	Stationary	71.4	7.8

tion of glucose is a well studied reaction [10] and the Atlas Powder Company in United States built a plant for the production of sorbitol as early as 1937 [11]. It has been found by Sanders and Hales [12] that the rate of reduction is constant in the glucose concentration range of 100 to 500 g/liter at a current density of 1 A/dm^2, and at concentrations less than 100 g/liter the rate of reduction has been reported to be proportional to the concentration of sugar. This clearly indicates that if the concentration of the glucose is high near the vicinity of the cathode, the rate of reduction will be high. The glucose molecule being neutral, the mass transfer of these molecules from the bulk of the solution to the surface of the cathode and be achieved by the rotation of the cathode. The reduction of glucose has been carried out on an amalgamated lead cathode by taking an initial concentration of 600 g/liter. The current efficiency for the reduction was more than 90% up to a current density of 15 A/dm^2, when the glucose concentration was reduced from 600 to 300 g/liter, but from 300 to 100 g/liter there was a sharp fall in current efficiency even at a current density of 5 A/dm^2. Thus it was clear that in the region of concentration (600 to 300 g/liter), where the reduction rate was independent of concentration, rotating of the cathode permitted high current density up to 15 A/dm^2, thus helping to bring about a fast reaction. In the lower concentration regions the reaction could not be speeded up even by the use of rotating cathodes because of the dependence of the reaction rate on the concentration of glucose. This also explained the results obtained by Pathy [13], who did not find much improvement in the current efficiency when the cathode was rotated at 1000 rpm compared to when it was kept stationary, because the concentration of glucose used for reduction was only 200 g/liter. The ratio of products of reduction of glucose, namely sorbitol and mannitol, is dependent on the alkalinity of the catholyte, and if the pH of the catholyte is maintained around 8, mostly sorbitol is formed. Increasing the glucose concentration in the catholyte back to 600 g/liter by adding fresh glucose after it has been reduced to 300 g/liter has resulted in good current efficiency. Since sorbitol is highly soluble in the catholyte, it was possible to add glucose to make the concentration 600 g/liter at least three times. However, the cell voltage and viscosity increased with repeated addition of glucose. In this way it has been possible to cathodically reduce 1.2 kg of glucose per liter of the initial catholyte volume per batch.

Reduction of Acetone

The reduction of acetone in alkaline media has been studied by Wilson and Wilson [14], who have explained the factors influencing the pinacol formation. In a study of the effect of rotation of the cathode [7] it was found that the yield of pinacol was much greater at rotating cathodes than at stationary cathodes. The results of the reduction of acetone are given in Table 8.6. The results obtained indicated that the values of r:

$$r = \frac{\text{no. of molecules of acetone converted to pinacol}}{\text{no. of molecules of acetone converted to isopropyl alcohol}}$$

for the rotating cathode are nearly double those for the stationary cathode. Further, it has been reported that with a rotating cathode, pinacol formation seems to increase even with as high a current density as 30 A/dm², as seen in Table 8.6.

Reduction of Aromatic Aldehydes

Aromatic aldehydes are more easily reduced than their aliphatic analogs. A comparative study on the reduction of aromatic aldehydes, such as benzaldehyde, salicylaldehyde, and anisaldehyde, at stationary and rotating amalgamated cathodes has been reported [15]. At low current densities the products of reduction of aldehydes have been predominantly alcohols, whereas at high current densities free radical formation and dimerization occur. The current density range in which reduction to alcohol occurs and the current efficiency for the same were increased by the rotation of the cathode. The results obtained by the reduction of the three aldehydes at a rotating amalgamated cathode are shown in Table 8.7.

Carboxylic Acids

Oxalic Acid

Oxalic acid has been reported to be reduced to glyoxylic acid at cathodes of high hydrogen overvoltages, the reduction process terminating at the glyoxylic acid stage at temperatures not exceeding 20°C because of stabilization of glyoxylic acid as a hydrate. Studies carried out to examine the effect of rotation of the cathode showed the current efficiency to be high only at low current densities [16, 17]. The current efficiency decreased slowly up to a

Table 8.6. Reduction of Acetone in Alkaline Medium

Cathode	Motion of Cathode	Current Density (A/dm^2)	Current Passed $(A \cdot hr)$	Pinacol Yield (g)	Isopropyl Alcohol (g)	Ratio r
Amalgamated lead	Rotation	3.26	34	9.2	28.8	0.325
	Stationary	3.26	68	5.5	36.1	0.157
Amalgamated Monel	Rotation	31	66	16.5	27.1	0.619
	Stationary	31	72	10.6	32.3	0.334
Amalgamated copper	Rotation	10.9	72	18.5	25.9	0.725
	Stationary	10.9	72	12.6	36.6	0.351

Table 8.7. Reduction of Benzaldehyde, Salicylaldehyde, and Anisaldehyde at Various Current Densities

Conditions: catholyte, 200 ml containing 20 g Na_2SO_4, 5–10 g $NaHCO_3$, and CO_2 passing through; anolyte, 20% H_2SO_4; wt. of aldehyde taken, 20 g; temp., 18°C.

Aldehyde Reduced	Current Density (A/dm^2)	Alcohol (g)	Dimer (g)	Aldehyde Recovered (g)	Yield (%)	Current Efficiency (%)
Benzaldehyde	5	18	—	1.2	95	88.2
	10	8.7	9.6	1.5	96	43.9
						23.9
	15	—	16.8	1.6	92	42
Salicylaldehyde	5	17.5	—	1.5	94	87.3
	10	8.5	9.5	2.0	97.5	41.8
						23.6
	15	—	16.1	2.4	92.3	40.0
p-Anisaldehyde	1.5	17.0	—	1.7	92.5	83.8
	2.5	7.4	8.3	1.7	94.5	36.5
						20.8
	15.0	—	14.4	4.2	92	36

current density of 8 A/dm², beyond which a sudden fall in current efficiency was observed. However, at stationary cathodes the current efficiency was high up to 2 A/dm². Thus rotation of the cathode helps in controlling the pH of the cathode surface during the reduction of oxalic acid, although rotation did not permit employing high current densities.

$$\begin{matrix} \text{COOH} \\ | \\ \text{COOH} \end{matrix} + 2H^+ + 2e \longrightarrow \begin{matrix} \text{COOH} \\ | \\ \text{CHO} \end{matrix} + H_2O$$

Salicylic Acid (o-Hydroxybenzoic Acid)

Salicylic acid is reduced to salicylaldehyde on mercury cathodes in the presence of boric acid at a pH of 5.4 to 5.7. Further reduction of aldehyde was generally prevented by trapping the aldehyde with $NaHSO_3$. Studies conducted to replace the stationary mercury cathode with rotating amalgamated cathodes showed that salicylaldehyde can be produced at yields of 40 to 43% at a rotating amalgamated monel cathode. At stationary cathodes, using an auxiliary stirrer, the yield never exceeded 10% [18, 19]. The rotation of the cathode not only permitted the use of high current density (up to 15 A/dm²), but also helped in controlling the pH of the catholyte at the desired level. The primary electrochemical reaction is the discharge of sodium ions that forms amalgam with the mercury on the cathode surface [20]. This reacts with the sodium salicylate present in the catholyte. The process has been scaled up employing rotating cathodes [21–27].

In the laboratory, as well as in scaled-up cells, the important factors for high efficiency include: catalytic activity of the electrode surface and the pH of the medium. In experiments of short duration, the yields have always been good, but on prolonged electrolysis, the yields have been poor because of fouling of the electrode surface. To some extent this drawback can be overcome by employing interrupted direct current. The current efficiency is high under these conditions and prolonged catalytic activity of the electrode surface is maintained. When a porous porcelain separator was employed, anolyte acidity increased in strength and the pH could be maintained only by the periodic external addition of acid, so that the variation of pH was not too great or too frequent. A separator of high porosity has to be used to overcome this difficulty.

Laboratory-Scale Study

Salicylic acid is reduced at an amalgamated monel cathode (0.5 dm^2) in a divided cell. Fifteen percent Na_2SO_4 (400 ml) solution containing 45 g of boric acid and 20% H_2SO_4 (60 ml) are used as catholyte and anolyte, respectively. A current of 6 A is passed for 55 min. A mixture of salicylic acid (13.8 g) and sodium sulfite (12.6 g) is dissolved in 30 ml of water and added in regular intervals during the reduction. After the reduction, the catholyte is acidified by adding 5 ml of conc. H_2SO_4 and is steam distilled. Five milliliters of salicylaldehyde separate out, corresponding to 50% yield.

Benzoic Acid

Benzoic acid has been reduced in an acid medium to benzyl alcohol at a lead cathode. Earlier processes made use of the addition of ethyl alcohol to keep the benzoic acid in solution. Studies [28, 29] on the use of rotating electrodes for this reduction showed that rotation helped not only in saturating the catholyte with benzoic acid, but also in favoring the mass transfer of the depolarizer to the cathode, even without addition of alcohol. Further, benzyl alcohol was quickly removed from the cathode surface as a result of the motion of the cathode. Influence of current density on the current efficiency is indicated in Table 8.8.

$$\underset{\text{COOH}}{\bigcirc} + 4H^+ + 4e \longrightarrow \underset{\text{CH}_2\text{OH}}{\bigcirc} + H_2O$$

Once free benzyl alcohol separated, the current efficiency dropped and it was necessary to clean the cathode to restore its effectiveness. Thorough cleaning of the cathode surface with water also made it very effective for further reduction.

Table 8.8.   Influence of Current Density on the Current Efficiency during the Reduction of Benzoic Acid on a Rotating Lead Cathode
Conditions: catholyte, 200 ml of 10% H_2SO_4; benzoic acid, 5 g; temp., 80–90°C; anolyte, 50 ml of 10% H_2SO_4.

Current Density (A/dm^2)	Current Efficiency (%)
10	92.8
20	91.4
30	93.0
40	93.2

In the reduction of benzoic acid, the purity of the lead cathode has been reported to be important [30, 31]. It has also been established that by keeping the temperature of the catholyte within 40 to 45°C the catalytic activity of the cathode could be retained. A lead cathode prepared by depositing lead from a lead fluoborate bath gave high efficiency for the reduction. In addition to benzyl alcohol, small quantities of dibenzyl were formed during the reduction of benzoic acid.

Nitro Compounds

The reduction of nitro compounds has been extensively studied using rotating electrodes. The products of reduction depend on the pH of the media, as well as on the electrode material used for this purpose. This is especially true in the case of aromatic nitro compounds.

Aliphatic Nitro Compounds

Among the aliphatic nitrocompounds, the reduction of nitrourea and nitroguanidine to their corresponding amines has been reported [32, 33].

$$O=C\begin{smallmatrix}NH_2\\\\NH-NO_2\end{smallmatrix} + 6H^+ + 6e \longrightarrow O=C\begin{smallmatrix}NH_2\\\\NH-NH_2\end{smallmatrix} + 2H_2O$$

$$HN=C\begin{smallmatrix}NH_2\\\\NH-NO_2\end{smallmatrix} + 6H^+ + 6e \longrightarrow HN=C\begin{smallmatrix}NH_2\\\\NH-NH_2\end{smallmatrix} + 2H_2O$$

In the reduction of nitroguanidine, it is possible to use either an ammoniacal or a sulfuric acid medium. According to Yamashita and Sugino [34], an ammoniacal medium is preferred because a pH of 8 to 9 is favorable for nitrosoguanidine, which is an intermediate in this reduction. Studies in ammoniacal medium revealed that the cathode rotation enabled the use of high current density (up to 15 A/dm^2) without the loss of current efficiency [35]. Amalgamated lead, lead, and zinc cathodes gave better yields (80 to 85%) than amalgamated copper cathode (51%). However, the surfaces of the lead or amalgamated lead cathodes were tarnished at the end of each experiment and required cleaning before the cathodes could be reused. In the case of zinc cathodes, the activity of the cathode was reproducible and the loss of zinc could be minimized by periodically adding nitroguanidine during the course of reduction [36, 37].

When pure zinc was used as cathode material corrosion of zinc was observed. However, while the loss of zinc was minimized by nitroguanidine additions, the shape of the electrode became deformed, resulting in wobbling of the rotating cathode. To overcome this, a mild-steel cathode was used with a thick deposit of zinc (0.125 to 0.625 mm). By controlling the quantity of nitroguanidine periodically added, the activity of the cathode could be maintained for reuse in subsequent batches. A 1000-A cell (Fig. 8.6) employing two rotating zinc deposited cylindrical mild steel cathodes (30-cm diameter, 40-dm^2 area each) has been in operation to produce 16 kg of aminoguanidine bicarbonate per 30-hr batch.

Aromatic Nitro Compounds

Nitrocompounds to Aminophenols

Rotating electrodes have been used in a number of cases involving both acidic and alkaline media. During the reduction of nitrobenzene and substituted nitrobenzenes in an acid media to the corresponding aminophenols, rotation of the amalgamated cathode caused an unexpected rise in the yield of p-aminophenol and a corresponding drop in the yield of aniline [2, 38–42]. Further, as the speed of rotation of the cathode was increased, the yield of p-aminophenol rose rapidly at first, but remained reasonably steady beyond

Fig. 8.6. Photograph of 1000-A aminoguanidine cell.

1000 rpm. The effect of rotation of an amalgamated monel cathode during the reduction of various nitrocompounds to their corresponding aminophenols is given [7] in Table 8.9. From the data given, it is clear that rotating cathodes helped in achieving a greater yield of aminophenols from nitrobenzene and other substituted nitrobenzenes. It is seen that the electrochemical step in this reaction is the formation of the corresponding phenylhydroxylamines, which undergo nucleophilic hydroxylamine rearrangement to give the corresponding p-aminophenols. If the phenylhydroxylamine is not removed as fast as it is formed, further reduction occurs to the corresponding anilines.

Table 8.9. Reduction of Aromatic Nitrocompounds to Corresponding Aminophenols Conditions: catholyte, 350 ml H_2SO_4; anolyte, 150 cc 20% H_2SO_4; temp., 90°C; current density, 2.7 A/dm^2; cathode, amalgamated copper; anode, lead.

Nitrocompound Reduced	Motion of Cathode	Yield (%)	
		Aminophenol	Aniline
Nitrobenzene	Rotation	71.7	6.6
	Stationary	56.5	10.6
2-Nitro-6-chlorotoluene	Rotation	82.8	8.5
	Stationary	45.6	27.2
2-Nitro-4-chlorotoluene	Rotation	81.7	4.8
	Stationary	43.8	26.4
o-Nitrotoluene	Rotation	73.5	7.7
	Stationary	55.5	10.3
o-Nitrochlorobenzene	Rotation	75.8	4.9
	Stationary	50.1	17.5
2,5-dichloronitrobenzene	Rotation	71.9	7.2
	Stationary	44.3	26.0

The effect of addition of stannous, thallous, and bismuth salts to the catholyte as an alternative to the amalgamated cathode on the reduction of nitrobenzene to p-aminophenol has been reported recently [43, 44]. The results of the study are given in Table 8.10.

Studies carried out by Rance and Coulson [45] revealed that the formation of p-aminophenol is maximum when the cathode potential is of the order of 100 to 150 mV and aniline formation is favored when the potential exceeds 200 mV. In a recent study using bismuth deposited rotating cathodes, the authors were able to arrive at the various optimum parameters by which the cathode potentials were within the limits indicated by Rance and Coulson even when working at a current density of 20 to 25 A/dm^2.

Laboratory-Scale Study

Fifty grams of nitrobenzene is reduced on an amalgamated monel cathode (0.5 dm^2) in a divided cell. Twenty percent H_2SO_4 is used as catholyte and anolyte. A current of 10 A is passed for 4.5 hr and the catholyte temperature is maintained between 90 and 95°C. After the electrolysis the catholyte is steam distilled to remove any unreduced nitrobenzene and after cooling it is neutralized with a mixture of sodium and ammonium carbonate to a pH of 6.8 to 7.0. The neutralized catholyte is steam distilled to remove the aniline and the contents in the still are cooled when p-aminophenol separates out. It is filtered, recrystallized from hot water after it is decolorized with animal charcoal and sodium bisulfite, and dried under vacuum. The yield is 31.5 g of p-aminophenol, which corresponds to an assay of 70%.

Nitrocompounds to Anilines

A number of nitrocompounds have been reduced to their corresponding anilines at rotating copper cathodes. Although copper is known to have a low hydrogen overvoltage, a fresh copper surface is able to bring about reduction

Table 8.10.   Effect of Addition Agents on the Reduction of Nitrobenzene
Conditions: catholyte, 500 ml aq. 20% H_2SO_4; anolyte, 170 ml aq. 20% H_2SO_4; temp., 90-95°C; nitrobenzene used, 60 g; current density, 20-25 A/dm^2; cathode, rotating cylindrical copper; anode, copper.

Addition Agent	*Yield of p-Aminophenol (%)*	*Yield of Aniline (%)*
Stannous chloride	55–65	10–20
Bismuth chloride	65–75	5–8
Thallous chloride	60–65	5–10

at high efficiencies. Hence, a small quantity of copper sulfate is generally added to the catholyte and electrodeposited on the cathode surface before the nitrocompounds are added. The results of the reduction of various nitrocompounds at rotating cathodes are given in Table 8.11.

Nitrocompounds to Hydrazo Compounds

The reduction of aromatic nitrocompounds in an alkaline medium leads to the corresponding hydrazobenzenes, which are commercially important compounds. The efficiency of reduction was found to be high only when the cathode had a spongy deposit of lead, zinc, or tin, all of which have high hydrogen overvoltages. In earlier studies, ethanolic sodium hydroxide was employed as catholyte to make a homogeneous catholyte. Sekine [70] reported higher yields of hydrazobenzene when the electrodes were oscillated above 1000 oscillations/min, corresponding to a Reynolds number of about 7000. Atanasiu and Dumitru [71] reported the reduction of nitrobenzene using a two-stage process in a cell consisting of a rotating iron cathode at 200 rpm. Studies [72-75] with a rotating cathode aimed at achieving high efficiency for the reduction of nitrocompounds present as an emulsion in a single stage using high current densities showed promising results for the reduction of nitrobenzene to hydrazobenzene/benzidine (see Table 8.12).

$$2\;\underset{R}{\underset{\diagdown}{\underset{\diagup}{\bigcirc}}}\!\!-NO_2 \;+\; 10H^+ \;+\; 10e \longrightarrow \underset{\diagup R}{\bigcirc}\!\!NH\!-\!NH\!\!\underset{R\diagdown}{\bigcirc} \;+\; 4H_2O$$

The study on the use of rotating cathodes has been extended to other substituted nitrocompounds [76, 77] (see Table 8.13 for comparative details).

4 ELECTROCHEMICAL OXIDATIONS

Indirect Electrochemical Oxidations

The oxidation of organic compounds can be brought about either by a direct electrochemical route or by an indirect route. The former process is based on direct electron transfer from the organic substrate to the anode, while the latter case employs an intermediate inorganic redox couple. The inorganic redox couple readily reacts at the electrode surface and can be present in concentrations sufficient to avoid mass transfer problems. Oxidation is brought about by homogeneous reaction with the organic compound present in a dissolved or emulsified state—the regeneration of the redox species and the reaction with the organic compound being carried out either *in situ* or in a two-stage process. These indirect electrochemical oxidations have been

Table 8.11. Reduction of Aromatic Nitrocompounds to the Corresponding Amines

No.	Nitrocompound	Cathode (Rotating)	Medium	Temp. (°C)	Current Density (A/dm²)	Products	Yield (%)	Ref.
1	Nitrobenzene	Copper (disc)	Aq. 25% H_2SO_4 contg. 0.6% TiO_2	40–50	25	Aniline	71	46–49
2	o-Nitrophenol	Copper (disc)	Aq. 10% H_2SO_4 contg. $CuSO_4$	60–65	20–30	o-Aminophenol	82	50, 51
3	p-Nitrophenol	Copper (disc)	Aq. 10% H_2SO_4	50–60	15	p-Aminophenol	76 (CE 85%)	52, 53
		Copper	Aq. 25% H_2SO_4 contg. 5–10g TiO_2	60–80			92 (CE)	54–56
4	o-Nitrotoluene	Tinned copper (cylindrical)	Aq. HCl	40–45	10–20	o-Toluidine	86–89	57, 58
5	m-Nitrotoluene	Copper (cylindrical)	Aq. 10% H_2SO_4 contg. 0.5% $CuSO_4$	70–80	2–20	m-Toluidine	80–86	59
6	p-Nitrotoluene	Tinned copper	Aq. 5% HCl	35–40	10–30	p-Toluidine	75–80	60, 61
7	p-Nitrobenzoic acid	Tinned copper (cylindrical)	Aq. HCl	65–70	10–20	p-Amino benzoic acid	90–95	62–64
8	3 Nitro-p-cresol	Copper (disc)	Aq. H_2SO_4	50–55	10–20	3 Amino p-cresol	80–95	65, 66
9	m-Nitrobenzene sulfonic acid	Copper (cylinder)	Aq. H_2SO_4	30–40	2–12	Metanilic acid	57–70	67, 68
10	p-Nitrophenetole	Copper (cylinder)	Aq. H_2SO_4	60–65	15–20	p-Phenetidine	70–75	69

Reaction Conditions

Table 8.12.   Effect of Current Densities on the Reduction of Nitrobenzene in Sodium Hydroxide Medium
Conditions: catholyte, 10% NaOH; anolyte, 30% NaOH; temp., 80–85°C; anode, nickel plated mild steel; separator, white asbestos cage.

Current Density (A/dm^2)	Mode of Cathode	Yield of Hydrazobenzene (%)
5	Rotation	86.5
5	Stationary	—
10	Rotation	87.1
10	Stationary	13.5
20	Rotation	88.0
20	Stationary	10.7
25	Rotation	88.5
30	Rotation	88.2

Table 8.13.   Products Formed during the Reduction of Nitrocompounds in Alkaline Medium
Conditions: catholyte, 10% NaOH; anolyte, 30% NaOH; temp., 80–85°C; anode, nickel plated mild steel; cathode, rotating lead deposited mild steel

Nitrocompound	Current Density (A/dm^2)	Yield (%)		
		Benzidines	Azo Compound	Amines
Nitrobenzene	30	80.0	7.3	6.1
o-Nitrochloroben-zene	25	82.2	—	9.8
o-Nitroanisole	30	70.2	19.7	7.8
o-Nitrotoluene	20	55.0	23.0	12.9

reported with various inorganic redox couples in which the regeneration of the redox couple is carried out employing rotating anodes.

In situ Oxidation Using Hypobromite

Oxidation of glucose to gluconic acid was one of the earliest reactions studied employing rotating anodes [78–82] in a single stage by anodically generated bromine and/or hypobromite formed by hydrolysis of bromine. The rotating anode permitted the use of current densities up to 25 A/dm² and controlled the loss of bromine to a close limit, as compared to high loss of bromine encountered in a stationary electrode cell. The discharge of bromide ions at the anode controls the anode potential for the controlled oxidation of glucose to gluconic acid without leading to higher oxidation products [83]. The oxidation of glucose takes place in the bulk of the electrolyte and for efficient oxidation, it is necessary to maintain a finite concentration of bromine or hypobromite in the bulk of the electrolyte. The gluconic acid reacts with calcium carbonate to form calcium gluconate. The reactions taking place at the anode and cathode and the overall reaction are as follows:

Anodic reactions

$$2Br^- \longrightarrow Br_2 + 2e$$
$$Br_2 + H_2O \longrightarrow BrO^- + HBr$$
$$C_6H_{12}O_6 + BrO^- \longrightarrow C_6H_{12}O_7 + Br^-$$
$$2C_6H_{12}O_7 + CaCO_3 \longrightarrow Ca(C_6H_{11}O_7)_2 + CO_2 + H_2O.$$

Cathodic reaction

$$2H_2O + 2e \longrightarrow 2OH^- + H_2$$

Overall reaction

$$2C_6H_{12}O_6 + CaCO_3 + H_2O \longrightarrow Ca(C_6H_{11}O_7)_2 + CO_2 + 2H_2.$$

The current efficiency for the oxidation of glucose was almost quantitative. The presence of excess of calcium carbonate in the electrolyte helped maintain the pH around 6.0. This minimizes the possibility of the formation of bromate and higher oxidation products.

In situ Oxidation Using Ceric Sulfate

Use of ceric sulfate as the oxygen carrier was widely described in many electroorganic oxidation processes in dilute sulfuric acid, giving high efficiency for oxidation. Anthracene was oxidized to anthraquinone at a rotating lead anode in solutions containing sulfuric acid and ceric sulfate [84]. Comparison of the current efficiency and yield at different current densities both at rotating and stationary anodes is given in Table 8.14.

The current efficiency is found to be high at rotating anodes, especially at higher current densities.

Two-Stage Oxidation Using Manganic Sulfate

Trivalent manganese was reported to be stable in sulfuric acid above a concentration of 9 N. However, oxidation of aromatic hydrocarbons carried out in 55% H_2SO_4 containing manganous sulfate as oxygen carrier always resulted in a small quantity of resinous matter because of the high concentrations of sulfuric acid. Furthermore, current efficiency decreased with time. To avoid these drawbacks a two-stage process was developed [85–87] in

Table 8.14. Efficiency of Oxidation of Anthracene to Anthraquinone with Ceric Sulfate

Conditons: electrolyte, 500 cc 30% H_2SO_4 containing 10 g ceric sulfate; temp., 85-90°C; anthracene, 5 g; cathode, lead strips; anode, cylindrical lead.

Current Density (A/dm^2)	Motion of Anode	Current Passed $(A \cdot hr)$	Yield of Anthraquinone $(\%)$	Current Efficiency $(\%)$
1	Rotation	8.25	95	53
1	Stationary	11.2	86	35
2	Rotation	10.5	90	38.4
2	Stationary	16.8	91.5	25.0
4	Rotation	10.5	90	38.4
4	Stationary	16.8	36.2	9.9
10	Rotation	12.6	81.0	28.8
10	Stationary	22.0	46.4	9.06

which manganic sulfate of high concentration was prepared in an electrochemical cell employing rotating lead anodes that was later used for the oxidation of various aromatic hydrocarbons. The rotating anode did not show much improvement in current efficiency over the stationary anodes. The current efficiency was found to remain more or less the same (70 to 77%) when a paste of $MnSO_4$ in 55% H_2SO_4 was oxidized at a lead anode. A concentration of 285 g $MnSO_4$/liter 55% H_2SO_4 was found to be the maximum concentration beyond which there was no stirring effect even though a rotating anode was employed.

Oxidations of a number of aromatic hydrocarbons employing trivalent manganese in 55% H_2SO_4 are summarized in Table 8.15.

Two-Stage Oxidation Using Hexavalent Chromium (Chromic Acid)

The use of chromic acid as an oxidant for the preparation of organic compounds is well known. This type of process can be improved if chromium sulfate present in the spent solution is recovered as chromic acid by electrolytic oxidation. A process has been developed [97, 98] in which the use of a separate cathode chamber is avoided. The influence of rotating the anode on the current efficiency was studied.

High yields were obtained with rotating anodes in divided cells at a current density of ~ 20 A/dm². The increased energy consumption as a consequence of operating at higher current densities is not as much with rotating anodes as with stationary ones.

Two-stage processes were developed using regenerated chromic acid for

Table 8.15. Efficiency of Oxidation of Aromatic Hydrocarbons with Manganic Sulfate

Hydrocarbon Oxidized	Product Formed	Oxidation Efficiency Based on Manganic Sulfate Used (%)	Overall Current Eff. for Aldehyde Formation (%)	Ref.
Toluene	Benzaldehyde	60–65	44–46	88–90
o-Xylene	o-Tolualdehyde	—	35–40	91–94
m-Xylene	m-Tolualdehyde	—	25–35	91–94
p-Xylene	p-Tolualdehyde	60	42–45	91–94
Mixed xylenes	Mixed tolualdehydes	55	51–57	91–94
m-Chlorotoluene	m-Chlorobenzal-dehyde	60.4	33.2	95
p-Cymene	Resinous product	—	. . .	95
Anethole	Anisaldehyde	—	25	95
Ethylbenzene	Acetophenone	25	25	96

the oxidation of various aromatic compounds, such as anthracene, p-nitrotoluene, and o-toluene sulfonamide. The oxidation of anthracene to anthraquinone occurs with a yield of 90% [99]. For the oxidation of p-nitrotoluene to p-nitrobenzoic acid [100, 101] the yield on the basis of p-nitrotoluene consumed was 88%, while the oxidation efficiency of chromic acid was 74.0 to 90%, depending on the number of regenerations of the chromic acid. For the oxidation of o-toluene sulfonamide to saccharin a yield of 80 to 85% was obtained [102, 103] based on the o-toluene sulfonamide consumed in the reaction.

5   EXPERIENCE WITH DESIGN OF HIGH-AMPERAGE CELLS EMPLOYING ROTATING ELECTRODES

In the preceding sections, it is shown that a number of reactions can be carried out preparatively using rotating electrodes. However, for larger scale production, it is necessary to develop higher amperage cells. Various factors must be taken into account in the design of such cells and our experience with the same is described below [104].

Rotating cylindrical or disc-type electrodes are used. They are fitted with a flange to the rotating assembly so that the electrodes can be easily removed as and when necessary for cleaning. If the dimensions of the flanges are standardized, one can make use of the rotating assembly to carry out any reaction by simply changing the type of electrodes required. These electrodes are suspended from the rotating assembly without any support at the bottom. This imposes restrictions on the length of the electrode, necessitating the use of multielectrode assembly in the scaled-up cells. Each electrode is connected to a mercury cup at the top through the flange placed in the rotating assembly. The direct current from the rectifier is fed through these mercury cups to the rotating electrodes. Figure 8.7 shows the arrangement of the 3000 A cell for the reduction of nitrobenzene to hydrazobenzene. There are two disc-type electrodes and each electrode carries a current of 1500 A. The permissible current density region for this reduction is 30 A/dm^2 and therefore each electrode is designed to have an area of 50 dm^2, within the permitted length. For

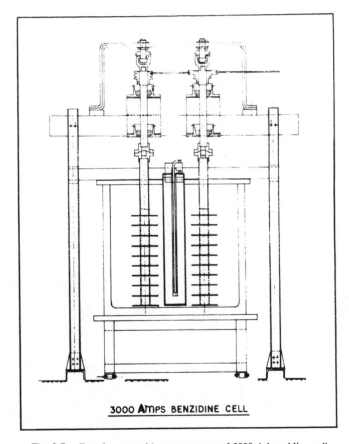

3000 AMPS BENZIDINE CELL

Fig. 8.7.   Rotating assembly arrangement of 3000-A benzidine cell.

the reduction of nitrobenzene to hydrazobenzene, a rotating mild steel cathode deposited with lead has been recommended. Since the electrical conductivity of mild steel is one-sixth that of copper, enough cross-sectional area is provided for the mild-steel cathode to carry a current of 1500 A. However, the weight of the cathode would be too heavy necessitating a motor with high horsepower. To minimize the weight of the cathode, a copper rod is used as the core for the outer mild-steel pipe riveted to it. The major current is carried by the copper rod and there is a little voltage drop along the length of the cathode. Figure 8.8 shows the 3000-A cell for hydrazobenzene, which can reduce 30 kg of nitrobenzene in 12 hr.

Another important factor in rotating electrode cells is the choice of suitable material for the cell container. If metallic containers are used they become anodic or cathodic depending on the polarity of the rotating electrode. This not only adversely affects the reaction, it may also bring about the corrosion of the electrode. It is therefore necessary that the cell container be a nonconductor—the commonly used materials being glass-lined, ceramic, PVC, pitch-lined, or epoxy-lined vessels or reinforced cement concrete tanks.

In cells requiring the use of separators, the separators have to be placed as close to the electrode as possible to minimize the cell voltage. This is possible only when the rotation of the electrode is quite smooth without any wobbling.

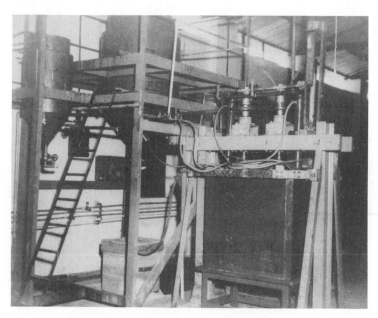

Fig. 8.8.   Photograph of 3000-A benzidine cell.

In cells having multielectrode assembly, the placement of counterelectrodes is important. It is better to consider each of the electrodes as an individual unit. Figure 8.9 shows the disposition of the anodes and separators around the two rotating cathodes in the 3000-A benzidine cell.

In reactions where there is no need to separate the anode and cathode compartments, the cell arrangement becomes simpler. The auxiliary electrodes are arranged surrounding the rotating working electrode in the form of strips or rods forming a regular polygon. This arrangement allows uniform current distribution at lower cell voltages and at high current densities, and free flow of the electrolyte around the rotating electrodes. Figure 8.10 shows the 500-A cell set up for the oxidation of glucose to calcium gluconate employing three rotating anodes, each surrounded by 10 graphite strips as cathodes. The top view of this cell showing the cathode assembly around each rotating graphite anode is given in Figure 8.11. The bromine evolved on the rotating anodes is removed as fast as it is formed, facilitating the bulk chemical reaction. The

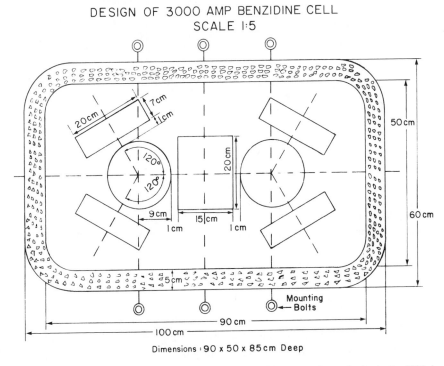

DESIGN OF 3000 AMP BENZIDINE CELL
SCALE 1:5

Dimensions : 90 x 50 x 85 cm Deep

Fig. 8.9. Disposition of anodes and separators around the two rotating cathodes in the 3000-A benzidine cell.

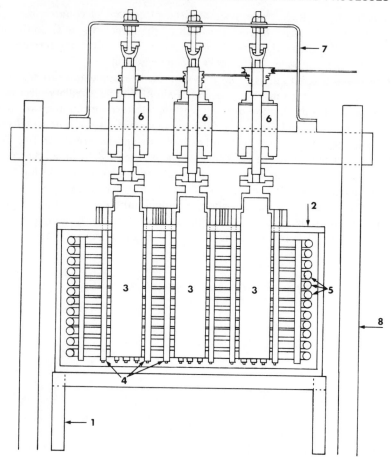

Fig. 8.10. Five-hundred ampere cell set up for the oxidation of glucose to calcium gluconate. (*1*) Wooden table, (*2*) PVC line wooden cell, (*3*) rotating graphite anodes, (*4*) stationary cathodes, (*5*) cooling coils, (*6*) core box assembly, (*7*) busbar, (*8*) R.C.C. support.

fast removal of the bromine from the surface of the anode also helps in improving the anode life. (No deterioration of the graphite anode has been noticed even after 3 years of plant operation*.) Figure 8.12 shows the photograph of the 500-A cell for the oxidation of glucose to calcium gluconate at a rate of 1.66 kg/hr in batch operations.

---

*Three plants each with a capacity of 200 kg/day are in operation and four more are in different stages of implementation based on the above technology.

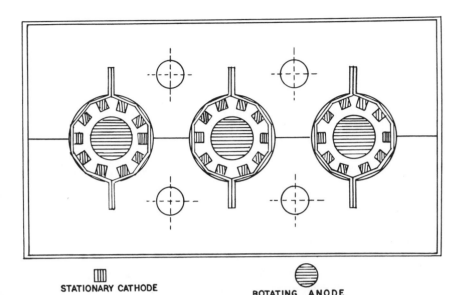

STATIONARY CATHODE   ROTATING ANODE

Fig. 8.11. Top view of the 500-A calcium gluconate cell.

Fig. 8.12. Photograph of 500-A calcium gluconate cell.

417

References

1. H. V. K. Udupa and B. B. Dey, *Proceedings of the Sixth Meeting of the International Committee of Electrochemical Thermodynamics and Kinetics,* Butterworths Scientific Publications, London, (1955), p. 87.
2. C. L. Wilson and H. V. K. Udupa, *J. Electrochem. Soc.,* **99,** 289 (1952).
3. C. L. Tobias, *J. Electrochem. Soc.,* **120,** 65C (1973).
4. R. Kanakam, M. S. V. Pathy, and H. V. K. Udupa, *Electrochim. Acta,* **12,** 329 (1967).
5. H. V. K. Udupa, M. S. V. Pathy, and R. Kanakam, Indian Patent 102,485 (1965), [*Chem. Abstr.,* **81,** 130199 (1974)].
6. E. Dineen, T. C. Schwan, and C. L. Wilson, *Trans. Electrochem. Soc.,* **96,** 226 (1949).
7. H. V. K. Udupa, Ph.D. Dissertation, Ohio State University (1950).
8. C. L. Wilson and K. B. Wilson, *Trans. Electrochem. Soc.,* **84,** 153 (1943).
9. K. S. Udupa, Ph.D. Thesis, Banaras Hindu University, Varanasi, 1971.
10. H. J. Creighton, *Trans. Electrochem. Soc.,* **75,** 289 (1939).
11. D. H. Killefer, *Ind. Eng. Chem. News Ed.,* **15,** 489 (1937).
12. M. T. Sanders and R. A. Hales, *Trans. Electrochem. Soc.,* **96,** 241 (1949).
13. M. S. V. Pathy, *Bull. Acad. Polon. Sci.,* **9,** 537 (1961).
14. C. L. Wilson and K. B. Wilson, *Trans. Electrochem. Soc.,* **80,** 151 (1941).
15. H. V. K. Udupa, G. S. Subramanian, K. S. Udupa, and K. Natarajan, *Electrochim. Acta,* **9,** 313 (1964).
16. S. Thangavelu, G. S. Subramanian, and H. V. K. Udupa, *Proc. 13th Seminar Electrochem.,* CECRI, Karaikudi, **1972,** 30.
17. H. V. K. Udupa, G. S. Subramanian, and S. Thangavelu, Indian Patent 132,716 (1971).
18. B. B. Dey and H. V. K. Udupa, *Curr. Sci.,* **22,** 371 (1953).
19. H. V. K. Udupa and B. B. Dey, *Proceedings of the Sixth Meeting of the International Committee of Electrochemical Thermodynamics and Kinetics,* Butterworths Scientific Publications, London, (1955), p. 465.
20. S. Satyanarayana and H. V. K. Udupa, *Bull. Acad. Polon. Sci.,* **7,** 625 (1959).
21. H. V. K. Udupa, *Bull. Acad. Polon. Sci.,* **9,** 51 (1961).
22. S. Ganesan, K. S. Udupa, G. S. Subramanian, and H. V. K. Udupa, *Chem. Age India,* **13,** 346 (1962).
23. K. S. Udupa, G. S. Subramanian, and H. V. K. Udupa, *Ind. Chem.,* **39,** 238 (1963).
24. T. D. Balakrishnan, K. S. Udupa, G. S. Subramanian, and H. V. K. Udupa, *Chem. Ind.,* 1622 (1970).
25. H. V. K. Udupa and B. B. Dey, Indian Patent 52,631 (1954).
26. H. V. K. Udupa and B. B. Dey, Indian Patent 60,864 (1957) [*Chem. Abstr.,* **54,** 7381 (1960)].
27. H. V. K. Udupa, G. S. Subramanian, K. S. Udupa, and T. D. Balakrishnan, Indian Patent 123,645 (1969) [*Chem. Abstr.,* **77,** 88092 (1972)].
28. K. Natarajan, K. S. Udupa, G. S. Subramanian, and H. V. K. Udupa, *J. Electrochem. Technol.,* **2,** 151 (1964).

29. H. V. K. Udupa, G. S. Subramanian, K. S. Udupa, and K. Natarajan, Indian Patent 88,851 (1963) [*Chem. Abstr.*, **81**, 25339 (1974)].

30. A. Muthukumaran, V. Krishnan, and H. V. K. Udupa, *Proc. 13th Seminar Electrochem.*, *CECRI, Karaikudi*, **1972**, 40.

31. H. V. K. Udupa, V. Krishnan, and A. Muthukumaran, Indian Patent 2350/Cal/73.

32. K. S. Udupa, G. S. Subramanian, and H. V. K. Udupa, *Bull. Acad. Polon. Sci.*, **14**, 849 (1966).

33. H. V. K. Udupa, G. S. Subramanian, and K. S. Udupa, Indian Patent 99,181 (1965).

34. M. Yamashita and K. Sugino, *J. Electrochem. Soc.*, **104**, 100 (1957).

35. T. D. Balakrishnan, K. S. Udupa, G. S. Subramanian, and H. V. K. Udupa, *Ind. Eng. Chem.*, **10**, 495 (1971).

36. H. V. K. Udupa, G. S. Subramanian, K. S. Udupa, and T. D. Balakrishnan, Indian Patent 127,059 (1970) [*Chem. Abstr.*, **77**, 121555 (1972)].

37. K. S. Udupa and H. V. K. Udupa, *Trans.*, *SAEST*, **11**, 153 (1976).

38. G. S. Krishnamurthy, H. V. K. Udupa, and B. B. Dey, *J. Sci. Ind. Res.*, **15B**, 47 (1956).

39. B. B. Dey, H. V. K. Udupa, and G. S. Krishnamurthy, *Bull. Cent. Electrochem. Res. Inst.*, **2**, 62 (1955).

40. T. D. Balakrishnan, K. S. Udupa, G. S. Subramanian, and H. V. K. Udupa, *Chem. Ing. Tech.*, **44**, 626 (1972).

41. G. S. Krishnamurthy, H. V. K. Udupa, and B. B. Dey, Indian Patent 53195 (1954).

42. H. V. K. Udupa, G. S. Subramanian, K. S. Udupa, and T. D. Balakrishnan, Indian Patent 126,677 (1970) [*Chem. Abstr.*, **82**, 36695 (1975)].

43. K. Jayaraman, K. S. Udupa, and H. V. K. Udupa, *Trans.*, *SAEST*, **12**, 143 (1977).

44. H. V. K. Udupa, K. S. Udupa, and K. Jayaraman, Indian Patent 2441/Cal/74.

45. H. C. Rance and J. M. Coulson, *Electrochim. Acta*, **14**, 283 (1969).

46. P. N. Anantharaman, G. S. Subramanian, and H. V. K. Udupa, *J. Appl. Chem.*, **2**, 169 (1972).

47. A. Pourasamy, P. N. Anantharaman, G. S. Subramanian, and H. V. K. Udupa. *Proc. 13th Seminar Electrochem.*, *CECRI, Karaikudi*, (1972), p 34.

48. H. V. K. Udupa, G. S. Subramanian, P. N. Anantharaman, and A. Pourasamy, Indian Patent 1207/72.

49. H. V. K. Udupa, G. S. Subramanian, and P. N. Anantharaman, Indian Patent 128412 (1973) [*Chem. Abstr.*, **82**, 23759 (1975)].

50. P. N. Anantharaman, G. S. Subramanian, and H. V. K. Udupa, *Indian J. Technol.*, **4**, 271 (1966).

51. H. V. K. Udupa, G. S. Subramanian, and P. N. Anantharaman Indian Patent 97,001 (1964) [*Chem. Abstr.*, **70**, 67893 (1969)].

52. H. V. K. Udupa and M. V. Rao, *Electrochim. Acta*, **12**, 353 (1967).

53. K. Swaminathan, P. N. Anantharaman, G. S. Subramanian, and H. V. K. Udupa, *Indian J. Technol.* **9**, 330 (1971).

54. A. Pourasamy, P. N. Anantharaman, G. S. Subramanian, and H. V. K. Udupa, *Proc. 14th Seminar Electrochem., CECRI, Karaikudi,* (1973), p 103.

55. H. V. K. Udupa, G. S. Subramanian, P. N. Anantharaman, and A. Pourasamy, Indian Patent 134,374 (1972).

56. H. V. K. Udupa, P. N. Anantharaman, and A. Pourasamy, Indian Patent 2436/Cal/73.

57. S. Chidambaram, M. S. V. Pathy, and H. V. K. Udupa, *Bull. Acad. Polon. Sci.,* **20,** 39 (1972).

58. H. V. K. Udupa, M. S. V. Pathy, and S. Chidambaram, Indian Patent 113,643 (1968) [*Chem. Abstr.,* **74,** 119466 (1971)].

59. T. D. Balakrishnan, P. N. Anantharaman, and G. S. Subramanian, *Indian J. Technol.,* **5,** 396 (1967).

60. R. Sridharan, S. Chidambaram, M. S. V. Pathy, and H. V. K. Udupa, *Proc. 13th Seminar Electrochem., CECRI, Karaikudi,* (1972), p 37.

61. H. V. K. Udupa, M. S. V. Pathy, S. Chidambaram, and R. Sridharan, Indian Patent 1491/72.

62. R. Kanakam, A. P. Shakuntala, S. Chidambaram, M. S. V. Pathy, and H. V. K. Udupa, *Electrochim. Acta,* **16,** 423 (1970).

63. H. V. K. Udupa, M. S. V. Pathy, S. Chidambaram, R. Kanakam Srinivasan, and S. Balagopal, *Proc. Int. Symp. Ind. Electrochem., SAEST, Karaikudi,* **1976,** 7.

64. H. V. K. Udupa, M. S. V. Pathy, A. P. Shakuntala, and R. Kanakam, Indian Patent 108,573 (1973).

65. P. N. Anantharaman, G. S. Subramanian, and H. V. K. Udupa, *Bull. Acad. Polon. Sci.,* **18,** 629 (1970).

66. H. V. K. Udupa, G. S. Subramanian, and P. N. Anantharaman, Indian Patent 110,852 (1967) [*Chem. Abstr.,* **72,** 106611 (1970)].

67. S. Thangavelu, G. S. Subramanian, and H. V. K. Udupa, *Denki Kagaku,* **38,** 5 (1971).

68. H. V. K. Udupa, G. S. Subramanian, and S. Thangavelu, Indian Patent 119,015 (1969) [*Chem. Abstr.,* **82,** 36697 (1975)].

69. H. V. K. Udupa et al., unpublished work.

70. T. Sekine, *J. Chem. Soc. Jap. (Pure Chem. Sec.),* **77,** 67 (1956).

71. I. Atanasiu and D. Dumitru, *Rev. Chim. (Buchar.),* **9,** 129 (1958) [*Chem. Abstr.,* **53,** 5916 (1959).

72. K. S. Udupa, G. S. Subramanian, and H. V. K. Udupa, *J. Electrochem. Soc.,* **108,** 373 (1961).

73. T. D. Balakrishnan, K. S. Udupa, G. S. Subramanian, and H. V. K. Udupa, *Chem. Ing. Tech.,* **41,** 776 (1969).

74. H. V. K. Udupa, G. S. Subramanian, K. S. Udupa, K. Jayaraman, T. D. Balakrishnan, M. Abdul Kadar, S. Krishnamurthy, and C. Seshadri, *Proc. 13th Seminar Electrochem., CECRI, Karaikudi,* (1972), p 57.

75. H. V. K. Udupa, G. S. Subramanian, and K. S. Udupa, Indian Patent 68,869 (1959) [*Chem. Abstr.,* **56,** 12670 (1962).

76. K. S. Udupa, G. S. Subramanian, and H. V. K. Udupa, *Bull. Chem. Soc. Jap.,* **35,** 1168 (1962).

77. K. S. Udupa, G. S. Subramanian, and H. V. K. Udupa, *Bull. Acad. Polon. Sci.*, **9**, 541 (1961).

78. V. Sarada Menon, H. V. K. Udupa, and B. B. Dey, *J. Sci. Ind. Res.*, **13B**, 746 (1954).

79. B. B. Dey, H. V. K. Udupa, and V. Sarada Menon, *Bull. Cent. Electrochem. Res. Inst.*, **2**, 59 (1955).

80. V. Sarada Menon, H. V. K. Udupa, and B. B. Dey Indian Patent 51,189 (1954).

81. H. V. K. Udupa, G. S. Subramanian, K. S. Udupa, T. D. Balakrishnan, M. Abdul Kadar, and S. Krishnamurthy, *Proc. 14th Seminar Electrochem., CECRI, Karaikudi,* (1973), p 79.

82. H. V. K. Udupa, G. S. Subramanian, K. S. Udupa, and T. D. Balakrishnan, Indian Patent 105/Del/77.

83. S. Satyanarayana and H. V. K. Udupa, *Bull. Acad. Polon. Sci.*, **6**, 493 (1958).

84. S. Krishnan, V. A. Vyas, M. S. V. Pathy, and H. V. K. Udupa, *J. Electrochem. Soc. (India)*, **14**, 32 (1965).

85. M. S. V. Pathy, R. Ramaswamy, and H. V. K. Udupa, *Curr. Sci.*, **28**, 63 (1959).

86. M. S. V. Pathy, R. Ramaswamy, and H. V. K. Udupa, *Proc. Symp. Electrolytic Cells, CECRI, Karaikudi,* (1961), p 147.

87. H. V. K. Udupa, M. S. V. Pathy, and R. Ramaswamy, Indian Patent 62,379 (1957) [*Chem. Abstr.*, **54**, 7384 (1960)].

88. M. S. V. Pathy, R. Ramaswamy, and H. V. K. Udupa, *Bull. Acad. Polon. Sci.*, **8**, 361 (1960).

89. M. S. V. Pathy, R. Ramaswamy, and H. V. K. Udupa, *Bull. Acad. Polon. Sci.*, **6**, 487 (1958).

90. H. V. K. Udupa, M. S. V. Pathy, and R. Ramaswamy, Indian Patent 62426 (1957).

91. M. S. V. Pathy and H. V. K. Udupa, *Proc. Electrolytic Cells, CECRI, Karaikudi,* **1961**, 204.

92. M. S. V. Pathy, R. Ramaswamy, and H. V. K. Udupa, *Bull. Acad. Polon. Sci.*, **7**, 629 (1959).

93. R. Ramaswamy, M. S. V. Pathy, and H. V. K. Udupa, *J. Electrochem. Soc.*, **110**, 202 (1963).

94. H. V. K. Udupa and M. S. V. Pathy, Indian Patent 66,175 (1958) [*Chem. Abstr.*, **55**, 15421 (1961)].

95. S. Chidambaram, M. S. V. Pathy, and H. V. K. Udupa, *J. Electrochem. Soc. (India)*, **17**, 95 (1968).

96. J. Alamelu, K. S. Lalitha, M. S. V. Pathy, and H. V. K. Udupa, *Proc. 12th Seminar Electrochem., CECRI, Karaikudi,* **1972**, 13.

97. S. Chidambaram, M. S. V. Pathy, and H. V. K. Udupa, *Indian J. Technol.*, **6**, 12 (1968).

98. H. V. K. Udupa, M. S. V. Pathy, and S. Chidambaram, Indian Patent 95 425 (1964) [*Chem. Abstr.*, **70**, 43435 (1969).

99. J. Alamelu, K. S. Lalitha, M. S. V. Pathy, and H. V. K. Udupa, *Tra. s., SAEST,* **5**, 148 (1970).

100.   S. Chidambaram, S. Balagopal, R. Kanakam, A. P. Shakuntala, M. S. V. Pathy, and H. V. K. Udupa, *Res. Ind.*, **15**, 215 (1970).
101.   S. Chidambaram, M. S. V. Pathy, and H. V. K. Udupa, *Indian J. Appl. Chem.*, **34**, 1 (1971).
102.   H. V. K. Udupa, M. S. V. Pathy, and S. Chidambaram, Indian Patent 116,488 (1968).
103.   H. V. K. Udupa, M. S. V. Pathy, and S. Chidambaram, Indian Patent 144,210 (1976).
104.   H. V. K. Udupa, *Electroorganic Synthesis Technol., Symp. Ser.*, No. 185, **75**, 26 (1979).

Chapter IX

ELECTROCHEMICAL ENGINEERING OF ELECTROORGANIC PROCESSES: ILLUSTRATED BY AN ENERGY ASSESSMENT OF SOME LARGE-TONNAGE PROCESS CHEMICALS

Theodore Beck
Electrochemical Technology Corporation,
Seattle, Washington

Richard Alkire
Department of Chemical Engineering,
University of Illinois,
Urbana, Illinois

Norman L. Weinberg
The Electrosynthesis Company,
East Amherst, New York

Robert Ruggeri
Electrochemical Technology Corporation
Seattle, Washington

and

Mark Stadtherr
Department of Chemical Engineering
University of Illinois
Urbana, Illinois

1 INTRODUCTION

The industrial synthesis of organic chemicals has always been practiced in the presence of uncertainty. Therefore it is necessary constantly to evaluate new candidate processes in the search for those that will be compatible with the future. Included among routes to the synthesis of many organic compounds are those involving electrolysis. The purpose of this chapter is to describe progress in various electrochemical fields related to industrial organic synthesis and to present new calculations of energy usage for both electrochemical and chemical routes of several high-tonnage organic chemicals.

The pace of electrochemical science and technology quickened during the very recent past. There have been major advances in the elucidation of electrochemical fundamentals in areas of reaction mechanisms, solvent effects, and the influence of the electrode on the course of reaction. A large number of experimental techniques have only recently become available for the study of electrochemical processes, many of which were triggered by concurrent developments in electronics, computer data processing, and catalytic science. Engineering methodologies have advanced rapidly as principles of current and potential distribution phenomena have been shaped into strategies for engineering design, scale-up, and optimization of electrolytic systems. New materials for electrodes, membranes, process sensors, and other cell components have appeared, often with extraordinary impact, as in the case of metal anodes and ion-selective membranes. Electrochemical science and engineering is coming of age, and those who practice electrochemistry are increasingly able to describe their knowledge in a manner that promotes rational engineering evaluation and design.

The field of electroorganic synthesis has seen steady growth since the earliest days of chemical synthesis. The annual publication of electroorganic synthesis documents during the past three decades is given in Fig. 9.1. These citations include scientific publications, patents, books, reviews, laboratory manuals, and dissertations [1]. It may be seen that the rate of publication has doubled every 8 years over the past 3 decades. Even *without any further increase* in the publication rate, the entire body of knowledge, which dates back to 1801, will have doubled again between 1975 and 1990.

By 1975 there were some 8000 organic compounds for which electrochemical synthesis routes were available [1, 2]. In view of electrochemical activities in general, and electroorganic synthesis in particular, it seems inevitable that new industrial technologies will emerge based on electrolytic routes for synthesis of organic compounds.

This chapter is divided into three main parts. In the first (Section 2), a historical perspective is presented by compiling the known industrial syntheses along with the advantages and disadvantages commonly associated with electroorganic process technologies. In the second part (Section 3), various specific areas of electrochemical science and engineering are discussed briefly, since knowledge from these areas will contribute to the emergence of new technology. The third part (Section 4) reports on engineering methodologies for predicting energy consumption from both electrochemical and chemical routes to synthesis of certain large-tonnage organic compounds [3] (see also Chapter V). These example evaluations are only preliminary in the sense that no process optimization has been attempted.

Not included in the comparison of electrochemical *versus* chemical routes (Section 4) are considerations of operating or investment costs, wastewater treatment credits for the electrochemical routes, and compounds produced at rates below 7000 tons/year. That is, the conclusions reached in Section 4 may not correspond to the more balanced perspective by which candidate processes are evaluated by the industrial sector. However, the purpose of this chapter is to contribute a general engineering perspective that, by further refinement and expansion, would lead to rational assessment of electro-

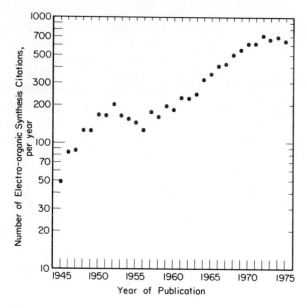

Fig. 9.1. Publication rate during the last three decades; data include papers, review, books, patents, and dissertations [1].

organic process routes, not only in the large-tonnage arena considered below, but also in other categories of intermediate and fine chemicals.

2 INDUSTRIAL ELECTROORGANIC PROCESSES

Current and Previously Used Processes

Industrial electroorganic synthesis has been practiced for many years and with various degrees of success. A measure of perspective is thus afforded by examining current and previously used processes. Table 9.1 provides a brief description of known processes. These include oxidation processes, reduction processes, and miscellaneous processes. The following symbols are used in the column marked "status" in Table 9.1.

P, pilot studies believed to be currently in progress.

EP, pilot studies that were begun and that are believed to have been abandoned.

C, commercial production currently in progress.

EC, commercial production that has been abandoned.

Table 9.1

Oxidations

Starting Material (Electrolyte)	Product(s)	Anode	Current Efficiency (%)	Energy Usage[a]	Status	Ref.
Anthracene (H$_2$O, Cr^{+6}, H$_2$SO$_4$)[b]	Anthraquinone	PbO$_2$	80	[4.8]	C—Holliday	4-7
Benzene (H$_2$O, H$_2$SO$_4$)	Hydroquinone (quinone)	PbO$_2$	40-60	15	P—Stavely, Carus, etc	8-10, 70
Butynediol (H$_2$O, H$_2$SO$_4$)	Acetylenedicarboxylic acid	PbO$_2$	75	[12.4]	P—BASF	11, 12, 69, 71
Cyanuric acid (H$_2$O, 20% KCl)	Pot. dichlorocyanurate	Pt/Ti	High	[3-5]	EP—Monsanto	13, 14
Dimethyldithiocarbamate (H$_2$O, sodium salt)	Tetramethylthiuram disulfide	Pt	90	1.7	P—Dupont	15, 16
Dimethyl sulfide (H$_2$O, DMSO, 5% H$_2$SO$_4$)	Dimethyl sulfoxide	Pt/Ti	100	0.7-1.3	P—Glanzstoff, Akzo?	17-20
Ethyl alcohol (H$_2$O, EtOH, NaOH, KI)	Iodoform	Pt	93-95	2	EC—Schering	21-23
Ethyl alcohol (H$_2$O, HCl)	Chloral	C	High	[8-10]	EC—Schering	24
Furan (CH$_3$OH, NaBr or NH$_4$Br)	Dimethoxydihydrofuran	C	85	2.5	C—BASF	25, 26, 69, 71
Glucose (H$_2$O, NaBr, CaCO$_3$)	Calcium gluconate	C	High	[2]	C—Sandoz, Chefaro?	27-29
Hydrogen cyanide (H$_2$O, NH$_4$Br)	Melamine	C	98	1.5-2	P—Sohio	30-31
Isobutanol (H$_2$O, 10% H$_2$SO$_4$)	Isobutyric acid	PbO$_2$	45	[14]	P—USSR	32-33
p-Methoxytoluene (H$_2$O, redox reagent, H$_2$SO$_4$)[b]	Anisaldehyde	PbO$_2$	High	[2.2]	P—Oxy	34
Methyl (ethyl) chloride (THF/diglyme, RMgCl)	Tetraalkyl lead	Pb	92	4-8	C—Nalco	35, 36
2-Methylnaphthalene (HOAc, NaOAc)	2-Methyl-a-naphthol acetate	C	Good	[1.9]	EP?—Socony EP—BASF	37, 69, 71
Monomethyl adipate (CH$_3$OH, sod. salt of acid)	Dimethylsebacate	Pt/Ti	75-85	5	EP—BASF	38-41, 69

427

Table 9.1. Continued

Oxidations

Starting Material (Electrolyte)	Product(s)	Anode	Current Efficiency (%)	Energy Usage[a]	Status	Ref.
Naphthalene	1-Naphthylacetate	C		[1.6]	P—BASF	69, 71
p-Nitrotoluene (H_2O, Cr^{+6}, H_2SO_4)[b]	p-Nitrobenzoic acid	PbO_2	98	7.5	P—India	42-44
Oleic acid (H_2O, Cr^{+6}, H_2SO_4)[b]	Azelaic acid, pelargonic acid	PbO_2	High	[4]	EC—Energy	45-46
Phenol (H_2O, 3% H_2SO_4)	Hydroquinone (quinone)	PbO_2	50	[8-12]	P? Union Carbide, Eastman Kodak	47
Propargyl alcohol	Propiolic acid			[8.5]	P—BASF	69
Potassium alkylaluminate (melt)	Tetraalkyl lead	Pb	100	1	P—MPI	48
Propylene (H_2O, NaCl or NaBr)	Propylene oxide	C, DSA^R	70-88	4-6	P—Bayer, Kellog, EP—BASF	49-52, 69
Quinoline (H_2O, H_2SO_4)	Nicotinic acid	PbO_2	40	25	?	53
Starch (H_2O, NaOH, periodate)[b]	Dialdehyde starch	PbO_2	High	[1.5-2]	C—Miles	54-58
Toluene (H_2O, H_2SO_4, Ce^{+4})[b]	Benzaldehyde	PbO_2, Pt	90	[6]	P—ETH	59, 60
o-Toluenesulfonamide (H_2O, Cr^{+6}, H_2SO_4)[b]	Saccharin	PbO_2	78	[6]	EP—Boots	61, 62
Various organics (HF, KF)	Perfluoroorganics	Ni	Low	High	C—3M	63-65
Various organics (KF,2HF melt)	Perfluorinated and partially fluorinated Organics[c]	C	Fair	High	EP—Phillips	166-68

Reductions

Starting Material (Electrolyte)	Product(s)	Cathode	Current Efficiency (%)	Energy Usage[a]	Status	Ref.
Acetone (H_2O, H_2SO_4, Cu^{2+})	Pinacol	Pb	70–75	3.3	EP—BASF, Bayer, Diamond Shamrock	69, 71–73
Acrylonitrile (H_2O, quat. salt)	Adiponitrile	Pb alloy	90–92	3–6	C—Monsanto, Asahi	74–77
Adiponitrile (H_2O, HCl)	1,6-Hexanediamine	Ni(Cu)	60	[15.8]	P—USSR	78–79
Ammonium phthalamate (H_2O, phthalamate salt)	Phthalide	Pb	70	[5.7]	EP—BASF	71, 80, 81
Anthranilic acid	o-Aminobenzyl alcohol				P—BASF	69
Benzene (NH_3 or CH_3NH_2, NaCl or LiCl)	1,4-Cyclohexadiene	C, Al	80	[4.2]	EP—Esso Res	82–84
Benzenediazonium chloride (H_2O, HCl)	Phenylhydrazine hydrochloride	Hg	70	5	?	85
Benzoic acid (H_2O, C_2H_5OH, H_2SO_4)	Benzyl alcohol	Pb	93	4.1	P?—India	86, 87
Benzyl cyanide (H_2O, HCl)	β-Phenethylamine hydrochloride	Pd	35–42	[12]	P—India	88, 89
Diazoaminobenzene (H_2O, CH_3OH, THF, NaOH)	Phenylhydrazine	C	82	[7]	EP—Hoechst	71, 90, 91
N,N-Dimethylaminoethyltetra-chlorophthalimide (H_2O, HOAc, H_2SO_4)	Corresponding isoindole	Hg	78	[2]	EP?—Ciba	92–94
m-Dimethylaminobenzoic acid	m-Dimethylaminobenzyl alcohol	Hg^b	60	[3.9]	P—BASF	69
m-Dinitrobenzene (H_2O, H_2SO_4)	2,4-Diaminophenol			[18]	P—India	95, 96
Dimethyl terephthalate	p-Carbomethoxybenzyl alcohol			[3.6]	P—Hoechst	70, 132

Table 9.1. Continued

		Reductions				
Starting Material (Electrolyte)	Product(s)	Anode	Current Efficiency (%)	Energy Usage[a]	Status	Ref.
Glucose (H_2O, Na_2SO_4)	Sorbitol	Amalgamated lead	90	1.5-2	EC—Atlas	97, 98
4-Methylimidazolecarboxylic acid	4-Methyl-5-hydroxymethyl imidazole			[5.3]	P—BASF	69
α-Methylindole	α-Methyldihydroindole	Pb		[2.3]	C—BASF	69, 71
N-Methylpyridinium chloride (H_2O, sodium carbonate)	Bipyridylium chloride		High	[1.2]	P—ICI	99
Nitrobenzene (H_2O, H_2SO_4)	p-Aminophenol	Hg, C, Cu etc.	90	[5.5]	P—Bayer, Miles, CJB, India etc.	100–102
Nitrobenzene (H_2O, H_2SO_4, Tl^{3+})	Aniline	Cu, Pb	97	12.4	P?—India	103, 104
Nitrobenzene (H_2O, NaOH)	Benzidine	NaHg, steel	Good	[4–5]	EC—I. G. Farben P—India	105, 106
Nitrobenzene	p-Anisidine			[4.8]	P—BASF	69
m-Nitrobenzenesulfonic acid (H_2O, H_2SO_4)	Metanilic acid	Pb	80	8–13	P—L. B. Holliday, CJB, India BASF	69, 107, 108
Nitronaphthalene	1-Amino-4-methoxynaphthalene			[3.5]	P—BASF	69
Nitroguanidine [H_2O, $(NH_4)_2SO_4$]	Aminoguanidine bicarbonate	Zn	89	[12]	P—India	109, 110
p-Nitrophenol (H_2O, H_2SO_4)	p-Aminophenol	Cu, Hg/Cu	60	5	P—India	111, 112
Nitrourea (H_2O, H_2SO_4)	Semicarbazide	Pb	Good	[13]	P—USSR	113, 114
Oxalic acid (H_2O, H_2SO_4)	Glyoxylic acid	Pb	97	[3.7]	P—Rhône Poulenc others	115–117
Phthalic acid (H_2O, dioxane, H_2SO_4)	Dihydrophthalic acid	Pb	90–95	1.6-4	EC—BASF	69, 118-120

Starting Material (Electrolyte)	Product(s)	Anode	Cathode	Current Efficiency (%)	Energy Usage[a]	Status	Refs.
Phthalimide (H_2O, HOAc, H_2SO_4)	Isoindole		Pb	50-55	[18]	P—Ciba	121, 122
Pyridine (H_2O, H_2SO_4)	Piperidine		Pb	90	[11]	C—Robinson Bros.	123
Quinoline (H_2O, H_2SO_4)	Tetrahydroquinoline		Pb	74	[5.4]	P?—Russ	124
Salicylic acid (H_2O, $NaHSO_3$, H_3BO_3)	Salicylaldehyde		Hg/Cu	50	[4.4]	C—USSR, India	125, 126
Tetrahydrocarbazole (H_2O, C_2H_5OH, H_2SO_4)	Hexahydrocarbazole		Pb	High	[1.8]	C—BASF	69, 71, 127–129

Miscellaneous Processes

Starting Material (Electrolyte)	Product(s)	Anode	Cathode	Current Efficiency (%)	Energy Usage[a]	Status	Refs.
Sodium citrate[d] (H_2O, sodium citrate)	Citric acid, NaOH			High		C?—Liquichimica Biosintesi	134
Cyanide waste stream (H_2O, various)	CO_2, N_2[e]	C	Steel		12-15	C—several	135-137
Methanol (CH_3OH, see footnote[f])	Sodium methoxide	DSA^R	Hg	High		C—Dynamit Nobel, Olin	138
Montan wax (H_2O, chromic acid)	Purified montan wax	PbO_2	?			C—Hoechst	139, 143
Tetramethylammonium chloride[d] (H_2O, quaternary chloride)	Tetramethylammonium hydroxide, chlorine	DSA^R C	Steel	30-60		EC—Monsanto	140-142

[a] Brackets indicate estimated value based on a cell voltage of 5 V and current efficiency of 90% unless otherwise given. Energy usage in KWh/kg.

[b] The process is a two-stage operation in which the redox reagent (Mn^{3+}, Ce^{4+}, Cr^{6+}, periodate, hypobromite, etc.) is regenerated electrochemically and passed into a second reactor, where the oxidant and the organic feed are contacted.

[c] Including perfluoropropane, perfluorobutyric acid, trifluoroacetic acid, and perfluoromethylether.

[d] Example of electrodialysis.

[e] The initial products are cyanate and cyanogen.

[f] The process utilizes the sodium amalgam from mercury cells as follows: $2Na(Hg) + 2ROH \rightarrow 2NaOR + H_2 + (Hg)$

Commercial status indicates that an electrochemically derived product is being sold, but suggests no measure of the size of the operation. Of course it is difficult to determine the status of present activity, and Table 9.1 may thus be incomplete and/or in error in this regard. Another published compilation is available in Ref. 144.

According to Table 9.1, electrochemical routes to some 70 organic compounds have undergone evaluation under industrial constraints. Of these about 15 are believed to be commercially viable, while another three dozen are currently under active evaluation at the pilot-plant stage.

Of the roughly 50 processes currently known [130, 131, 133] to be under evaluation, well over half were known in the laboratory *at the turn of the century*. What has occurred in many of these cases is achievement of economical configuration through the development and adaptation of new cell materials (see Chapter IV) and designs.

Characteristics of Electroorganic Synthesis Processes

Experience to date with electroorganic processes has led to recognition of several advantages of electrolytic process routes:

1. Electroorganic syntheses usually require cheaper and thus more readily available feedstocks than alternate chemical process routes.

2. Electrochemical processes usually exhibit mild process conditions and low temperatures throughout so that entropy is conserved.

3. Electrochemical processes exhibit high chemical yields so that downstream recovery and purification is thus simplified.

4. Electricity is cleaner and easier to transport and handle than corrosive chemical reagents needed in many organic syntheses, so that the cost of containment materials is lower.

5. Wastewater treatment of effluents from electrolytic processes is minor owing to the absence of spent redox agents.

On the other hand, the disadvantages are as follows:

1. Operating costs are high because electricity is an expensive form of energy.

2. Investment costs are high.

These characteristics have been discussed at length by Fitzjohn [145] and Krumpelt et al. [146], who compared electrochemical and chemical processes over a range of production rates up to 500 million lb/year and concluded that investment costs per pound of product *were similar*.

In the field of industrial electrolysis, rules of thumb have come into use largely through experience with chlor-alkali, aluminum, and copper refining systems. It would be thoughtless, however, to apply such empirical rules to the evaluation of prospective electroorganic processes without some added measure of engineering insight. Rational evaluation of new electroorganic processes must be based on sound economic assessment, not rules of thumb developed from experience in other fields of electrochemical technology.

Uncertainties in process development must also be taken into account in the evaluation of new technologies. Systematic methodologies for obtaining electrochemical data are becoming available. Improvements in methods for transferring technology from laboratory cell to pilot plant are being reported with increased frequency. As understanding of the interdisciplinary nature of the field develops, exploration of process candidates has been carried out with greater perception and with more cost-effective development programs. The material in the next section has been selected to provide access to the background literature required for the engineering development of new electroorganic process technologies.

3 DEVELOPMENTS IN ELECTROCHEMICAL SCIENCE AND ENGINEERING

New techniques for the elucidation of mechanisms have been developed within the last decade and have been vigorously applied to the study of electroorganic synthesis reactions. During the same period, a large number of excellent books and reviews containing much of the fundamental information and techniques have become available. The purpose of this section is to draw together the components needed for engineering evaluation and development of electroorganic processes intended for commercial exploitation. The chapters that make up the remainder of this volume provide more thorough coverage of many specific aspects of the broad and interdisciplinary literature presented in this section.

Background Material

The electroorganic synthesis literature through 1975 is contained in the bibliographies of Fichter [2] and of Swann and Alkire [1]. These collections include scientific publications, reviews, books, laboratory manuals, dissertations, and patents. In addition, the polarographic and other properties of many electroorganic systems have been tabulated by Meites and Zuman [147]. A selected bibliography of source materials appears in Appendix I of Ref. 148 and includes books, monograph series, journals, proceedings and special issues, reviews, abstracting services, and bibliographies.

A description of past industrial electroorganic activity is available [149, 150], while more recent activity has been the subject of a symposium publication [146] containing over 20 papers. A perspective of progress during the past 25 years was recently offered by Baizer [151].

Electrochemical Synthesis and Mechanistic Studies

Although major reviews of electroorganic chemistry have appeared for more than 50 years [152], the publications of Swann [153] and Allen [154] were particularly timely and influential. During the past decade, a remarkable number of books and major reviews have appeared of which an incomplete, but representative, list would include those by Wawzonek [155], Eberson and Schäfer [156], Tomilov et al. [157], Fry [158], Baizer [148], F. Beck [159], Rifi and Covitz [160], Weinberg [161], Eberson and Nybert [162], Bard [163], and Langer and Sakellaropoulos [164].

Laboratory Procedures, Apparatus, Techniques

One advantage of electroorganic synthesis is the high selectivity often achieved for the desired product, a result of proper control of electrode potential and reaction environment (electrode, solvent, electrolyte, etc.). Procedures for laboratory investigations naturally depend on the goals at hand. Organic chemists and electrochemists tend to emphasize mechanistic aspects that would lead to a fundamental understanding of the mechanism, from which it may be possible to predict optimum conditions for high yield.

On the other hand, technologists often require information before there is understanding. Important initial considerations in the design of industrial processes include reaction stoichiometry, cell voltage, current density, current efficiency, and specification of the potential range over which high yields are possible. Of subsequent importance is knowledge of influence on yield of operating conditions such as temperature, solvent, electrolyte composition and concentration, reactant concentration, hydrodynamics, electrode material, and impurities in the solution phase. Knowledge of side reactions is critically important in order to design around them or, better yet, to avoid them. The decision of whether or not a separator is needed between anode and cathode can exert a major impact on process economics.

Engineering insight (see Chapter I) usually comes from knowledge of what phenomena control the behavior of the process, typical rate-limiting phenomena being ohmic resistance, mass transfer, and charge-transfer kinetics. It is invariably easier to obtain such insight from experiments conducted under controlled and reproducible conditions in continuous flow cells.

The book of Gileadi et al. [165] provides an excellent introduction to numerous electrochemical laboratory techniques. The chapter by Goodridge and King in Part I of this series [166] describes not only experimental aspects, but also more fundamental considerations involved in the choice of electrode, solvent, and electrolyte materials. Additional sources for laboratory methods for electroorganic investigations include Swann [153], Adams [167], Wawzonek [155], Eberson and Schäfer [156], Fry [158], Tomilov et al. [157], Baizer (Chpt. II by Cauquis and Parker) [148], Rifi and Covitz [160], and F. Beck [159].

The elucidation of mass transfer processes often relies on electrochemical measurements by the limiting-current method for which an extensive review is available [168].

Cell Component Materials: Electrodes, Solvent, Electrolyte, Separator

Qualitative aspects of specification of electrode material (see Chapter IV) have been widely reported, but fundamental understanding is meager. It may be expected that substantial progress will be made as understanding of surface science and catalysis is applied to electrochemical systems. Even applied studies, however, should be carried out with an appreciation for the importance of metallurgical factors on the electrochemical performance of an electrode material, since, for example, the temperature at which a particular electrode is cast can influence the product spectrum obtained.

General discussion of materials suitable for various electrochemical purposes appears in a book by Kuhn [169], in the aforementioned review of Goodridge and King [166] in Part I of this series, and in Chapter IV of a book of Baizer [148]. Additional discussions appear in works by Adams [167], F. Beck [159], Eberson and Nybert [162], and Danly [144].

The use of a separator between anode and cathode is often required when one electrode is corroded or poisoned by components in the feed to the other electrode; separate flow systems for catholyte and anolyte are thus required. Separators are also needed when the product of one electrode would react if it were to pass into the vicinity of the other electrode. For example [155], reduction of benzene in methylamine containing lithium chloride gives cyclohexene in a divided cell; in an undivided cell, however, the lithium methylamide in the catholyte reacts with the methylamine hydrochloride in the anolyte to yield dihydrobenzene. In simpler systems where homogeneous reactions are absent, careful hydrodynamic design can be used to ensure that reactants are removed from the cell before they diffuse across to the counterelectrode.

Source materials on separator materials for electrolytic cells, including ion-exchange membranes, may be found in Chapter IV of a book by Baizer [148] and in reviews by Goodridge and King [166], F. Beck [159], Eberson and Schäfer [156], and Danly [144]. More general coverage of membranes and membrane processes is provided by several recent books [170–176].

The choice of solvent, electrolyte, and solution additives is discussed in the works of Mann and Barnes [177], Goodridge and King [166], Baizer [148, Chapter IV], F. Beck [159], and Danly [144]. On the practical side, the attainment of high solubility of the organic, good mass transfer, and high ionic conductivity is sought in the presence of low cost, easy downstream recovery, and noncorrosive properties. The use of quaternary ammonium salts and their analogs is widely practiced in aqueous systems since these salts enhance both ionic conductivity and solubility of organic species, while, at the same time, contribute toward improved yields by hindering electrolytic breakdown of the solvent.

New Design Concepts for Electroorganic Cells

The attainment of high volumetric current density is important in order to keep investment costs at a minimum. Low power consumption is requisite to keeping operating costs down, and uniform potential distribution is needed for achieving high yields. A large number of design concepts have come forth that seek to achieve one or more of these goals. Reviews have been published by Houghton and Kuhn [178], Tomilov and Fioshin [179], Gallone [180], Goodridge and King [166], and Danly [144, 181]. The design concepts appear to fall into two general categories as described below.

The first category consists of cells wherein enhancement of mass transfer to the electrode surface is sought by manipulating electrode shape and fluid flow configurations. For example, flow-through porous electrodes exhibit high internal surface area and good mass transfer rates; recent reviews by Newman and Tiedemann [182, 183] survey the literature. There are essentially two flow configurations in use, the first having parallel directions of current and electrolyte flow [184], and the second having essentially perpendicular directions of current and electrolyte flow [185]. The latter configuration appears to be emerging as superior for applications involving electroorganic synthesis and has been used, for example, in the Braithwaite cell for tetraalkyl lead synthesis [186], for the "Swiss-roll" cell configuration of Ibl and co-workers [187], and for the "ESE" cells described by Keating and Sutlic [188]. Other methods for achieving high mass transfer rates at electrode surfaces include placement of stationary "turbulence promoting" grids on or near the electrode surface [189], use of continuously moving solid electrodes [190, 191], use of wipers that periodically disrupt the mass transfer layer [192], and various types of fluidized [193, 194] and tumbling [195] bed systems. In many electroorganic systems, the reactants include organic vapors and liquids that are only modestly soluble in polar solvents; thus various bipolar [196] and trickle-bed [197] cell designs have been suggested for enhancing gas/liquid contact, while emulsion cells [198] can be used to achieve enhanced liquid/liquid extraction rates.

The second general category of design concept emphasizes manipulation of the potential field distribution. Minimization of ohmic resistance losses has been achieved with the use of capillary-gap cells, developed originally by

Beck and Guthke [199], in which the anode/cathode gap spacing is very small; such cells have been extensively developed [200] and have been employed in pilot-plant operations in both divided and undivided configurations. Hydrodynamic and mass transfer conditions within capillary-gap cells have been modeled [201] in order to predict operating conditions under which such cells may operate in undivided configurations [202]. Along different lines, manipulation of the potential field to achieve bipolarity along an electrode surface [203] has been used for a variety of applications to electroorganic synthesis processes [195, 204].

To select a reactor concept appropriate for a given application it is important to bear in mind several elements of overall perspective. Electrochemical cells, for example, are always carefully designed to match the process chemistry. Adaptation of a concept to actual hardware involves countless decisions that must be based on the process chemistry. Also, the relative importance of investment capital, electric power cost, and chemical yield depends greatly on the selling price. Appropriate design concepts must therefore be responsive to the economic constraints under which the process will eventually be judged. In addition, the economic evaluation of various reactor configurations must incorporate the entire array of peripheral equipment, including electrical distribution, flow devices, and cooling equipment; thus seeming advantages in cell operation may be offset by nonelectrochemical considerations.

Engineering Methods, Modeling, and Economic Assessment*

The papers of MacMullin provide general discussions of engineering procedures and considerations involved in scale-up of electroorganic and other electrolytic processes, as well as of the design and materials of construction of cells, of innovation, and of the use of computer simulation in design [204]. The use of analytical and numerical methods for simulation of cell processes based on current and potential distribution phenomena represented an important development in the mathematical modeling of electrochemical processes [205].

The elucidation of fundamental principles of electrochemical engineering has led to improved ability to predict electrochemical behavior from knowledge of operating parameters. Principles of thermodynamics, double-layer phenomena, electrode kinetics, ohmic resistance, fluid mechanics, heat, and mass transport form the underlying concepts needed for electrochemical engineering [206]. While mathematical complications may be formidable for highly exact results, the deft use of simple limiting cases can often result in cost-effective improvements. In many cases it is not necessary to solve equa-

*See Chapter I for details.

tions, but only to recognize the dimensionless variables that dominate behavior of the process. That is, modeling electrochemical processes provides a methodology for dealing with the unknown. Perhaps the greatest value of such endeavors is the quantitative and rational thought processes thereby promoted.

A book by Hine [207] provides an excellent general introduction, while that by Newman [206] is rigorous and detailed. Concepts involved in electrochemical reactor design are described from a chemical engineering viewpoint by Pickett [208].

The application of current and potential distribution concepts to electroorganic synthesis cells promises to provide added leverage needed to design and optimize reactors. For example, one major advantage of electroorganic syntheses is the high degree of selectivity for a desired reaction product that can be achieved through careful control of the potential along the electrode surface. However, to achieve large reaction rates per unit volume, it is necessary to optimize around competing needs for large surface area, adequate mass transfer, and low ohmic resistance, in addition to uniform potential distribution. Defining the optimum process conditions that achieve both high selectivity and high reaction rate is clearly an important engineering design consideration [209]. In electroorganic syntheses, where multiple reactions occur, the use of modeling techniques appears to be especially important to avoid loss of intuition.

In the near future it seems reasonable to expect that modeling activities will continue to expand in scope so that entire cells will be included within the model. Examples to date include nonelectroorganic processes [204, 210–212], although the extension of such methods to electroorganic systems is clearly within grasp. The consequence of this development would be to greatly enhance the capability for accurate engineering evaluation and design.

Use of simple models in optimization of electrochemical systems has been extensively reviewed by Beck [213], who provides many references and examples of specific design trade-offs. Fitzjohn [145, 214] and Danly [144] provide discussion and references on additional economic considerations involving investment and operating costs.

It is generally agreed that investment costs for electrolytic cells scale with a power between 0.9 and 1.0. Other (nonelectrolytic) components of the process scale with a 0.5 to 0.6 power and can often represent a significant fraction of total investment [144, 145, 214], so that the overall process may exhibit substantial economies of scale. While the exponents cited above are generally accepted, the literature is nearly devoid of data on actual investments required. The examples reported by Fitzjohn [145, 214], Danly [144], and Keating and Sutlic [188] therefore represent particularly important contributions. Additional data on methods for investment cost estimation would be extremely beneficial in promoting the accuracy with which electroorganic process economics may be evaluated.

4 ENERGY USAGE OF LARGE-TONNAGE ORGANIC ELECTROLYTIC PROCESSES

The objective of this section is to demonstrate the use of chemical engineering methods for evaluation of energy consumption to be expected of electroorganic processes in comparison with chemical processes. Additional details may be found in Ref. 3.

No attempt was made to iterate the process calculations to achieve optimized electrochemical designs. The results reported below could undoubtedly be improved upon by simple optimization procedures and by better pilot-plant data. Only synthesis of large-tonnage organic chemicals was evaluated. Many additional low-tonnage process candidates exist. Other important engineering considerations not introduced in the discussion below include economics of operating and investment costs, wastewater treatment, and synthesis of lower-tonnage chemicals for which electrochemical routes are also available.

There should be no question whatsoever that additional and broader engineering studies are needed before arriving at firm conclusions on the question of electrochemical versus chemical process selection. The chemical engineering methods described below can easily be expanded to embrace a wider range of considerations, as well as a larger population of process candidates. By such refinement and expansion, a more realistic engineering assessment of electroorganic process routes would surely emerge.

Organic chemicals produced and sold at greater than 7000 tons/year in the United States number about 220, of which 95 were found to have electrochemical synthesis routes. The literature was examined and a group of compounds was chosen for which engineering estimates were made on process reaction conditions, yields, current efficiencies, and so on. Of these, the nine chemicals for which the most complete data were available were then chosen for detailed evaluation as reported below. Material and energy balances were made for both chemical and electrochemical routes for each compound; the energy consumption (kcal/kg) for both process routes was compared. Also, the potential impact that large-tonnage electrochemical processes might have on the U.S. petrochemical industry was simulated using a mathematical model of that industry.

Indentification of Candidate Electroorganic Reactions

Candidates for evaluation were selected from a list of organic chemicals produced and sold in the United States in amounts exceeding 7000 tons/year in 1975 [215]. This list does not contain organic intermediates and, because of the difficulties in obtaining this often proprietary information, organic intermediates were excluded from evaluation except in the case of adiponitrile. For each compound, two bibliographic collections [1,2] were examined to obtain literature citations on the electrolytic synthesis. The bibliographic data base covered the period 1801 to 1975 and included the patent literature. Some of the chemicals, however, were generic in nature (e.g., tar) and were not listed as such in the bibliographic sources. Any literature reference that involved obviously unattractive feedstock material or untenable reaction conditions was discarded from further evaluation. The compounds that survived this preliminary screening step are shown in Table 9.2. *Electrochemical synthesis routes were found for 95 compounds, or 57% of the nongeneric largetonnage compounds.*

To narrow the list of candidate processes, additional selection rules were established as follows:

1.   There should be sufficient published electroorganic literature to make reasonable estimates of process conditions. Thus promising but speculative process routes were not considered.

2.   Hydrocarbons currently synthesized by efficient catalytic routes should not be evaluated.

3.   Polymers should not be evaluated.

4.   Pollutants presently being phased out should not be evaluated (e.g., tetraalkyl lead compounds).

5.   The product molecular weight should be high, and there should be fewer than eight electrons transferred per molecule.

6.   The product cost should be greater than the cost of raw materials.

The application of these rules to the 95 compounds in Table 9.2 led to the selection of 18 compounds listed in Table 9.3. Of these, the 9 compounds given in capital letters were selected for detailed evaluation reported below, since they had the most complete and accurate process data files.

The 9 compounds selected for detailed evaluation represent several types of electrochemical reactions and include oxidations, halogenations, and reductions. The EPA numbers cited in Table 9.3 refer to information on chemical process routes [216] that, along with information in the open literature, was used in making process calculations.

Methods Used for Making Calculations

The basic flowsheet for a typical electroorganic process is shown in Fig. 9.2. The cells and the separation and purification steps are most often energy intensive. Electricity is generally the only energy applied to the cells, and cooling is usually required. Thermal energy is used for the separation and

Table 9.2. Large-Tonnage^a Organic Chemicals Having Electrochemical Synthesis Routes

Acetic acid	2,2'-Dithiobis (benzothiazole)	Phenol
Acetic acid salts	Ethane	Phenol salts
Acetone	Ethyl acetate	Phenyl hydrazine
Acrylic acid	Ethyl acrylate	Phosgene
Acrylonitrile	Ethyl alcohol	Phthalic anhydride
Adipic acid	Ethylbenzene	Phthalide
o-Aminobenzyl alcohol	Ethylene	Polyacrylamide
Aniline	Ethylene glycol	Polyacrylonitrile and acrylonitrile copolymers
Benzene	Ethylene oxide	Polyethylene terephthalate
Benzoic acid	Formaldehyde	Propane
Biphenyl	Formic acid, Na$^+$	Propionic acid
n-Butane	Fumaric acid	Propyl alcohol
2-Butanone	Glycerol	Propylene
1-Butene and 2-butene	Heptanes	Propylene glycol
n-Butyl alcohol	Hexane	Propylene oxide
Butylamines	1,6-Hexanediamine	Quinone (intermediate for hydroquinone)
Butyraldehyde	Heptenes	Salicylic acid
Carbon tetrachloride	Hydroquinone	Semicarbazide
Chlorobenzene	Isobutyl alcohol	Sorbitol
Chlorodifluoromethane	Isobutyraldehyde	Stearic acid salts
Chloroethane	Isopropyl alcohol	Styrene
Chloroform	Isobutane	Terephthalic acid
Cyclohexanone	Isobutylene, 2-butene, and mixed butylenes	Tetrachloroethylene
Chloromethane	Isoprene	Tetrafluoroethylene
Cresols	Maleic anhydride	Toluene
1,4-Cyclohexadiene	Methanol	Toluene-2,4-diamine
Cyclohexane	Monoethylamine	Toluenesulfonic acid, K$^+$, and Na$^+$
1,2-Dibromoethane	Monoisopropylamine	Trichlorethylene
o-Dichlorobenzene	Naphthalene	Trimethylamine
Dichlorodifluoromethane	Nitrobenzene	Urea
1,2-Dichloroethane	Pentaerythritol	Xylene
Dichloromethane		
Dimethylamine		

^{a} 14,163,000 lb/year and above.

Table 9.3.   Candidate Electrochemical Processes

Class	Product Compound	EPA Number
Reduction	ADIPONITRILE	
	ANILINE	31
	Toluene-2,4-diamine	343
	SORBITOL	
Oxidation	TEREPHTHALIC ACID	362A,B
	PHENOL	37,295
	Propylene oxide	286,328
	METHYL ETHYL KETONE (2-BUTANONE)	55,56
	MELAMINE	
	Salicyclic acid	303A,B
	QUINONE (then to HYDROQUINONE)	16
	Dithiobis(2,2-dithiobisbenzothiazole)	
Halogenation	DICHLOROETHANE	131,132
	Chlorobenzene	4
	1,2-Dibromoethane	
	1,2-Dichloropropane	
	Tetrafluoroethane	
	2,4-Dichlorophenol	

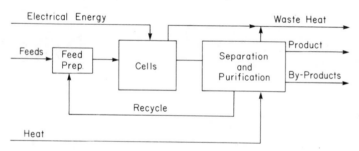

Fig. 9.2.   Electrochemical process material and energy streams.

purification steps downstream from the cells. Feed preparation usually requires a minor amount of energy, such as for mixing, and therefore was usually neglected. In the case of chemical processes, chemical reactors are used for which thermal energy is most often required.

To give consistent treatment for all processes, the set of design rules shown in Table 9.4 was used. According to rule 1, current chemical routes were based on Ref. 216. According to rule 7, it was assumed that 3 moles of water per hydrogen ion passed through a membrane in acid electrolyte [217].

Table 9.4.   Process Design Rules

1. Current chemical routes used for comparison with each electrochemical route
2. Electrochemical and current chemical routes are chosen that are as similar as possible
3. Electrochemical processes are based on largest-scale data available
4. All electrochemical processes are continuous
5. Cells are based on state-of-the-art technology, for example, metal anodes, membranes
6. Membranes or other separators are not used unless essential
7. Cation membranes are preferred
8. Atmospheric pressure cells are used unless otherwise specified
9. Hydrogen by-product is credited at $\Delta H_{combust}$ thermal
10. No credit is given for oxygen by-product
11. Product separation energy requirements are based on key component binary separation
12. When thermodynamics dictates that a reaction does not go to completion, 100% of the equilibrium conversion is assumed
13. All compressors and pumps are electrically driven
14. Only consider pumps and compressors for high-pressure reaction; no transfer pumps are to be considered
15. Neglect energy cost of vacuum for stills and evaporators

Following rule 11, product separation energy requirements were calculated for each separation process; energy requirements for liquid/liquid extraction and for phase separations were assumed to be negligible; and energy requirements for distillation were based on heat required for boiling the bottoms for a column with constant reflux ratio.

In some cases the chemical and electrochemical routes used different raw materials. It was then necessary to assign energy costs to the various raw materials to compare both routes on a consistent basis. Energy costs for common raw materials are given in Table 9.5. These data are based on published process energy data [218–220], and calculations are available in Ref. 3. Table 9.5 gives energy costs for organic raw materials and products as calculated from data given in Ref. 221. By-product hydrocarbons were credited at an energy cost equal to the heat of combustion [222], since they could have been burned under a boiler to produce steam.

An energy balance for a continuous electrochemical reactor at steady state is given in Fig. 9.3. The algebraic sum of the enthalpies of formation of reactants and products, and the electrical and thermal energy is zero.

$$q_r + q_p + \text{EI} + Q = 0 \tag{1}$$

Table 9.5. Thermal Energy Costs Assigned to Chemical Raw Materials

Chemical	$\Delta H_{combustion}$ (kcal/kg)
Hydrocarbons	
Carbon	7,824
Hydrogen	33,944
Methane	13,266
Ethane	12,400
Propane	12,034
n-Butane	11,838
Ethylene	12,024
Propylene	11,689
n-Butene	11,578
Isobutene	11,517
Benzene	10,117
Toluene	10,244
Xylene	10,361
Fuel oil	10,430
Hydrogen	
Steam reforming of hydrocarbons	49,000
Water electrolysis (at 1.5 V)	105,770
Ammonia	9,000
Hydrogen cyanide	18,000
Nitric acid	3,100
Urea	6,400
Oxygen from air	964
Chlorine (prorated 50% with NaOH)	5,875
Sodium hydroxide (prorated 50% with Cl_2)	5,875
Sulfuric acid	29
Iron	4,050
Air	Zero
Water	Zero
Manganese dioxide	Zero

The heat of reaction is the sum of the products of the component flow rates and their heats of formation.

$$q = \sum_{i=1}^{n} N_i \Delta H_{fi} \qquad \begin{matrix} N_{ri} \text{ is } + \\ N_{pi} \text{ is } - \end{matrix} \qquad (2)$$

where

$$\Delta H_f = \Delta H_{fg} - \Delta H_v \qquad (3)$$

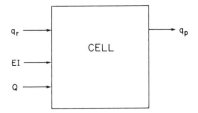

Fig. 9.3.   Energy balance for continuous reactor at steady state.

Usually a cell reaction is endothermic; the electrical potential applied drives a reaction that would otherwise require a higher temperature. Generally the electrical energy delivered to the cell exceeds the endothermic heat of reaction owing to irreversible effects, so that heat is generated and must be removed by exchange or by evaporation of electrolyte.

Heats of formation of most organic compounds studied below are not experimentally known. The ASTM program (CHETAH: Chemical Thermodynamic and Energy Release Evaluation Program) [223] was used to estimate values of H_f for the ideal gas state.

Additional thermodynamic data were obtained from the following reference sources which are listed in decreasing order of preference of use [224-234]. Additional sources were used for critical properties [233, 234], heat of vaporization estimates [235], and solubility data [224-232, 236].

Phase equilibria between multicomponent mixtures of aqueous and organic phases are generally unknown. Where binary data were available, they were used without consideration of other components or ternary effects. In liquid phases, ideal vapor/liquid equilibria were assumed to be valid. Solid phases were assumed to be present as pure compounds.

Several separation methods were employed in the flowsheets discussed below, including filtering, centrifuging, settling, liquid/liquid extraction, and distillation. Of these, the distillations were usually the most energy intensive. All distillations were handled as binary separations [237] for which a reflux ratio of 1.25 times the minimum was used.

Yield was used in the calculations below to indicate the fraction of primary reactant ultimately converted to the desired product and refers to the overall plant process independent of recycle streams. Conversion, on the other hand, represents the fraction of limiting reactant converted to product on each pass through the reactor or cell.

While the foregoing procedures were used in the example process evaluations that follow, it is possible to make very quick estimates of energy consumption of the electrolysis cells. Based on Faraday's law, the power consumption in engineering units of an electrochemical cell can be determined with the formula

$$\frac{12.6\, n\, V_{\text{cell}}}{M_w\, \eta_c} \text{ kWh/lb product}$$

where

n = number of electrons per g·mole of product
V_{cell} = cell voltage, V
M_w = molecular weight of product
η_c = current efficiency.

These results may be connected to equivalent fossil fuel values with the factor 0.325. Thus an electroorganic process with an equivalent weight (M_w/n) of 50, $V_{\text{cell}} = 5$ V, $\eta_c = 0.90$, would consume 1.36 kWh/lb product in electricity or 1.36/0.325 = 4.18 kWh/lb in fossil fuel at the generating station.

Energy Calculations for Specific Chemicals

Descriptions of electrochemical processes are available [3] for those chemicals listed in Table 9.2. Still more detailed evaluation is given below for the nine chemicals given in capital letters in Table 9.3. For each capitalized chemical, flowsheets, stream conditions, and energy calculations for *both* chemical and electrochemical process routes are available [3] and are summarized below. The comparison of energy consumptions for various processes is given in Table 9.18.

Adiponitrile

Adiponitrile is a chemical intermediate and therefore is not included in the production figures cited in Ref. 216. However, its production rate should be closely related to that of nylon 66 which, in the United States, is expected to be about 2.7 billion lb in 1979 [238, 239].

CHEMICAL PROCESS (ADN FROM BUTADIENE)

The energy required for the chemical route to adiponitrile from butadiene is based on data reported by Rudd et al [221]. The figure reported in Table 9.7 (65,808 kcal/kg) includes the energy required to make butadiene, as well as the gross heating value of the butylene feedstock. A more recent chemical process route, commercialized in 1972, does not use chlorine [220] and has superceded the older process for which data were available. Avoiding the use of chlorine would save 6668 kcal/kg and would probably therefore decrease the energy for the chemical route to 59,140 kcal/kg. Detailed calculations for the newer process are not possible owing to lack of sufficient information.

ELECTROCHEMICAL PROCESS (ADN FROM ACRYLONITRILE)

The electrochemical production of adiponitrile from acrylonitrile is commerically practiced by Monsanto and has been described in numerous publications [240–247] (see also Chapter VI). The process chosen for evaluation in this study corresponds to the early Monsanto patents; it is to be expected that even lower energy consumption would be exhibited by more recent processes. Electrolysis is carried out in a cation-exchange membrane cell in a plate-and-frame type assembly. The catholyte consists of an aqueous mixture of adiponitrile, acrylonitrile, and a quaternary ammonium salt. The following reactions occur:

		Current Efficiency
Cathode reactions:		
$2C_3H_3N(\ell) + 2H^+ + 2e \longrightarrow C_6H_8N_2(\ell)$		89%
$C_3H_3N(\ell) + 2H^+ + 2e \longrightarrow C_3H_5N(\ell)$		11%
Anode reactions		
$H_2O(\ell) \longrightarrow \frac{1}{2}O_2(g) + 2H^+(aq) + 2e$		100%
Primary overall reaction		
$2C_3H_3N(\ell) + H_2O(\ell) \longrightarrow C_6H_8N_2(\ell) + \frac{1}{2}O_2(g)$		

The quaternary ammonium salt, assumed to be tetraammonium-p-toluene sulfonate, increases both the solubility of organic compounds and the conductivity of the solution. The cells are assumed to operate at 6 V.

The flowsheet for the electrochemical process, shown in Fig. 9.4, has associated with it Table 9.6, in which stream conditions at each numbered position are provided. From the reactor, the catholyte is pumped to an extractor where acrylonitrile is used to strip adiponitrile from the aqueous phase. The resulting organic phase contains appreciable QAS, which is removed using fresh water in the QAS extractor. The resultant organic stream containing adiponitrile, propionitrile, and acrylonitrile is sent to the first of two distillation towers, where the adiponitrile is separated as the bottom stream. The distillate stream is split, with part being recycled to the solvent extractor and part entering the propionitrile distillation column where the by-product, propionitrile, is separated. The acrylonitrile is the distillate and is recycled to the cell.

The energy requirement calculated for the electrochemical route agreed with the data of Rudd et al. [221] for fuel oil equivalent energy to within 8%. To put the electrochemical route on a comparable basis with the chemical route, the energy required to manufacture acrylonitrile and the gross heating value of the propylene feedstock are added to give a total of 43,177 kcal/kg.

Energy data for various process routes are compiled in Table 9.7. The electrochemical route has a significant saving in energy as compared to the chemical route. The electrochemical process gives about 12% by-product

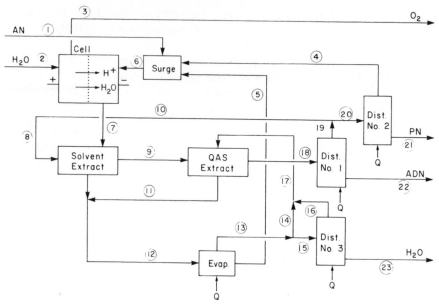

Fig. 9.4.   Electrochemical route to adiponitrile.

propionitrile, which is presumably a saleable product. No information was obtained on by-products of the chemical route, but hydrogen cyanide, a hazardous chemical to transport, is a chemical raw material.

Methyl Ethyl Ketone

CHEMICAL ROUTES (MEK FROM BUTENE)

The thermochemical production of methyl ethyl keton (MEK) from butene is discussed in the *EPA 600 Report* [216] and by Lowenheim and Moran [232]. The process flowsheet is shown in Fig. 9.5.

The first reaction is the absorption of butene in concentrated (75%) sulfuric acid:

$$C_4H_8(g) + H_2SO_4(aq) \longrightarrow C_4H_{10}SO_4(aq)$$

A conversion of 100% was assumed. The second reaction is the hydrolysis of the butyl hydrogen sulfate to form butanol,

$$C_4H_{10}SO_4(aq) + H_2O(\ell) \longrightarrow C_4H_{10}O(aq) + H_2SO_4(aq)$$

A conversion and yield of 84% was used.

The butanol and most of the water are separated from the sulfuric acid by distillation in the first column. The acid bottoms is sent to a second column, where more water is removed as overhead product. The concentrated acid

Table 9.6. Stream Conditions for Adiponitrile Process Electrochemical Route (Flow Units are g·mole/sec)

Stream No.

Flows	1	2	3	4	5	6
C_3H_3N	19.62			18.69		38.31
H_2O		85.36			50.00	50.00
O_2			5.18			
QAS					8.60	8.60

Flows	7	8	9	10	11	12
C_3H_3N	18.69	119.50	133.69	4.5	1.50	6.00
H_2O	125.00			125.0	58.41	183.41
QAS	8.60		1.01	7.59	1.01	8.60
$C_6H_8N_2$	9.26		9.26			
C_3H_5N	1.10	7.03	8.13			

Flows	13	14	15	16	17	18
C_3H_3N	6.00	2.40	3.60	3.60	6.00	138.19
H_2O	133.41	53.36	80.05	5.05	58.41	
$C_6H_8N_2$						9.26
C_3H_5N						8.13

Flows	19	20	21	22	23
C_3H_3N	138.19	18.69			
H_2O					75.00
$C_6H_8N_2$				9.26	
C_3H_5N	8.13	1.10	1.10		

Equipment Condition	Cell	Solvent Extractor	QAS Extractor	Dist. 1	Dist. 2	Dist. 3	H_2O Evap.
T (°C)	50	25	25	87-200	77-97	75-100	112
P (atm)	1	1	1	0.07	1	1	1
VI (kcal/sec)	2868						
Q(kcal/sec)	-2288			1229	358	773	1730

449

Table 9.7. Adiponitrile Energy Requirements (kcal) [Basis: 1 kg (9.26 g·moles)]

Chemical Route (Refs. 221 and 3 p. C24)

Adiponitrile from butadiene (duPont)	65,808

Electrochemical Route

1. Process Energy
    a. As computed in Ref. 3

Cell	8,825
Evaporator (H_2O)	1,730
Distillation no. 1	1,229
Distillation no. 2	358
Distillation no. 3	773
Total	12,915
b. As reported in Ref. 221; see also Ref. 3, Table C-O	11,890
Average process energy from parts a and b	12,402

2. Energy to normalize feedstock to common basis as chemical route above

a. Acrylonitrile from propylene (Sohio)	31,287
Total for Electrochemical route	43,690

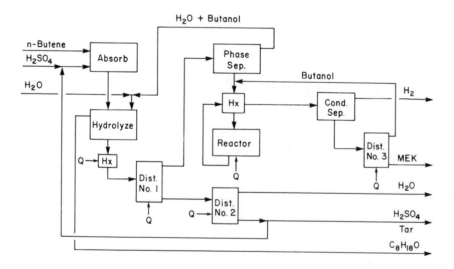

Fig. 9.5. Methyl ethyl ketone from *n*-butene, chemical.

stream contains the tar, which must be purged from the process. The remaining acid with dissolved tar is recycled to the butene absorber.

The butanol and water mix from the first distillation column is separated in the liquid phase; the butanol proceeds first to a heat exchanger, where it is boiled, and then to a gas phase catalytic reactor. The aqueous phase is recycled to the hydrolysis reactor.

The gas phase reactor operates at 450°C and 1 atm according to the reaction

$$C_4H_{10}O(g) \longrightarrow C_4H_8O(g) + H_2(g) \quad \Delta H_{700°K} = 13.81 \text{ kcal/mole MEK}$$

The reactor conversion is 85% and the yield is assumed to be 100%.

The exit stream from the gas phase reactor is cooled and the hydrogen gas is separated off. The MEK–butanol mix is separated in a third distillation column. The light butanol is recycled to the gas reactor feed. The final product, MEK, makes up the column bottoms. The overall process yield is 84%.

The largest energy requirement is for the first distillation column, which vaporizes a large quantity of water and butanol. The gas phase reactor requires a small amount of heat, but the temperature is moderately high. Part of this heat is recovered by preheating the reactants with the hot product stream.

Process energy calculations for the chemical route to MEK [3] are compared with those for the electrochemical route in Table 9.8.

ELECTROCHEMICAL ROUTE (MEK FROM BUTENE)

The electrochemical production of 2-butanone (MEK) from butene is discussed in a series of patents by Worsham [248] and by Griffin and Worsham [249]. Other related citations include Refs. 250 to 254. The process flowsheet is given in Fig. 9.6, with stream conditions specified in Table 9.9.

The cation membrane electrochemical cell operates at about 80°C and at 1 atm pressure. The reaction sequence is as follows:

Anode reaction

$$C_4H_8(g) + H_2O(\ell) \longrightarrow C_4H_8O(\ell) + 2H^+(aq) + 2e \quad \text{C.E.} = 49\%$$

$$C_4H_8(g) + 8H_2O(\ell) \longrightarrow 4CO_2(g) + 24H^+(aq) + 24e \quad \text{C.E.} = 51\%$$

Cathode reaction

$$\frac{1}{2}O_2(g) + 2H^+(aq) + 2e \longrightarrow H_2O(\ell) \quad \text{C.E.} = 100\%$$

Primary overall

$$C_4H_8(g) + \frac{1}{2}O_2(g) \rightarrow C_4H_8O(\ell) \quad \Delta H_{350°} = -64.4 \text{ kcal/mole MEK}$$

The conversion per pass through the cell was assumed to be 81% based on butene.

Chemical Route

Process calculations in Ref. 3

Heat exchanger	623
Reactor	277
Distillation no. 1	4907
Distillation no. 2	480
Distillation no. 3	1352
H_2 credit	−949
Total	6690
Data reported in Ref. 221; see also Ref. 3 App. C-12	3233[a]

Electrochemical Route

Cell (short-circuited fuel cell)	—
Vacuum distillation	3300
Distillation no.1	1710
Distillation no. 2	47
Total	5057

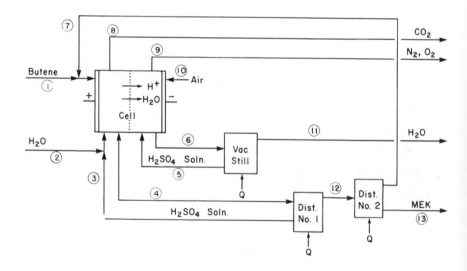

Fig. 9.6. Methyl ethyl ketone from butene, electrochemical.

The aqueous reaction products pass out of the cell into the first distillation column, where the lighter organic products form the distillate, and the organic-free bottoms are recycled to the anode compartment. The organic phase is further separated in a second column, with the butene distillate being recycled to the cell while the bottoms stream constitutes the product MEK.

The water passing through the membrane with the hydrogen ions is purged from the system after separation from the acid solution by vacuum distillation.

The solubility of butene in sulfuric acid has been assumed to be 0.5 mole butene for each mole of acid. Since the overall chemical reaction is exothermic, the cell should be able to produce energy. No energy credit has been given, that is, a short-circuited cell has been assumed. Based on these assumptions, the vacuum still requires more energy than any other single unit, about 65% of the total process energy.

Energy requirements for the chemical and electrochemical routes to methyl ethyl ketone are compared in Table 9.9. Only the process energies are listed because both processes use the same raw materials. The electrochemical route, based on the study calculations, has a lower energy requirement than the chemical route. It has the further potential advantage of generating a net amount of energy in the electrogenerative mode rather than the short-circuit mode.

The chemical route produces waste sulfuric acid, tar, and dibutyl ethers. The electrochemical route oxidizes a part of the butene feed to CO_2.

Hydroquinone

CHEMICAL PROCESS (HYDROQUINONE FROM BENZOQUINONE)

This process is described in the *EPA 600 Report* [216] and in Volumes 2, 11, 16, and 18 of Ref. 231. The flowsheet and process conditions are described in Fig. 9.7. The reaction takes place at 70°C and 1 atm with the consumption of iron.

$$C_6H_4O_2(c) + H_2O(\ell) + \tfrac{2}{3} Fe(c) \longrightarrow C_6H_6O_2(c) + \tfrac{1}{3} Fe_2O_3(c)$$

$$\Delta H_{298} = -40.32 \text{ kcal/mole hydroquinone}$$

The reaction proceeds with a conversion per pass of 79% and a yield of 100%.

It is characterized by a large volume of water because of the low solubility of hydroquinone. The solubility of hydroquinone at 70°C was assumed to be 9% by weight [229].

ELECTROCHEMICAL ROUTE (HYDROQUINONE FROM BENZENE)

The electrochemical production of hydroquinone from benzene has been described by Fremery et al. [255]. Further, the process was evaluated [3,

Table 9.9. Stream Conditions for 2-Butanone Process Electrochemical Route (Flow Units are g·moles/sec)

Stream No.

	1	2	3	4	5	6	7
T (°C)	25	80	151	80	170	80	
P (atm)	1	1	1	1	1	1	1
Flows							
C_4H_8	15.07			3.47			3.47
H_2O		227.00	40.79	40.79	17.92	184.24	
H_2SO_4			13.87	13.87	62.65	62.65	
C_4H_8O				13.87			

	8	9	10	11	12	13
T (°C)	80	80	25	80	170	80
P (atm)	1	1	1	1	1	1
Flows						
C_4H_8					3.47	
H_2O	0.32	8.51		166.32		
N_2		112.86	112.86			
O_2		15.83	30.00			
C_4H_8O					13.87	13.87
CO_2	4.82					

Equipment Condition	Cell	Vac. Still	Dist. 1	Dist. 2
T (°C)	80	80	25	80
P (atm)	1	14 mm	1	3
VI (kcal/sec)	0			
Q (kcal/sec)	0	3300	1710	47

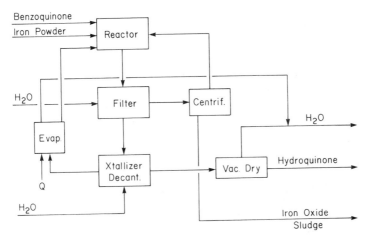

Fig. 9.7. Hydroquinone from benzoquinone, chemical.

App. B] according to suggestions based on Refs. 256 to 262. The process flowsheet is shown in Fig. 9.8 and stream conditions appear in Table 9.10.

The reaction of benzene to produce hydroquinone proceeds by the following reaction sequence:

Anode reaction

$$C_6H_6(\ell) + 2H_2O(\ell) \longrightarrow C_6H_4O_2(c) + 6H^+(aq) + 6e \qquad C.E. = 40\%$$

$$C_6H_6(\ell) + 12H_2O(\ell) \longrightarrow 6CO_2(g) + 30H^+(aq) + 30e \qquad C.E. = 50\%$$

$$2H_2O(\ell) \longrightarrow O_2(g) + 4H^+(aq) + 4e \qquad C.E. = 10\%$$

Cathode reaction

$$C_6H_4O_2(c) + 2H^+(aq) + 2e \longrightarrow C_6H_6O_2(c) \qquad C.E. = 13\%$$

$$2H^+(aq) + 2e \longrightarrow H_2(g) \qquad C.E. = 87\%$$

Basic overall reaction

$$C_6H_6(\ell) + 2H_2O(\ell) \longrightarrow C_6H_6O_2(c) + 2H_2(g)$$
$$\Delta H_{298} = 139.39 \text{ kcal/mole hydroquinone}$$

The cell for this process has been assumed to contain a cation-exchange membrane. As the chemical reactions above indicate, benzoquinone is formed at the anode along with hydrogen ions. The benzoquinone is pumped to the cathode compartment, where it is reduced to hydroquinone. The hydrogen ions traverse the ion-exchange membrane and replace those that are consumed at the cathode surface.

After leaving the cell, the hydroquinone is pumped to a phase separator

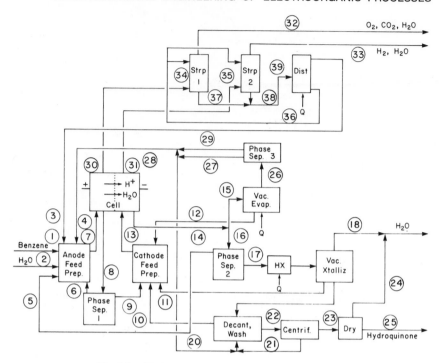

Fig. 9.8. Hydroquinone from benzene, electrochemical.

where the benzene floats and is recycled to the anode compartment. The aqueous phase containing hydroquinone is heated and flashed to a vacuum crystallizer, where the solid hydroquinone is centrifugally filtered out. The solid hydroquinone is then washed, centrifuged again, and dried to produce the final product.

The water that is passed through the membrane to the catholyte with the hydrogen ions is removed by vacuum evaporation and the resulting benzene/water mixture is recycled to the anode compartment.

The off-gases, hydrogen, oxygen, and carbon dioxide, contain benzene, which is absorbed in oil in the gas plant. The benzene is then stripped out of the heavy oil and recycled to the cell feed.

The water evaporator in this process consumes about one-fourth the energy of the cell. The high cost of recycling the water carried across the membrane with hydrogen ions is typical of membrane cell systems.

An additional large energy-consuming step not mentioned in the patents or publications on the process is purification of the sulfuric acid electrolyte [263]. Soluble organic impurities build up in the electrolyte and cause a pro-

Table 9.10. Stream Conditions for Hydroquinone Process Electrochemical Route (Flow Units are g·moles/sec)

Stream No.

Flows	1	2	3	4	5	6	7
C_6H_6	11.35		35.69	17.49	494.79		559.32
H_2O		133.56		270.84		936.69	1341.09
H_2SO_4						43.03	43.03
$C_6H_4O_2$					2.64		2.64

Flows	8	9	10	11	12	13	14
C_6H_6	540.00	540.00	40.58		212.54	540.00	512.28
H_2O	936.69			14,514.68		14,767.80	15,109.55
H_2SO_4	43.03		0.44	472.70	16.73	489.87	489.87
$C_6H_4O_2$	11.72	11.72			0.09	11.81	2.73
$C_6H_6O_2$				214.33	7.90	222.23	231.31

Flows	15	16	17	18	19	20	21
C_6H_6	17.49	494.79			12.76		
H_2O	515.83	14,593.72	14,593.72	66.28		32.45	9.28
H_2SO_4	16.73	473.14	473.14		0.44		
$C_6H_4O_2$	0.09	2.64					
$C_6H_6O_2$	7.90	223.41	223.41		9.08		

Flows	22	23	24	25	26	27	28
C_6H_6	13.91	4.63	4.63		17.49		
H_2O					303.29	303.29	270.84
$C_6H_6O_2$	9.08	9.08		9.08			

Table 9.10. Continued

Stream No.

Flows	29	30	31	32	33	34	35
C_6H_6	17.49	7.97	27.72		8.16		
H_2O		2.26	8.16	2.26			
O_2		3.39		3.39			
CO_2		13.64		13.64			
H_2			59.05		59.05		
Oil							

Flows	36	37	38	39	40
C_6H_6		9.97	27.72	35.69	
H_2O					349.91
Oil					
H^+					136.25

Equipment Conditions	Cell	Phase Sep. 1	Phase Sep. 2	Phase Sep. 3	Vac. Evap.	Vac. Cryst.	Dryer	Strip 1	Strip 2	Dist.
T (°C)	45		40			20	100	25	25	
P (atm)	1	1	1	1	0.309	0.309	1	1	1	1
VI (kcal/sec)	12,600									
Q (kcal/sec)					3244	1430	93			233

gressive decline in current efficiency to the point where sulfuric acid must either be replaced or purified. Not enough technical information is available at present to make an estimate of the energy requirement for sulfuric acid purification.

COMPARISON (HYDROQUINONE SYNTHESIS)

Energy requirements for the chemical and electrochemical routes to hydroquinone are compared in Table 9.11. The chemical route has a lower energy requirement according to these calculations, but the case is not clearcut, because the calculations were based on rather sketchy process information and the results are sensitive to the assumptions made. The following factors can increase the energy consumption for the chemical process [263]:

1. Refrigeration may be used to operate the reactor for oxidation of aniline to benzoquinone below ambient temperature.

2. Evaporation of water from the salt solution effluent before disposal or sale of the salts for fertilizer use would require 3816 kcal/kg of hydroquinone.

3. The amount of water used in the process may be larger than that used in the calculations, thus increasing the heat loads for evaporation.

It has already been pointed out that there is an additional energy cost to the electrochemical process for sulfuric acid purification. On the other hand, the current efficiency might be increased from the 40% used in the calculations to a higher figure through more research and development. The 12,600 kcal/kg of electrical energy for the cell in Table 9.10 is in good agreement with the 14 kWh/kg or 12,900 kcal/kg given in Ref. 264.

Millington's calculations [264] show that the electrochemical process is more favorable on an energy basis than the chemical process. His data show 43,950 kWh (thermal)/ton (41,600 kcal/kg) for the electrochemical route and 69,900 kWh (thermal)/ton (66,100 kcal/kg) for the chemical route; these numbers are exclusive of the heat of combustion of the benzene feed to each process. These results clearly show an advantage for the electrochemical route. The overall energy requirement (41,600 kcal/kg) for the electrochemical route is in reasonable agreement with the 43,769 kcal/kg (exclusive of benzene heating value) calculated here. The former exhibit a higher overall energy requirement for the chemical route because:

1. An obsolete iron-reduction method was used to make the aniline intermediate.

2. A higher energy was assigned to the iron used for reduction of nitrobenzene to aniline and benzoquinone.

3. A higher steam usage was employed in the hydroquinone part of the process, presumably for a greater water evaporation load.

Table 9.11.   Hydroquinone Energy Requirements (kcal) [Basis: 1 kg (9.08 g·moles)]

Chemical Route	
H_2O evaporation	2,979
Dryer	147
Iron	2,633
Benzoquinone (9.08 g moles)	
Steam stripper	11,050
Aniline from benzene	14,152
Total	30,961
Electrochemical Route	
Cell	38,769
Vacuum evaporation	3,244
Vacuum crystallizer	1,430
Dryer	93
Distillation	233
Benzene	8,970
Total	52,739

The chemical route to hydroquinone given in the EPA report [216] and used as a basis in this study is also old technology. Two newer processes [263] are chemical oxidation of phenol to hydroquinone and catechol, and hydroperoxidation of *p*-diisopropylbenzene, leading to hydroquinone and acetone.

The chemical reduction of benzoquinone to hydroquinone by iron gives an iron oxide sludge that must be discarded. The chemical oxidation of aniline to benzoquinone gives calcium sulfate sludge, which must be discarded. The electrochemical route is relatively clean, although it is understood that the process is prone to produce tars.

Dichloroethane

CHEMICAL ROUTE (DICHLOROETHANE FROM ETHYLENE)

The gas phase production of dichloroethane (DCE) from ethylene and hydrogen chloride (HCl) is discussed in the *EPA 600 Report* [216] and *Hydrocarbon Processing* [265].

The catalytic reactor operates at 300°C and 5.5 atm pressure. The reaction is

$$C_2H_4(g) + \tfrac{1}{2} O_2(g) + 2HCl(g) \longrightarrow C_2H_4Cl_2(g) + H_2O(g)$$

$$\Delta H_{573} = -57.18 \text{ kcal/mole DCE}$$

The only energy demands for this process are made by the DCE stripper and the air compressor. The reactor operates at the highest temperature in the process, and the heat of reaction could be used in the DCE stripper. With the exception of the small requirement for the air compressor, this process is energy self-sufficient.

A second chemical route based on Cl_2 is also known. Energy data for both process routes are given in Table 9.12.

ELECTROCHEMICAL ROUTE (DICHLOROETHANE FROM ETHYLENE)

The electrochemical production of dichloroethane (DCE) from ethylene and hydrogen chloride [266–272] has led [1] to the flowsheet given in Fig. 9.9.

The diaphragm electrochemical cell used for this process converts hydrochloric acid to chlorine gas and hydrogen as shown in the following sequence of reactions:

Anode reaction

$$2Cl^-(aq) \longrightarrow Cl_2(g) + 2e \qquad C.E. = 98\%$$

$$2H_2O(\ell) \longrightarrow O_2 + 4H^+(aq) + 4e \qquad C.E. = 2\%$$

Cathode reaction

$$2H^+(aq) + 2e \longrightarrow H_2(g) \qquad C.E. = 100\%$$

The overall electrochemical reaction can thus be written as

$$2H^+(aq) + 2Cl^-(aq) \longrightarrow H_2(g) + Cl_2(g)$$

The chlorine gas is absorbed into solution, where it undergoes reaction with ethylene to form several compounds, with DCE constituting the primary product. About 75% of the ethylene is ultimately converted to DCE, and the conversion per pass is about 64% based on ethylene. Ethylene chlorohydrin is the primary by-product. The cells are assumed to operate at 2.2 V, which is similar to the voltage of commercial HCl electrolysis cells.

The chlorohydrin formed in the cell is absorbed primarily into the aqueous phase. Part of this phase is withdrawn from the cell and passed to a chlorohydrin still. This still is a very large energy user, primarily because of the small concentration of chlorohydrin assumed for the water phase. If the concentration of chlorohydrin is allowed to increase beyond the value assumed here, a significant energy savings will be possible. Water and HCl pass overhead from the chlorohydrin still and are recycled to the cell. The primary energy requirements for this process are for the chlorohydrin still and the electrochemical cell.

COMPARISON (DICHLOROETHANE SYNTHESIS)

Energy requirements for the chemical and electrochemical routes to dichloroethane are compared in Table 9.12. Although both the HCl and chlo-

Table 9.12.   Dichloroethane Energy Requirements (kcal) [Basis: 1 kg (10.10 g·moles)]

Chemical Route	
From HCl	
Compressors and refrigerator	116
Process heat	—
Ethylene	6,014
Total	6,130
From Cl₂ (see Ref. 3, App C-12)	14,819
Electrochemical Route	
Cell	4,148
Compressor 1	9
Compressor 2	30
Refrigeration	73
Chlorohydrin distillation	5,753
Ethylene	7,760
HCl—no cost	—
Total	17,773

rine chemical routes are used commercially to make DCE, the HCl route has a clearly lower energy requirement. Over 95% of the DCE produced in the United States in 1974 was made by the "balanced process," which is a balance of these two processes with pyrolysis of DCE to produce vinyl choride monomer [216]. The by-product HCl from the vinyl chloride process is used in the DCE process so that there is no net production of HCl. The electrochemical route has a 20% greater energy requirement than the chemical chlorine route, but this energy requirement can be decreased below that of the chemical route if formation of the by-product chlorohydrin can be suppressed. This is because conventional NaCl brine diaphragm cells that supply the chlorine for the chemical process operate at 3.5 V versus 2.2 V for HCl electrolysis. The electrochemical process produces about 25% chlorohydrin, which is a saleable product.

Melamine

CHEMICAL ROUTE (MELAMINE FROM UREA)

The process producing melamine from urea is described by Kirk-Othmer [273], Lowenheim and Moran [232], and Mackay [274].

The primary reaction produces cyanamide, ammonia, and carbon dioxide

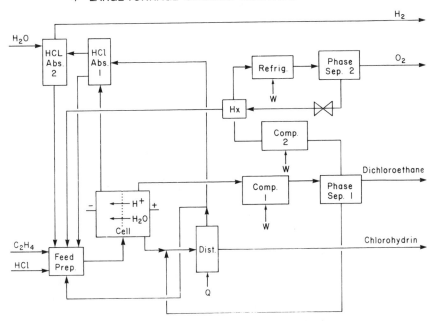

Fig. 9.9. Dichloroethane from ethylene and HCl, electrochemical.

from urea. The cyanamide trimerizes in excess ammonia to yield melamine. The overall reaction is

$$6CH_4N_2O(c) \xrightarrow{\text{NH}_3} 3CO_2(g) + 6NH_3(g) + C_3H_6N_6(c)$$

$$\Delta H_{298} = 114.75 \text{ kcal/mole melamine}$$

The aqueous phase containing the melamine as crystals passes to a vacuum thickener where some water and most of the ammonia and carbon dioxide are evaporated off. The thickener bottoms then move to a steam stripper where the residual ammonia and carbon dioxide are removed. The melamine passes, as an aqueous slurry, to the crystallizer, where water is driven off and the solid product is separated by a centrifugal filter.

The ammonia/carbon dioxide stream is recycled to a urea plant, where it is ultimately recycled to the urea feed. A credit has therefore been given for the off-gases.

The primary energy demand for this process is the reactor. The overall reaction is highly endothermic and requires a large amount of thermal energy.

ELECTROCHEMICAL ROUTE (MELAMINE FROM HYDROGEN CYANIDE)

The electrochemical production of melamine from cyanide is reviewed [1] based on work by Forman et al. [275, 276] and by Sprague et al. [277]. The process discussed below is discussed more thoroughly in Ref. 3.

The reactions in the electrochemical cell are as follows:

Anode reaction

$$2Br^-(aq) \longrightarrow Br_2(g) + 2e \qquad C.E. = 100\%$$

Cathode reaction

$$H^+(aq) + e \longrightarrow \frac{1}{2}H_2(g) \qquad C.E. = 50\%$$

$$NH_4^+(aq) + e \longrightarrow \frac{1}{2}H_2(g) + NH_3(g) \qquad C.E. = 50\%$$

A chemical reaction takes place in the anolyte:

$$HCN(\ell) + Br_2(g) \longrightarrow CNBr(c) + H^+(aq) + Br^-(aq)$$

The primary overall reaction can thus be written as

$$HCN(\ell) + NH_4^+(aq) + Br^-(aq) \longrightarrow CNBr(c) + H_2(g) + NH_3(g)$$

$$\Delta H_{298} = 48.62 \text{ kcal/mole cyanogen bromide}$$

The cell is operated at 3.1 V and 25°C at 100% conversion efficiency for HCN.

The product stream leaves the reactor as a slurry and enters the cyanogen bromide stripping tower where CNBr is extracted with tetrahydrofuran (THF). The THF stream is pumped to a chemical reactor, where cyanamid is formed,

$$CNBr(c) + 2NH_3(g) \longrightarrow CH_2N_2(c) + NH_4Br(c)$$

$$\Delta H_{298} = -60.62 \text{ kcal/mole cyanamide}$$

The product ammonium bromide is precipitated and separated from the THF stream in a centrifuge. The centrifuge overflow is pumped to a second chemical reactor, where polymerization to melamine occurs:

$$3CH_2N_2(c) \longrightarrow C_3H_6N_6(c)$$

$$\Delta H_{298} = -59.30 \text{ kcal/mole melamine}$$

This process contains two large energy sinks, the electrochemical cell, and the cyanogen bromide stripper. The stripper energy requirement is high because of the moderate solubility of CNBr in water.

COMPARISON (MELAMINE SYNTHESIS)

Energy requirements for the chemical and electrochemical routes to melamine are compared in Table 9.13. The electrochemical energy requirements

Table 9.13. Melamine Energy Requirements (kcal) [Basis: 1 kg (7.93 g·moles)]

Chemical Route	
Reactor	1,378
Thickener	1,009
Steam stripper	69
Crystallizer	618
Dryer	88
Total	3,162
Credit (heat exchange)	− 410
Net	2,752
Urea	19,876
NH_3 credit	−7,156
Total	15,472
Electrochemical Route	
Cell	10,462
CNBr stripper	4,932
NH_3 still	183
THF dryer	15
H_2O evaporator	88
Total	15,680
Credits (heat exchange)	− 736
Net	14,944
HCN	11,570
NH_3	3,645
Total	30,159

are significantly greater than those of the chemical route largely because of the cell and the CNBr stripper.

The chemical process gives about 20% biuret as a by-product, which is a saleable product.

Aniline

CHEMICAL ROUTE (ANILINE FROM NITROBENZENE)

The production of aniline from nitrobenzene is outlined by Lowenheim [232]. Other sources include Kirk-Othmer [231], *Hydrocarbon Processing* [265], and *EPA 600 Report* [216].

The process is a simple one with one phase separator, an absorber, and a distillation column. The primary reaction is

$$C_6H_5NO_2(g) + 3H_2(g) \longrightarrow C_6H_7N(g) + 2H_2O(g)$$

$$\Delta H_{298} = -111.34 \text{ kcal/mole aniline}$$

The equipment for this process presented no serious calculation problems. The heat, both sensible and latent, in the reactor product stream is sufficient to run the downstream separation equipment and preheat the nitrobenzene feed. Since the overall process is highly exothermic, this energy would have to be sold to a consumer outside the plant.

ELECTROCHEMICAL ROUTE (ANILINE FROM NITROBENZENE)

The electrochemical production of aniline is described in the literature [3, 278–284].

The cell operates at 25°C without a membrane and produces aniline by the following reactions:

Anode reaction

$$2H_2O(\ell) \longrightarrow O_2(g) + 4H^+(aq) + 4e \qquad C.E. = 100\%$$

Cathode reaction

$$C_6H_5NO_2(\ell) + 6H^+(aq) + 6e \longrightarrow \quad C_6H_7N(\ell) \;+\; 2H_2O(\ell) \qquad C.E. = 95\%$$

$$C_6H_5NO_2(\ell) + 4H^+(aq) + 4e \longrightarrow C_6H_7ON(c) + H_2O(\ell) \qquad C.E. = 4\%$$

$$2H^+(aq) + 2e \longrightarrow H_2(g) \qquad C.E. = 1\%$$

Primary overall reaction

$$C_6H_5NO_2(\ell) + H_2O(\ell) \longrightarrow C_6H_7N(\ell) + {}^3\!/_2\, O_2(g)$$

$$\Delta H_{298} = 71.94 \text{ kcal/mole aniline}$$

By far, the greatest energy cost in this process is the electrical energy going to the cell, which was assumed to operate at 5 V. Nearly all this energy is degraded to low-quality heat.

COMPARISON (ANILINE SYNTHESIS)

Energy requirements for the chemical and electrochemical routes to aniline are compared in Table 9.14. The electrochemical route has a decided energy disadvantage compared to the chemical route, primarily because of the high energy requirement for the cells.

The electrochemical route gives about 7% *p*-amino phenol by-product, which is a saleable product. Information on the corresponding chemical routes is given in Ref. 216.

Table 9.14. Aniline Energy Requirements

Chemical Route

(Basis: 1 kg = 10.75 g·moles)

Nitrobenzene (10.75 g·moles)	10,733
Hydrogen (32.5 moles)	3,186
No thermal credit taken	—
Total	13,919

Electrochemical Route

Cell	24,120
Crystallizer	441
Aniline evaporator	179
Process energy	24,740
Nitrobenzene (11.45 g·moles)	11,432
Total	36,172

Sorbitol

CHEMICAL ROUTE (SORBITOL FROM GLUCOSE)

The production of sorbitol from glucose is described by Lowenheim and Moran [232] and Fedor et al. [285].

The reaction is carried out in aqueous solution at 200°C and 140 atm of pressure over a nickel catalyst. The reaction is

$$C_6H_{12}O_6(c) + H_2(g) \longrightarrow C_6H_{14}O_6(c)$$

$$\Delta H_{298} = -16.15 \text{ kcal/mole sorbitol}$$

As with most highly hydrated processes, the primary energy demand is that required to evaporate the water.

ELECTROCHEMICAL ROUTE (SORBITOL FROM GLUCOSE)

The electrochemical reduction of glucose to sorbitol is described by various authors [286–291]. The chemical reaction sequence is as follows:

Anode reaction
$$2H_2O(\ell) \longrightarrow O_2(g) + 4H^+(aq) + 4e \quad \text{C.E.} = 100\%$$

Cathode reaction
$$C_6H_{12}O_6(c) + 2H^+(aq) + 2e \longrightarrow C_6H_{14}O_6(c) \quad \text{C.E.} = 90\%$$

$$2H^+(aq) + 2e \longrightarrow H_2(g) \quad \text{C.E. } 10\%$$

Primary overall reaction

$$C_6H_{12}O_6(c) + H_2O(\ell) \longrightarrow C_6H_{14}O_6(c) + \tfrac{1}{2}O_2(g)$$

$$\Delta H_{298} = 52.16 \text{ kcal/mole sorbitol}$$

This process is quite straightforward and the primary energy demand, other than that of the electrochemical cell, involves the removal of water. No allowance has been made for the heat of solution of glucose in water or of sorbitol in ethanol.

COMPARISON (SORBITOL SYNTHESIS)

Energy requirements for the chemical and electrochemical routes to sorbitol are compared in Table 9.15. The electrochemical route is at a substantial energy disadvantage to the chemical route because of the higher energy requirement for the cells and of the difficulty of separation of the product sorbitol from the sodium sulfate electrolyte. The chemical and electrochemical routes should be similar in terms of by-products.

Terephthalic Acid

CHEMICAL ROUTE (TEREPHTHALIC ACID FROM XYLENE)

The production of terephthalic acid (TPA) from p-xylene is discussed in the *EPA 600 Report* [216] and *Hydrocarbon Processing* [292], as well as by Lowenheim and Moran [232] and Sherwood [293].

The reaction involved is the air oxidation of p-xylene to TPA:

$$C_8H_{10}(\ell) + 3O_2(g) \longrightarrow C_8H_6O_4(c) + 2H_2O(\ell)$$

$$\Delta H_{298} = -319.78 \text{ kcal/mole TPA}$$

The largest energy requirements for this process are for distillation and the air compressor, for which three stages have been assumed.

ELECTROCHEMICAL ROUTE (TEREPHTHALIC ACID FROM P-XYLENE)

The electrochemical production of terephthalic acid (TPA) is discussed by various authors [3, 291, 294–300].

The chemical reactions comprising the heart of this process take place in both an electrochemical cell and an accompanying reactor. The reduction of dichromate to chromic ions drives the second reaction. Dichromate ions are regenerated in the cell. The cell reactions are as follows:

Anode reaction

$$Cr^{+3}(aq) + 3H_2O(\ell) \longrightarrow CrO_3(aq) + 6H^+(aq) + 3e \qquad \text{C.E.} = 90\%$$

$$2H_2O(\ell) \longrightarrow O_2(g) + 4H^+(aq) + 4e \qquad \text{C.E.} = 10\%$$

Cathode reaction

$$2H^+(aq) + 2e \longrightarrow H_2(g) \qquad \text{C.E.} = 100\%$$

The heat of reaction for the primary reaction in the cell is

$$\Delta H_{298} = 126.1 \text{ kcal/mole chromic acid}$$

The cell operates at 55°C and atmospheric pressure without a membrane at 3.3 V.

The cell effluent, a sulfuric acid/chromic acid mix, is pumped to the second reactor, where it is mixed with p-xylene, and the following reaction takes place:

$$C_8H_{10}(\ell) + 4CrO_3(aq) + 12H^+(aq) \longrightarrow C_8H_6O_4(c)$$
$$+ 4Cr^{3+}(aq) + 8H_2O(\ell)$$

$$\Delta H_{298} = -417.8 \text{ kcal/mole TPA}$$

The primary overall reaction for the entire process is

$$C_8H_{10}(\ell) + 4H_2O(\ell) \longrightarrow C_8H_6O_4(c) + 6H_2(g)$$
$$\Delta H_{298} = 86.68 \text{ kcal/mole TPA}$$

The largest energy consumer in this process is the cell. A small amount of energy is consumed by the boiler and ether dryer. Both the cell and the reactor are highly exothermic, but they both operate at such low temperatures that this heat is unavailable for use elsewhere.

Table 9.15. Sorbitol Energy Requirements (kcal) [Basis: 1 kg (5.49 g·moles)]

Chemical Route	
Compressor 1	63
Compressor 2	1
H$_2$O evaporation	352
H$_2$	542
Total	958
Electrochemical Route	
Cell	5194
Ethanol evaporation	2539
Ethanol distillation	1795
Dryer	121
Total	9649

COMPARISON (TEREPHTHALIC ACID SYNTHESIS)

Energy requirements for the chemical and electrochemical routes to tere-phthalic acid are compared in Table 9.16. Both processes use the same raw materials. The energy requirement calculated for the chemical process is considerably smaller than the 4172 kcal/kg given in Ref. 221 based on plant data. The reason for the high actual value is that terephthalic acid must be very pure for polymer production; insufficient information was available for the present calculations to make an accurate estimate of distillation energy. In either case, the chemical process is overwhelmingly favored in terms of en-ergy requirement. The catalytic air-oxidation step is exothermic and only re-quires a small amount of energy for compression. The electrolytic cells re-quire a large amount of energy for the overall 12-electron reaction. One of the process selection rules was violated in this case in order to include an ex-ample of an indirect oxidation from the processes in Table 9.3. An overall one-electron process for an oxidation would have a chance of being much more competitive, but a high-yield catalytic air oxidation is very difficult competition.

The electrochemical route uses toxic sodium dichromate as an intermedi-ate oxidizing agent and high recovery is required. The chemical route uses acetic acid as a solvent, part of which is lost in effluent streams.

Table 9.16. Terephthalic Acid Energy Requirements (kcal) [Basis: 1 kg (6.02 g·moles)]

Chemical Route	
Calculated in Ref. 3	
Compressor	206
Dryer	64
Distillation 1	411
Distillation 2	19
Total	700
Electrochemical Route	
Cell	19,015
Boiler	160
Ether evaporation	48
Refrigeration	1,120
Dryer	60
Hydrogen credit	−3,021
Total	17,382

Phenol

CHEMICAL ROUTE (PHENOL FROM CUMENE)

The energy requirements for phenol synthesis by way of cumene are available from the compilation of Rudd et al. [221].

ELECTROCHEMICAL ROUTE (PHENOL FROM BENZENE)

The electrochemical conversion of benzene to phenol is reviewed in Refs. 3, 301, and 302, which describe other fundamental studies [303–311].

The cation-membrane electrochemical cell operates at 50°C. The following reaction sequence is assumed:

Anode

$$2H_2O(\ell) \longrightarrow O_2(g) + 4H^+(aq) + 4e \qquad C.E. = 100\%$$

Cathode

$$Fe^{+3}(aq) + e \longrightarrow Fe^{2+}(aq)$$
$$O_2(g) + 2H^+(aq) + 2e \longrightarrow H_2O_2(aq) \Bigg\} \quad C.E. = 85\%$$

$$O_2(g) + 4H^+(aq) + 4e \longrightarrow 2H_2O(aq) \qquad C.E. = 15\%$$

The hydrogen peroxide then reacts in solution with the ferrous ions:

$$Fe^{+2}(aq) + H_2O_2(aq) \longrightarrow Fe^{3+}(aq) + OH^-(aq) + OH \cdot (aq)$$
$$Fe^{+2}(aq) + OH \cdot (aq) \longrightarrow Fe^{3+}(aq) + OH^-(aq)$$

The hydroxyl free radical also can react directly with benzene:

$$C_6H_6(\ell) + 2OH \cdot (aq) \longrightarrow C_6H_6O(c) + H_2O(\ell)$$

Thus the overall chemical reaction can be written

$$C_6H_6(\ell) + \tfrac{1}{2}O_2(g) \longrightarrow C_6H_6O(c)$$

$$\Delta H_{298} = -51.16 \text{ kcal/mole phenol}$$

The cell, assumed to operate at 4 V, is by far the largest energy-consuming unit in this process; the water evaporator consumes about two-thirds as much energy as the cell.

COMPARISON (PHENOL SYNTHESIS)

Energy requirements for the chemical and electrochemical routes to phenol are compared in Table 9.17. The chemical route is based on Ref. 221. Both processes include the heat of combustion of the raw materials to put them on an equal basis. The chemical route has a sizable energy advantage because of the energy credit for readily marketable acetone by-product and because of the sizable cell energy requirement for an overall six-electron reaction.

Table 9.17.   Phenol Energy Requirements (kcal) [Basis: 1 kg (10.64 g·moles)]

Chemical Route by way of Cumene (Ref. 3, App. C-12)	12,251
Electrochemical Route	
Cell	21,317
H_2O evaporation	4,610
Crystallizer-centrifuge	1,083
Dryer	8
Benzene	8,409
Oxygen	165
Total	35,592

Biphenyl is a by-product of the electrochemical route and is presumably saleable. Acetone is a desired by-product in the chemical route.

Impact of Electrochemical Routes on Process Industry

A linear-programming model of the U.S. petrochemical industry was used to evaluate the energy efficiency of the electrochemical processes reported above and to assess their potential impact on the industry as a whole. The model was formulated and described in detail by Stadtherr and Rudd [312]. It was used to determine the optimal structure of the industry with respect to resource use. That is, the model was used to determine the combination of process alternatives that can meet the demands of the economy while consuming a minimum of feedstock and energy resources. The version of the model described by Stadtherr and Rudd [312] accounted for feedstock consumption only. For purposes of this study, the process energy data of Rudd et al. [221] were also incorporated in the model. Thus the objective function in the version of the model used here was minimum energy consumption, counting energy consumed both as utilities and as feedstock (feedstock energy consumption is measured using its gross heating values).

The energy efficiency of the electrochemical processes was evaluated by adding the processes to the model and noting which processes, if any, were used in the energy-optimal industry. Since the model accounts for all commercially proven processes, not just those that predominate today, use of the model in this way, in effect, provided comparison of the electrochemical routes *to all available chemical routes.* Since the various segments of the industry are highly interactive, the introduction of a more energy-efficient route in one part of the industry, say in the manufacture of some chemical intermediate, may affect other parts of the industry, as other route changes may occur in order to make use of the more energy-efficient intermediate. The overall effect on the industry of adding the electrochemical processes may be assessed by noting whether any other route changes occur in the energy-optimal industry as a result of its adoption of an electrochemical process.

The study outlined above was performed by adding to the model industry six of the electrochemical processes, those for adiponitrile, aniline, ethylene dichloride (dichloroethane), methyl ethyl ketone, phenol, and terephthalic acid. The processes for hydroquinone, melamine, and sorbitol are not considered here since these chemicals are not manufactured on a large enough scale for their production to be accounted for explicitly in the model. Of the six processes added, two were used in the energy-optimal industry model: the electrochemical routes to adiponitrile and to methyl ethyl ketone. Addition of the adiponitrile process results in a savings of 5260 kcal/kg adiponitrile; addition of the MEK process results in a savings of 1960 kcal/kg methyl ethyl ketone.

The savings in adiponitrile manufacture are less than indicated in Table 9.18 because the chemical route used by the energy-optimal industry before the addition of the electrochemical processes is butane dehydrogenation, while the figure for the chemical route in Table 9.18 is based on butylene dehydrogenation. The butane process is more energy efficient, but is used to a lesser extent than the butylene process.

The savings in methyl ethyl ketone manufacture are greater in the present estimates than in Table 9.18 because those figures account for process energy only. Since the electrochemical process has a high yield of butylene compared to the chemical route, there is a significant savings in feedstock energy.

The adoption of this process does not result in any other route change in the energy-optimal industry. Of course, this does not rule out the development of new routes not now included in the model. For instance, it has been suggested that a route might be developed from adiponitrile to adipic acid, which is currently derived from benzene by way of a cyclohexane intermediate. Thus both monomers for nylon 66 could be derived from propylene by way of acrylonitrile and adiponitrile intermediates, using the energy-efficient electrochemical technology.

The main effect indicated by the model on the overall industry is a tightening in the supply of propylene. Since there is a trend toward increased use of heavier feedstocks (naphtha or gas oil) in the manufacture of ethylene, and this results in increased production of by-product propylene, the effect of the electrochemical technology on propylene supplies does not seem significant.

Table 9.18.   Summary of Process Energy Requirements (Electrochemical Processes Are Not Optimized)

Chemical	Energy Requirement (kcal/kg)	
	Electro-chemical	Chemical
Adiponitrile	43,177[a]	65,808[b]
Aniline		
Nitrobenzene route	36,172[a]	13,919
Phenol route	—[a]	16,736[b]
Sorbitol	9,649	958
Terephthalic acid	17,382	700
Phenol	35,592[a]	12,251[b]
Methyl ethyl ketone	5,057	6,690
		3,233[b]
Melamine	30,159[a]	15,472
Hydroquinone	52,739[a]	30,961
Dichloroethane		
HCl route	17,773[a]	6,130
Cl$_2$ route	—[a]	14,819[b]

[a] Energy charged for hydrocarbon raw materials (different compounds).
[b] Chemical route energy from Reference [221].

Summary

The objective of Section 4 is to review the commerical status of electroorganic process technology and to provide methods for estimating whether energy savings might be realized by the introduction of electroorganic processes for production of high-tonnage organic chemicals.

It is important to recognize that severe restraints were set on the scope of this section. The conclusions reached here therefore may not correspond to the more balanced perspective with which candidate processes are normally evaluated by the industrial sector. In particular, only the energy consumption of large-tonnage organic processes is evaluated, without consideration of waste-water treatment or economics of operating and capital costs.

This section provides a previously unavailable data base of electroorganic synthesis processes, along with engineering methodologies needed for process evaluation. Although the constraints on this section are limited to energy-efficiency conditions of existing process candidates alone, the data base and the methods of evaluation can be readily expanded to include a wider range of criteria and process candidates. By these additional refinements, a more realistic assessment of electroorganic process routes surely will emerge, not only in the large-tonnage arena considered here, but also in the lower-tonnage categories, where product compounds tend to be more complex, to have high molecular weight, and to be polyfunctional.

The nine chemicals for which detailed estimations of process energy requirements were made in this study were selected either because they illustrated a particular class of electrochemical reaction or because they were considered to have a reasonable chance of a more-energy-efficient electrochemical route. Within the constraints specified on this study of nine compounds, the electrochemical routes offered energy savings in two processes and came close to an even match in two other processes. In many of the cases the chemical and electrochemical processes started with different raw materials. It was therefore necessary to link more than one process in series back to the same raw materials or to basic building blocks.

A summary of the nonoptimized process energy requirements for the nine chemicals is given in Table 9.18. The energy requirements include those for chemical and electrochemical reactors and for separation processes. Electrical energy inputs to cells, compressors, and refrigerators were divided by 0.325, the U.S. average fossil fuel-to-electricity efficiency, to convert to a common fossil-fuel basis. The gross heating values of alkane and benzene feedstocks were assigned to each process when different feedstocks were used in the chemical and electrochemical routes. By-product hydrogen was credited its gross heating value and other organic by-products were credited their manufacturing energy requirements for their main synthesis route. Energy data for some of the chemical products with several series processes were obtained from the literature and are so noted in Table 9.18. Detailed calculations are available in Ref. 3.

Although every effort is made to make the comparisons based on state of the art technology, the information available was not always complete. Chemical process flowsheets are for the most part based on a 1977 survey of industrial organic processes by the Environmental Protection Agency, but even this source for reasons of industrial secrecy was not completely current. The electrochemical process calculations were based on data from bench to industrial scale, from recent patents and papers, as well as from a 40-year-old process. The electrochemical process flowsheets are *not* optimized to minimum energy consumption conditions. Thus the "balance sheet" in Table 9.18 is not for a true instant in time, but spans a number of years for the various processes.

The electrochemical route to adiponitrile shows a clear energy advantage over the chemical route. This is one of the more accurate comparisons because the electrochemical route calculations were based on a published plant flowsheet and the chemical route result is based on production data. Qualifications are that a new chemical route introduced in 1973 probably has a lower energy requirement of at least 6600 kcal/kg, and that improvements have been made in the electrochemical process.

Production of methyl ethyl ketone in an electrochemical cell appears to be attractive on an energy basis in comparison with the current chemical route. In addition, replacement of the chemical route by the electrochemical route would offer the further advantage of conserving feedstock material by higher efficiency utilization. The cell uses an air cathode and oxidizes butene at a catalytic anode. For the calculations it was assumed that the cell operated in the short-circuited mode although it could generate electricity.

The electrochemical synthesis of hydroquinone begins with the oxidation of benzene to the quinone intermediate. While the electrochemical process involves a single step in contrast to a multistep chemical route to this intermediate, it is reported that current efficiencies are only 40%. However, if research and development activity would focus on increasing the current efficiency toward 100%, then the energy required for the electrochemical route would be competitive with that required for the existing chemical route.

In-cell halogenations, as used to make dichloroethane, are also close to breakeven with chemical routes, all of which use electrolytically produced chlorine. More detailed analyses and optimization calculations would clearly be justified to provide a better definition of the relative benefits of the two routes.

The chemical synthesis of melamine is less energy intensive than the electrochemical route owing, to an appreciable extent, to power consumption by the electrolysis cell.

The two electrochemical hydrogenation processes (aniline and sorbitol syntheses) are considerably more energy intensive than high-yield catalytic hydrogenations. Electrochemical generation of hydrogen is more energy intensive than generation from hydrocarbons by steam reforming, and separation of products from electrolytes requires more energy than that from corresponding chemical process streams. While economic considerations are not taken into account in this section, it is worth mentioning that electrochemical hydrogenations occur under very mild process conditions in comparison with some chemical routes that require very high pressure. Other types of electrochemical reductions deserve evaluation, including hydrodimerizations and electrode-specific reductions.

Terephthalic acid synthesis represents an example of indirect oxidation using the chromic/dichromate redox couple. However, one of the process selection rules (Table 9.4) was broken, since the electrochemical route requires 12 electrons per mole of product. While the resulting comparison may thus be foreordained to favor the chemical route, it is also true that high-yield catalytic air oxidations are difficult competition for electrochemical processes. Evaluation of competing electrochemical routes would be worthwhile in cases of difficult chemical oxidation processes, which often exhibit low yields and strenuous reaction conditions.

If the electrochemical route to phenol is to become attractive on an energy basis, research must be directed to reducing the cell voltage and increasing the current density.

This section provides the first extensive comparison of energy requirements for a significant number of processes (electrochemical versus chemical) treated in a consistent manner. The overall errors of the calculations for each process are judged to be less than ±50%, although the electrochemical processes are not optimal. Because only one iterative step is made toward optimization for each electrochemical process route, it seems undeniably clear that the results given here represent a *preliminary assessment* against which future improvements in process technologies may be measured. That is, considerable research and development effort has already been devoted to current large-tonnage chemical processes to make them economically competitive. With even modest development efforts, there is no question but that the energy efficiencies for electroorganic processes estimated here could be substantially improved.

5   CONCLUSIONS AND RECOMMENDATIONS

Many of the critical components that make up the field of electroorganic synthesis technology have recently grown into place. Chemists, for example, can synthesize many thousands of organic compounds by electrolysis and have thereby come to a deeper understanding of the molecular basis underlying many reactions. Advances in surface science and catalysis are leading to controlled study of electrode properties, which should further enhance ability to synthesize chemicals in high yields. Further, electrochemical engineers have developed reactor designs along with modeling techniques needed to guide scale-up and optimization. Chemical engineers are therefore increasingly able to assess economic aspects of process routes involving electrochemistry. It seems inevitable that consolidation of these interdisciplinary elements will continue and will promote increased effectiveness in the development of electroorganic process technologies.

Generally, electroorganic processes can be favorable when the product molecular weight is high, when there are fewer than four electrons in the electrolysis step, when the yield is high compared to a corresponding chemical route, and when the raw materials are less expensive than those used in the chemical synthesis. In some cases the electroorganic route may be the only way to make some desirable products in reasonable yield; often there are fewer processing steps and fewer pollutants.

The task of identifying process candidates may involve widespread searching of general categories, such as the high-tonnage search described in Section 4. On the other hand, most industrial producers carry out evaluation studies based on a specific chemical of interest. In such situations, the consideration of potential electrochemical synthesis routes should be included as a matter of routine, especially if the chemical routes are cumbersome from the viewpoint of energy, feedstock, or environment. That is, electrochemical methods should be regarded as one of several tools available for the purpose of synthesis. It is no longer realistic to ignore electrochemical candidates, because the methods of evaluation and implementation of electrolytic routes have so vastly improved.

With the recent development of new electrode and membrane materials, the prospects for imaginative new process chemistry have been greatly expanded. There can be no question that the synthesis of organics from nonpetroleum feedstock is extremely important; electrolytic routes based on CO and CO_2 have been known since the mid-1800s and have been vigorously pursued in the laboratory. Similarly, the possibility of paired synthesis (where useful chemical products are made at *both* electrodes) could offer highly competitive process economics.

In the development of an electrochemical process route, three viewpoints are critically important, namely, those of the *organic electrochemist, the electrochemical engineer*, and the *process engineer*. For example, the prospect of using a nonaqueous solvent may intrigue the chemist since wholly new avenues of chemistry would be available in comparison with aqueous chemistry. The electrochemical engineer, however, may express concern over low electrolyte solubility, consequent high power consumption, and poor in-cell mass transfer. The chemical engineer might recognize still other nonelectrochemical implications, such as easier downstream solvent recovery, higher investment cost of solvent inventory, and environmental concern. In the early stages of process development, it is therefore especially critical to incorporate different viewpoints to ensure cost-effective workup.

Many important engineering questions cannot be adequately answered with data from batch-type cells. The early deployment of continuous flow cells is often an important step in obtaining an improved assessment of technical problems and realities. The early identification of gaps in needed process data can also be important. Measurement of heat capacity, enthalpy, solubility, and conductivity, for example, are often required for early evaluation and are often unavailable from literature sources.

The application of computer-based process design simulators to electrochemical process studies would bring a major new engineering methodology into the industrial electrolytic field. More work is needed to identify the type of laboratory and pilot-plant data required to scale-up and design optimum cell configurations and operating conditions. Of particular importance is the need for more investment cost data and estimation procedures for various cell configurations.

The evaluation of competing process routes depends to a great extent on the relative costs of power, feedstock, capital, wastewater treatment, labor, and materials as projected over the lifetime of the process. It would therefore be helpful to examine the effect of such projected future trends on the viability of selected electroorganic processes versus chemical counterparts. Of special importance would be assessment of the effect of decoupling electricity costs from petroleum costs owing to the advent of nuclear or solar generating capability.

This chapter reviews a number of major trends that, taken together, indicate electroorganic synthesis technology has come of age. Our understanding of electroorganic chemistry has advanced toward levels that exhibit more rigorous and more scientific approaches to explaining observed behavior on the basis of molecular events. Electrochemical engineering has become established as a quantitative discipline for the design of cells and processes. As a consequence, chemical engineers are increasingly able to assess electrochemical options for organic synthesis. It is yet to be seen whether these trends will lead to substantial expansion of industrial electroorganic synthesis. It is abundantly clear, however, that the essential groundwork has been laid and the field is ripe for industrial development.

Acknowledgment

The survey of organic electrolytic processes reported in Sections 2 and 4 of this chapter was carried out under Argonne National Laboratory Contract No. 31-109-38-4209 at Electrochemical Technology Corporation. Portions of the work were subcontracted to the University of Illinois and to the Electrosynthesis Company.

References

1. S. Swann, Jr., and R. C. Alkire, *Bibliography of Electro-Organic Syntheses, 1801–1975,* The Electrochemical Society, Princeton, NJ, 1980.
2. F. Fichter, *Organische Elektrochemie,* Steinkopff, Dresden, 1942; available from University Microfilms International, Ann Arbor, MI.
3. T. R. Beck, R. T. Ruggeri, R. C. Alkire, M. A. Stadtherr, and N. L. Weinberg, "A Survey of Organic Electrolytic Processes," Report #ANL/OEPM-78-5, Argonne National Laboratory, Argonne, IL, 1979.
4. E. Opperman, U.S. Patent 323,435 (1906).
5. C. J. Thatcher, Canadian Patent 222,264 (1922).

6. S. Krishan, V. A. Vyas, M. S. Y. Pathy, and H. V. K. Udupa, *Bull. India Sect. Electrochem. Soc.*, **14**, 32 (1964).
7. K. Niki, T. Sekine, and Sugino, *J. Electrochem. Soc. Jap.*, **37**, 74 (1969).
8. J. P. Millington, *Chem. Ind. (Lond.)*, **1975** 780; British Patent 1,377,681 (1974).
9. M. Fremery, H. Hoever, and G. Schwarzlose, *Chem. Ing. Tech.*, **46**, 635 (1974); German Patent 2,204,351 (1973).
10. F. J. Anderson and J. L. Kessler, U.S. Patent 3,758,391 (1973).
11. V. E. Wolf, *Chem. Ber.*, **87**, 668 (1954).
12. U.S. Patent 2,786,022 (1955).
13. J. C. Marchon and B. Rosset, *Bull. Chem. Soc. Fr.*, **1965** 2776.
14. E. J. Matzner, U.S. Patent 3,449,225 (1969).
15. L. H. Cutler, U.S. Patent 4,032,416 (1977).
16. L. H. Cutler, Abstracts, AICHE Meeting, Atlanta, Georgia, 1978.
17. British Patent 737,567; 737,577 (1968).
18. British Patent 1,177,308.
19. C. Thibault and P. Mathieu, U.S. Patent 3,808,112 (1974).
20. S. J. Reddy and V. R. Krishnan, *J. Electroanal. Chem.*, **27**, 474 (1970).
21. K. Elbs and W. Herz, *Z. Elektrochem.*, **4**, 113 (1897).
22. J. Billiter, *Die Technische Elektrolyze der Nichtmetalle*, Springer-Verlag, Wien, 1954.
23. R. Ramaswamy, M. S. Venkatachalapathy, and H. V. K. Udupa, *J. Electrochem. Soc.*, **110**, 294 (1963).
24. E. Schering, *Z. Elektrochem.*, **1**, 70 (1894).
25. N. Clauson-Kaas and F. Limborg, U.S. Patent 2,714,576 (1955); U.S. Patent 2,801,252 (1957).
26. N. Clauson-Kaas, F. Limborg, and K. Glens, *Acta Chem. Scand.*, **6**, 531 (1952).
27. H. S. Isbell, *Natl. Bur. Std. (U.S.)*, **914**, 331 (1936).
28. C. G. Fink and D. B. Summers, *Trans. Electrochem. Soc.*, **74**, 625 (1938).
29. L. Mezynski and M. Pawelczak, *Przem. Chem.*, **51**, 28 (1972).
30. R. W. Foreman and J. W. Sprague, *Ind. Eng. Chem. Prod. Res. Dev.*, **2**, 303 (1963).
31. J. W. Sprague, U.S. Patent 3,294,657 (1966); 3,300,398 (1967).
32. E. A. Dzhafarov, V. G. Khomyakov, N. G. Bakhchirats'yan, M. Y. A. Fioshin, G. A. Kokarev, and M. A. Khrizolitova, USSR Patent 143,391 (1962) [*Chem. Abstr.*, **57**, 9668g].
33. K. I. Kryshchenko et al., *Khim. Prom.*, **45**, 496 (1969).
34. O. R. Brown, S. Chandra, and J. A. Harrison, *J. Electroanal Chem.*, **34**, 505 (1972).
35. D. G. Braithwaite, U.S. Patent 3,256,161 (1966); 3,391,066, and 3,391,067 (1968).
36. L. L. Bott, *Hydrocarbon Proc.*, **44**, 115 (1965).
37. W. J. Koehl, Jr., U.S. Patent 3,252,876 (1966).
38. W. Himmele, N. Kutepow, and W. Schwab, German Patent 1,162,347 (1964).
39. F. Beck, H. Nohe, and J. Haufe, German Offen., Patent 2,014,985 (1971).

40. E. P. Kovsman, G. N. Greidlin, and Y. M. Tyurin, *Khim. Prom.*, **49**, 16 (1973).

41. N. T. Kobzeva, E. P. Kovsman, Yu. B. Vasil'ev, G. A. Tarkhanov, and G. N. Freidlin, *Sov. Electrochem.*, **11**, 667 (1975).

42. S. Chidambaram, M. S. V. Pathy, and H. V. K. Udupa, *Res. Ind. (New Delhi)*, **15**, 215 (1970).

43. R. Kanakam, A. P. Shakunthala, S. Chidambaram, M. S. V. Pathy, and H. V. K. Udupa, *Electrochim. Acta*, **16**, 423 (1971).

44. S. Chidambaram, M. S. V. Pathy, and H. V. K. Udupa, *Indian J. Appl. Chem.*, **34**, 1 (1971).

45. J. D. Fitzpatrick and L. D. Myers, U.S. Patent 2,450,858 (1943).

46. M. Kappel, *Chem. Ing. Tech.*, **35**, 386 (1963).

47. F. Covitz, U.S. Patents 3,509,031 (1970) 3,616,323 (1971), and 3,616,324 (1971).

48. K. Ziegler and H. Lehmkuhl, U.S. Patent 3,372,097 (1968).

49. J. A. M. Leduc, U.S. Patents 3,288,692 (1966), 3,379,627 (1968), 3,342,717 (1967), and 3,427,235 (1969).

50. Farhenfabriken Boyer, French Patent 1,540,800 (1968).

51. T. D. Binns and D. C. G. Cattiker, U.S. Patent 3,635,803 (1972).

52. K. H. Simmrock and G. Hellemanns, German Patent 2,658,189.

53. N. A. Dschabanowski and L. D. Borkhi, Russian Patent 166,654 and 172,330 (1963).

54. H. F. Conway and V. E. Sohns, *Ind. Eng. Chem.*, **51**, 637 (1959).

55. H. F. Conway, E. B. Lancaster, and V. E. Sohns, *Electr. Technol.*, **2**, 43-46 (1964).

56. M. S. V. Pathy and H. V. K. Udupa, *Central Electrochemical Research Institute (India)*, **1964**, 20.

57. C. L. Mehltretter, U.S. Patent 2,713,553 (1955); 2,830,941 (1958);

58. J. E. Slager, U.S. Patent 3,086,969 (1963).

59. B. B. Dey and R. K. Maller, *J. Sci. Ind. Res. (India)*, **12B**, 255 (1953).

60. N. Ibl, J. C. Puippe, and H. Angerer, Abstracts of Papers, AICHE Meeting, Atlanta, Georgia, 1978.

61. Netherlands Patent. 41,338 (1937) [*Chem. Abstr.*, **31**, 8399 (1937)].

62. S. Mizuno and Y. Kohtoku, *Chem. Abstr.*, **69**, 15324g (1968).

63. J. H. Simons et al., *J. Electrochem. Soc.*, **95**, 47, 53, 55, 59, 64 (1949).

64. J. H. Simons, U.S. Patent 2,519,983 (1950).

65. E. A. Kauck and J. H. Simons, U.S. Patent 2,594,272 (1952); British Patent 666,733 (1952).

66. H. M. Fox, F. N. Ruehlen, and W. V. Childs, *J. Electrochem. Soc.*, **118**, 1246 (1971).

67. W. V. Childs, U.S. Patents 3,558,449 (1971), 3,617,453 (1971), 3,655,548 (1972), and 3,853,737 (1974).

68. F. N. Ruehlen, U.S. Patent 3,620,941 (1971).

69. Private communication from BASF (Ludwigshofen) from Drs. Reif and Pape, January 29, 1979.

70. Private communication from Hoechst (Frankfurt) from Drs. Jensen and Lohaus, January 17, 1979.

71. Private communication from Prof. F. Beck, Gesamthochschule Duisburg, April 10, 1979.

72. N. Shuster, K. J. O'Leary, and A. D. Babinsky, Abstracts of Papers, Electrochemical Society Meeting, Washington, DC May 2-7, 1976, pp. 694-695.

73. H. Nohe and F. Beck, German Patent 2,343,054 (1975).

74. M. M. Baizer, *J. Electrochem. Soc.*, **111**, 215 (1964); U.S. Patents 3,193,480 (1965) 3,193,481 (1965), and 3,440,154 (1969).

75. Asahi Chemical, French Patents 1,503,244 (1965) and 1,533,472 (1968).

76. F. Beck and H. Guthke, *Chem. Ing. Tech.*, **41**, 943 (1969).

77. W. V. Childs and H. C. Walters, Abstracts of Papers, AICHE Meeting, Atlanta, Georgia, Feb. 1978.

78. M. Y. Fioshin and A. P. Tomilov, *Usp. Elektrochim. Org., Soedin.*, **1966** 256.

79. Russian Patents 137,924 (1961), and 445,647 (1974).

80. F. Beck and F. P. Woerner, German Offen. Patent 2,144,419 (1971).

81. B. Sakurai, *Bull. Chem. Soc. Jap.*, **7**, 127 (1932).

82. R. A. Benkesser, E. M. Kaiser, and R. F. Lambert, *J. Am. Chem. Soc.*, **86**, 5272 (1964).

83. U.S. Patent 2,182,284 (1938).

84. *Chem. Eng.*, 78B (1971), Esso Res.: see F. Beck, Ref 90.

85. P. Ruetschi and G. Trumpler, *Helv. Chim. Acta*, **36**, 1649 (1953).

86. K. Natarajan, K. S. Udupa, G. S. Subramanian, and H. V. K. Udupa, *Electrochem. Technol.*, **2**, 151 (1964).

87. G. Schwarzlose, German Offen. Patent 2,237,612 (1974).

88. V. Krishnan, K. Ragupathy, and H. V. K. Udupa, *Trans. SAEST,* **11**, 509 (1976).

89. V. Krishnan, K. Ragupathy and H. V. K. Udupa, *Electrochim. Acta,* **21**, 449 (1976).

90. F. Beck, *Elektroorganische Chemie,* Verlag Chemie, Berlin 1974, p 17.

91. H. Alt and H. Clasen, German Offen. Patent 2,157,608 (1973) [*Chem. Abstr.,* **79**, 48729c].

92. M. J. Allen and J. Ocampo, *J. Electrochem. Soc.,* **103**, 452 (1956).

93. M. J. Allen, *Can. J. Chem.,* **37**, 257 (1959); British Patent 817,097 (1959).

94. J. W. Drew and G. J. Moll, *Ind. Eng. Chem.,* **53**, 948A (1961).

95. Indian Patent 60,865 (1957).

96. H. V. K. Udupa, G. S. Subramanian, and K. S. Udupa, *Res. Ind. (India),* **5**, 309 (1960).

97. H. J. Creighton, U.S. Patents 1,612,361 (1926), 1,653,004 (1927), 1,712,952 (1929), and 1,990,582 (1935).

98. M. T. Sanders and R. A. Hales, *J. Electrochem. Soc.,* **96**, 241, (1949).

99. J. E. Colchester and J. H. Entwistle, U.S. Patent 2,478,042 (1966); British Patent (ICI), 1,913,149 (1968).

100. H. C. Rance and J. M. Coulson, *Electrochim. Acta,* **14**, 283 (1969).

101. H. V. K. Udupa, G. S. Subramanian, and S. Thangavelu, Indian Patent 130,295 (1974).

102. D. W. Lawson and D. S. Saltera, German Offen. Patent 2,026,039 (1970) [*Chem. Abstr.,* **74**, 42154 (1971)].

103. C. N. Otin, *Z. Elektrochem.,* **16**, 674 (1910).

104. P. N. Anantharam and H. V. K. Udupa, Abstracts of Papers, Electrochemical Society Meeting, Washington, DC, May 1976, pp. 708–709.

105. German Patents (Rohner) 181,116 (1903) and 94,736 (1896).

106. T. D. Balakrishnan, K. S. Udupa, G. S. Subramanian, and H. V. K. Udupa, *Chem. Ing. Tech.*, **41**, 776 (1969).

107. S. Thangavelu, G. S. Subramanian, and H. V. K. Udupa, *Denki Kagaku*, **39**, 5 (1971).

108. H. V. K. Udupa et al., Indian Patent 119,015 (1974).

109. M. Yamashita and K. Sugino, *J. Electrochem. Soc.*, **104**, 100 (1957).

110. T. D. Balakrishnan, K. S. Udupa, G. S. Subramanian, and H. V. K. Udupa, *Ind. Eng. Chem. Proc. Des. Dev.*, **10**, 495 (1971); Indian Patent 127,059 (1972).

111. R. H. McKee et al., *Trans. Am. Elect. Soc.*, **62**, 203 (1932).

112. H. V. K. Udupa and M. V. Rao, *Electrochim. Acta*, **12**, 353 (1967).

113. M. Y. Fioshin and A. P. Tomilov, *Khim. Prom.*, **43**, 243 (1967).

114. R. Kanakam et al., *Electrochim. Acta*, **16**, 423 (1971).

115. Rhone-Poulenc, French Patent 2,151,150 (1973) [*Chem. Abstr.*, **79**, 60911P (1973)].

116. D. J. Pickett and K. S. Yap, *J. Appl. Electrochem.*, **4**, 17 (1974).

117. F. Goodridge and K. Lister, British Patent 1,411,371 (1975) [*Chem. Abstr.*, **84**, 66955f (1976)].

118. P. C. Condit, *Ind. Eng. Chem.*, **48**, 1252 (1956).

119. H. Suter, H. Nohe, F. Beck, W. Bruegel, and H. Aschenbrenner, U.S. Patent 3,542,656 (1970).

120. H. Nohe, *Chem. Ing. Tech.*, **46**, 594 (1974).

121. B. Sakurai, *Bull. Chem. Soc. Jap.*, **5**, 184 (1930).

122. N. Ibl, *Chem. Ing. Tech.*, **33**, 69 (1961); **35**, 353 (1963); **36**, 601 (1964).

123. Robinson Bros. and D. W. Parkes, British Patent 395,741 (1933).

124. N. A. Dschabanowski, N. E. Khomutov and V. V. Tsodikov, *Zh. Prikl. Khim*, **38**, 2720 (1965).

125. U.S. Army Technical Translation FSTC-HT-23-258-68: "Industrial Electrosynthesis of Organic Compounds," by M. Y. Fioshin and A. P. Tomilov, 1966.

126. T. D. Balakrishnan et al., *Chem. Ind. (Lond.)*, **1970** 1969.

127. M. Lukes, *Chem. Listy*, **27**, 392 (1933).

128. W. H. Perkin and S. G. Plant, *J. Chem. Soc.*, **125**, 1503 (1924).

129. F. Beck (BASF), private communication.

130. Private communication from BASF (Ludwigs Hafen) from Drs. Reif and Paper, January 29, 1979.

131. Private communication from Hoechst (Frankfurt) from Drs. Jensen and Lohaus, January 17, 1979.

132. DT-OS 2,428,878 (1974); Euchem Conference on Organic Electrochemistry, Pitlochry, Scotland May 27, 1977.

133. Private communication from Prof. F. Beck, Gesamthochschule Duisburg, April 10, 1979.

134. Brochure #EC-2, *Chemomat*, published 1976, Ionics Inc., 65 Grove Street, Watertown, MA 02172.

135. J. K. Easton, *J. Wat. Pollut. Control Fed.*, **39**, 1621 (1967).

136. M. Tarjanyi and M. P. Strier, U.S. Patents 3,764,497 and 3,764,498 (1971).
137. Report, 1974, Oxy Effluent Control Ltd., England.
138. R. B. Macmullin, *Chem. Eng. Prog.*, **46**, 440 (1950).
139. M. Kappel, *Chem. Ing. Tech.*, **35**, 386 (1963).
140. Monsanto Co., British Patent 1,066,930 (1967); Netherlands Appl. 6,510,357 (1966); Netherlands Appl. 6,603,164 (1966).
141. G. S. Supin, USSR Patent 132,205 (1960).
142. S. Swann, Jr., and R. C. Alkire, *Bibliography* of *Electo-Organic Syntheses, 1801–1975,* The Electrochemical Society, Princeton, NJ, 1980.
143. Private communication from Hoechst (Frankfurt) from Drs. Jensen and Lohaus, January 17, 1979.
144. D. Danly, *Kirk-Othmer Encyclopedia of Chemical Technology*, Vol. 8, 3rd ed., Wiley, New York, 1979 pp. 696–720.
145. J. L. Fitzjohn, *Chem. Eng. Prog.*, **71**, 85 (1975).
146. M. Krumpelt, E. Y. Weissman, and R. C. Alkire, *Electroorganic Synthesis Technology, AIChE Symp. Ser.*, No. 185, **75** (1979).
147. L. Meites and P. Zuman, *Electrochemical Data, Part I: Organic, Organometallic and Biochemical Substances*, Wiley, New York, 1974.
148. M. M. Baizer, *Organic Electrochemistry: An Introduction and a Guide*, Dekker, New York, 1973.
149. C. L. Mantell, "Electro-Organic Chemical Processing," Noyes Development Corp., Park Ridge, NJ, 1968.
150. A. T. Kuhn, "Industrial Electrochemical Processes," Elsevier, New York, 1971, Chap. 13.
151. M. M. Baizer, *J. Electrochem. Soc.*, **124**, 185C (1977).
152. C. J. Brockman, *Electroorganic Chemistry*, Wiley, New York, 1926.
153. S. Swann Jr., "Electrolytic Reactions," in *Technique of Organic Chemistry*, Vol. 2, A. Weissberger, Ed., Interscience, New York, 1956.
154. M. J. Allen, *Organic Electrode Processes*, Chapman & Hall, London, 1958.
155. Stanley Wawzonek, *Synthesis, p.* **285** (1971).
156. L. Eberson and H. Schafer, "Organic Electrochemistry," in *Fortschrette der chemischen Forschung*, Springer-Verlag, Berlin, 1971. p. 21.
157. A. P. Tomilov, S. G. Mairanovskii, M. Ya. Fioshin, and V. A. Smirnov, *Electrochemistry of Organic Compounds*, Halsted Press (a division of John Wiley & Sons, Inc.), New York, 1972.
158. Albert J. Fry, *Synthetic Organic Electrochemistry*, Harper & Row, New York, 1972.
159. Fritz Beck, *Elektroorganische Chemie: Grundlagen and Anwendungen*, Verlag Chemie, Weinbein, 1974.
160. M. R. Rifi and Frank H. Covitz, *Introduction to Organic Electrochemistry*, Dekker, New York, 1974.
161. Norman L. Weinberg, Ed., *Technique of Electroorganic Synthesis, Part II*, Techniques of Chemistry Series, Vol. V, A. Weissberger, Ed., Wiley-Interscience, New York, 1975.
162. L. Eberson and K. Nyberg, "Structure and Mechanism in Organic Electrochemistry," in *Advances in Physical Organic Chemistry*, Vol. 12, V. Gold and D. Bethell, Eds., Academic Press, New York, 1976.

163. A. J. Bard, Ed., *Encyclopedia of Electrochemistry of the Elements*, Organic Sections Vols. XI (1978), XII (1978), XIII and XIV, Dekker, New York.

164. S. H. Langer and G. P. Sakellaropoulos, *Ind. Eng. Chem. Proc. Des. Dev.*, 18(4), 567 (1979).

165. E. Gileadi, E. Kirowa-Eisner, and J. Penciner, *Interfacial Electrochemistry— An Experimental Approach*, Addison-Wesley, Reading, MA, 1975.

166. F. Goodridge and C. J. H. King, in *Technique of Electroorganic Synthesis*, N. L. Weinberg, Ed., Techniques of Chemistry Series, Vol. V Part I, A. Weissberger, Ed., Wiley, New York, 1974.

167. Ralph N. Adams, *Electrochemistry at Solid Electrodes*, Dekker, New York, 1969.

168. J. R. Selman and C. W. Tobias, in *Adv. Chem. Eng.*, 10, 211 (1978).

169. A. T. Kuhn, Ed., *Industrial Electrochemical Processes*, Elsevier, New York, 1971.

170. R. G. Kesting, *Synthetic Polymeric Membrane*, McGraw-Hill, New York, 1971.

171. R. E. Lacey and S. Loeb, *Industrial Processing with Membranes*, Wiley, New York, 1972.

172. N. Lakshminarayaniah, *Transport Phenomena in Membranes*, Academic Press, New York 1969.

173. P. Meares, Ed., *Membranes Separation Processes*, Elsevier Scientific Pub. Co., New York, 1976.

174. H. S. Kaufman and J. J. Tralcetta, *Introduction to Polymer Science and Technology*, Interscience, New York, 1977.

175. S. Sourirajan, Ed., *Reverse Osmosis and Synthetic Membranes: Theory, Technology, Practice*, National Research Council of Canada, Ottawa, 1977.

176. R. F. Madsen, *Hyperfiltration and Ultrafiltration in Plate-and-Frame Systems*, Elsevier, Amsterdam, 1977.

177. C. K. Mann and K. K. Barnes, *Electrochemical Reactions in Non-aqueous Systems*, Dekker, New York, 1970.

178. R. W. Houghton and A. T. Kuhn, *J. Appl. Electrochem.*, 4, 73 (1974).

179. A. P. Tomilov and M. Ya. Fioshin, *Br. Chem. Eng. 16*, 154 (1971).

180. P. Gallone, *Electrochim. Acta*, 22 913 (1977).

181. *Hydrocarbon Process*, 56(11), 127 (1977).

182. J. S. Newman and W. Tiedemann, *AIChE J.*, 21, 25 (1975).

183. J. S. Newman and W. Tiedemann, *Adv. Electrochem. Electrochem. Eng.*, 11, 353 (1978).

184. D. N. Bennion and J. S. Newman, *J. Appl. Electrochem.*, 2, 113 (1972).

185. R. C. Alkire and P. K. Ng, *J. Electrochem. Soc.*, 121, 95 (1974); 124, 1220 (1977).

186. D. A. Braithwaite, (Nalco Chem. Co.), *U.S. Patent 3,391,067 (July 2, 1968)*.

187. N. Ibl, K. Kramer, L. Ponto and P. Robertson, *J. Appl. Electrochem.*, 7, 323 (1977); *Electroorganic Synthesis Technology, AIChE Symp. Ser.*, No. 185, 75, 45 (1979).

188. K. B. Keating and V. D. Sutlic, *Electroorganic Synthesis Technology, AIC·E Symp. Ser.*, No. 185, 75, 76 (1979).

189. F. B. Leitz and L. Marincic, *J. Appl. Electrochem.*, 7, 473 (1977).

486   ELECTROCHEMICAL ENGINEERING OF ELECTROORGANIC PROCESSES

190. H. V. K. Udupa, et al., *Chem. Age India*, **15**, 501 (1964).
191. J. Ghoroghchian, R. E. W. Jansson, and R. J. Marshall, *Electrochim. Acta* **24**, 1175 (1979).
192. M. Farooque and T. Z. Fahidy, *J. Electrochem. Soc.*, **125**, 1777 (1978).
193. M. Fleischmann and F. E. W. Jansson, *Chem. Ing. Technol.*, **49**, 283 (1977).
194. F. Goodridge and C. J. Vance, *Electrochim. Acta*, **24**, 1247 (1979).
195. R. P. Tison, *J. Electrochem. Soc.*, **127**, 122C (1980).
196. F. Goodridge, C. J. H. King, and A. Wright, *Electrochim. Acta*, **22**, 347 (1977).
197. C. Oloman, *J. Electrochem. Soc.*, **126**, 1885 (1979).
198. R. Dworak, H. Fees, and H. Wendt, *Electroorganic Synthesis Technology*, *AIChE Symp. Ser.*, No. 185, **75**, 38 (1979).
199. F. Beck and H. Guthke, *Chem. Ing. Techn.*, **41**, 943 (1969).
200. F. Wenisch, H. Nohe, H. Hannebaum, R. Horn, M. Stroezel, and D. Degner, *Electroorganic Synthesis Technology, AIChE Symp. Ser.*, No. 185, **75**, 14 (1979).
201. R. K. Horn, *Electroorganic Synthesis Technology, AIChE Symp. Ser.*, No. 185, **75**, 125 (1979).
202. R. Dworak and H. Wendt, *Ber. Bunsenges. Phys. Chem.*, **80**, 77 (1976).
203. R. C. Alkire, *J. Electrochem. Soc.*, **120**, 900 (1973).
204. M. Fleischmann, I. N. Justinijanovic, *J. Appl Electrochem.*, **10**, 143, 151, 157, 169 (1980).
205. R. B. MacMullin, *Electrochem. Technol.*, **1**, 5 (1963); **2**, 106 (1964); *Denki Kagaku*, **38**, 570 (1970); *J. Electrochem. Soc.*, **120**, 135C (1973).
206. J. S. Newman, *Electrochemical Systems*, Prentice-Hall, Englewood Cliffs, NJ, 1973.
207. F. Hine, *Denkikagaku-hannososa to Denkaiso-kogaku*, Kagaku-dojin Publishers, Kyoto, 1978.
208. D. J. Pickett, *Electrochemical Reactor Design*, Elsevier Scientific, New York, 1977.
209. R. C. Alkire and R. M. Gould, *J. Electrochem. Soc.*, **123**, 1842 (1976); **126**, 2125 (1979); **127**, 605 (1980).
210. C. W. Tobias, *J. Electrochem. Soc.*, **120**, 65c (1973).
211. K. W. Choi, D. N. Bennion, and J. Newman, *J. Electrochem. Soc.*, **123**, 1616, 1628 (1976).
212. J. B. Riggs, "Modeling of the Electrochemical Machining Process," Ph. D. Thesis, University of California, Berkeley (1977).
213. T. R. Beck, in *Techniques of Electrochemistry*, Vol. 3, E. Yeager and A. Salkind, Eds., Wiley-Interscience, NY, (1978).
214. J. Fitzjohn, *Electroorganic Synthesis Technology, AIChE Symp. Ser.*, No. 185, **75**, 64 (1979).
215. United States International Trade Commission, "Synthetic Organic Chemicals, United States Production and Sales, 1975," USITC Publications 804, U.S. Government Printing Office, Washington, DC, 1977.
216. R. Liepins, F. Mixon, C. Hudak and T. B. Parsons, "Industrial Profiles for Environmental Use, Chapter 6, The Industrial Organic Chemicals Industry,"

EPA Contract No. 68-02-1319, Document No. EPA-600/2-77-023f, February 1977.

217. D. E. Danly, Monsanto Co., Pensacola, FL, personal communication.
218. R. N. Shreve and J. A. Brink, Jr., *Chemical Process Industries*, 4th ed., McGraw-Hill, New York, 1977.
219. Kirk-Othmer, *Encyclopedia of Chemical Technology*, Interscience, New York, 1963.
220. F. A. Lowenheim and M. K. Moran, *Industrial Chemicals*, Wiley-Interscience, New York, 1975 (previous authors: W. L. Faith, D. B. Keyes, and R. L. Clark).
221. D. F. Rudd, S. Fathi-Afshar, A. A. Trevino, and M. A. Stadtherr, *Chemical Technology Assessment*, by Wiley-Interscience, New York, 1981.
222. R. H. Perry and C. H. Chilton, *Chemical Engineers Handbook*, 5th ed., McGraw-Hill, New York, 1973.
223. W. H. Seaton, E. Freedman, and D. N. Treweak, "CHETAH-The ASTM Chemical Thermodynamic and Energy Release Evaluation Program," ASTM DS 51, Philadelphia, PA, 1974.
224. D. R. Stull, and H. Prophet, *JANAF Thermochemical Tables*, 2nd ed., National Standard Reference Data Service, National Bureau of Standards (U.S.), July 1970, p. 37.
225. D. R. Stull, E. F. Westrum, and G. C. Sinke, *The Chemical Thermodynamics of Organic Compounds*, Wiley, New York, 1969.
226. R. C. Reid, J. M. Prausnitz, and T. K. Sherwood, *The Properties of Gases and Liquids*, 3rd ed., McGraw-Hill, New York, 1977.
227. R. H. Perry and C. H. Chilton, Ed., *Chemical Engineers' Handbook*, 5th ed., McGraw-Hill, New York, 1973.
228. C. L. Yaws, *Physical Properties*, McGraw-Hill, New York, 1977.
229. N. A. Lange, Ed., *Handbook of Chemistry*, 10th ed., McGraw-Hill, New York, 1969.
230. M. Windholz, Ed., *The Merck Index*, 9th ed., Merck and Co., Inc., Rahway, NJ, 1976.
231. Kirk-Othmer, Encyclopedia of Chemical Technology, 2nd ed., Vol 2, Interscience, New York, 1963.
232. F. A. Lowenheim and M. K. Moran, *Industrial Chemicals*, 4th ed., Wiley-Interscience, New York, 1975 (previous authors: W. L. Faith, D. B. Keyes, and R. L. Clark).
233. G. V. Michael, and G. Thodos, *Critical Temperatures and Pressures of Hydrocarbons, Chem. Eng. Prog. Symp. Ser.*, **49**(7), 1313 (1953).
234. A. P. Kudchadker, G. H. Alani, and B. J. Zwolinski, *Chem. Rev.*, **68**, 659 (1968).
235. D. R. Stull and H. Prophet, *JANAF Thermochemical Tables*, 2nd ed., National Standard Reference Data Service, National Bureau of Standards (U.S.)., July 1970, p. 208.
236. A. Seidell, and W. F. Linke, *Solubilities, Organic and Metal-Organic Compounds*, 4th ed., American Chemical Society, Washington, DC, 1965.
237. R. E. Treybal, *Mass-Transfer Operations*, 2nd ed., McGraw-Hill, New York, 1968, Chap. 9.

488   ELECTROCHEMICAL ENGINEERING OF ELECTROORGANIC PROCESSES

238. Anon., *Chem. Eng. News,* 11 (December 4, 1978).
239. D. E. Danly, private communication.
240. M. M. Baizer, Ed., *Organic Electrochemistry,* Dekker, New York 1973, pp. 936-939.
241. M. M. Baizer, *J. Electrochem. Soc.,* **111,** 215 (1964).
242. J. H. Prescott, *Chem. Eng.,* **Nov. 8,** 238 (1965).
243. Anon., *Chem. Eng. News,* **Oct. 2,** 58 (1967).
244. F. Beck and H. Gutheke, *Chem. Ing. Tech.,* **41,** 943 (1969).
245. I. L. Knunyants and N. P. Gambaryan, *U. Khim.,* **23,** 781 (1958).
246. Anon., *Eur. Chem. News,* **51,** Dec. 19 26 (1969).
247. W. V. Childs and H. C. Walters, Symposium on Electroorganic Synthesis Technology, AIChE National Meeting, Atlanta, February/March 1978.
248. C. H. Worsham (Esso Research), U.S. Patents 3,219,562 (1965), 3,247,084 (1966), 3,247,085 (1966); British Patent 935,236 (1963).
249. L. I. Griffin and C. H. Worsham, U.S. Patent 3,329,593 (1967).
250. G. E. Nelson, U.S. Patent 3,419,483 (1968).
251. *Eur. Chem. News,* **1969** 15.
252. L. I. Griffin, Jr., J. V. Clarke, Jr., and C. H. Worsham, (Esso Research) U.S. Patent 3,324,593 (1967).
253. A. F. Maclean and A. L. Stantzenberger, U.S. Patent (Celanese) 3,479,262 (1969).
254. Farbwerke Hoechst, Belgium Patent 648,304 (1964) [*Chem. Abstr.,* **63,** 13082i (1965)].
255. M. Fremery, H. Hoever, and G. Schwarzlose, *Chem. Ing. Tech.,* **46,** 645, (1974).
256. J. P. Millington, British Patent 1,377,681 (1974).
257. J. P. Millington, *Chem. Ind. (Lond.),* **1975,** 780.
258. J. P. Millington and M. Trotman, Abstracts of Papers, Electrochemical Society Meeting, Washington, DC, May 2-7 1976, pp 700-702.
259. G. Schwarzlose, M. Fremery, and K. Erdt, German Patent 2,205,351 (1973); German Patent 2,108,623 (1972).
260. F. J. Anderson and J. L. Kessler, U.S. Patent 3,758,391 (1973).
261. F. Covitz, U.S. Patent 3,616,324 (1971); U.S. Patent 3,509,031.
262. F. A. Keidel, U.S. Patent 3,884,776 (1975).
263. A. H. Reidies, personal communication.
264. J. P. Millington, *Chem. Ind. (Lond.),* **1975** 783.
265. *1975 Petrochemical Handbook Issue, Hydrocarbon Processing,* November 1975, p. 114.
    1977 Petrochemical Handbook Issue, Hydrocarbon Process., November 1977, p. 131.
266. P Askenasy, *Z. Electrochem.,* **15,** 773, (1909).
267. M. A. Kalinin and V. V. Stender, *J. Appl. Chem. USSR,* **19,** 1045, (1946) [*Chem. Abstr.,* **41,** 4767c].
268. G. S. Tedoradze, V. A. Paprotskaya, and A. P. Tomilov, *Boinet Electrochem.* **10,** 1047 (1974); *ibid.,* **10,** 1239 (1974); *ibid.,* **10,** 612 (1974); *ibid.,* **12,** 206 (1976).

269. B. A. Tedoradze, V. A. Paprotskaya, N. B. Vladimirova, and Yu. A. Yuzbekov, *Elektrokhimiya,* **11,** 502, (1975).
270. Belgian Patent 648997 (1964).
271. U.S. Patents 1,253,615 and 1,253,616 (1918); 1,295,339 and 1,308,797 (1919).
272. A. P. Tomilov, Uy. D. Smirnov, and M. I. Kolitma, *Zh. Prikl. Khim.,* **38,** 2123 (1965).
273. Ref. 219, 3rd ed., Vol 2, p. 443.
274. J. S. Mackay, U.S. Patent No. 2,760,961 (1956).
275. R. W. Foreman and J. W. Sprague, *Ind. Eng. Chem., Prod. Res. Dev.,* **2,** 303 (1963).
276. R. W. Foreman and F. Veatch, U.S. Patent 3,119,760 (1964).
277. J. W. Sprague, U.S. Patent 3,294,657 (1966); U.S. Patent 3,300,398 (1967).
278. M. Smialowski and H. Jarmolowica, *Bull. Acad. Pol. Sci., Classe III.,* **3,** 107 (1955).
279. H. V. K. Udupa et al., *Chem. Age India,* **15,** 501 (1964).
280. P. N. Anantharaman, G. S. Subramanian, and H. V. K. Udupa, *Ind. J. Technol.,* **4,** 271 (1966).
281. P. N. Anantharaman, G. S. Subramanian, and H. V. K. Udupa, *Bull. Acad. Pol. Sci. Ser. Chim.,* **18,** 629 (1970).
282. K. Swaminathan, P. N. Anantharaman, G. S. Subramanian, and H. V. K. Udupa, *J. Appl. Electrochem.,* **2,** 169 (1972).
283. A. Pourassamy, P. N. Anantharaman, G. S. Subramanian, and H. V. K. Udupa, *Prep. Semin. Electrochem.,* **13,** 34–36 (1972).
284. H. V. Udupa, S. G. Subramanian, and N. Payyalur, Indian Patent 128,412 (1974).
285. W. S. Fedor, J. Millar, and A. J. Accola, Ed., *Ind. Eng. Chem.,* **52**
286. H. J. Creighton (Atlas Powder Co.), U.S. Patents 1,612,361 (1926), 1,653,004 (1926), 1,712,952 (1929), and 1,990,582 (1935).
287. R. L. Taylor, *Chem. Met. Eng.,* **44,** 588 (1937).
288. D. H. Killeffer, *Ind. Eng. Chem. News, Ed.,* **15,** 489 (1939).
289. H. J. Creighton, *Trans. Am. Electrochem. Soc.,* **75,** 289 (1939).
290. M. T. Sanders and R. A. Hales, *J. Electrochem. Soc.,* **96,** 241 (1949).
291. C. L. Mantell, *Electro-Organic Chemical Processing,* Noyes Development Corp., NJ, 1968, pp. 147–150.
292. Ref 181, November 1977, p. 229.
293. P. W. Sherwood, *Chem. Ind. (Lond.),* **1960** 1096.
294. J. W. Shipley and M. T. Rogers, *Can. J. Res.,* **17B,** 147 (1939).
295. H. Hayashida, *Chem. Abstr.,* **75,** 36984m (1971).
296. A. J. Allmand and A. Puttick, *Trans. Faraday Soc.,* **23,** 641 (1927).
297. F. Fichter and C. Simon, *Helv. Chim. Acta,* **17,** 717 (1934).
298. R. M. Englebrecht, J. C. Hill, and R. N. Moore, U.S. Patent 3,714,003 (1973).
299. E. G. Bondarenk and A. P. Tomilov, *Zh. Prikl. Khim.,* **42,** 2033 (1969); USSR Patent 193,490 (1967).
300. R. H. McKee and S. T. Leo, *J. Ind. Eng. Chem.,* **12**(1), 16 (1920).
301. R. Tomat and E. Vecchi, *J. Appl. Electrochem.,* **1,** 185 (1971).
302. E. Steckhan and J. Wellmann, *Angew. Chem., Int. Ed.,* **15,** 1076 (1976).

303. C. M. Wright, R. F. Heck, and W. Yokoyama, Electrochemical Society Abstracts of Meeting, New York, May 1969, p. 36.
304. N. L. Weinberg and C. N. Wu, *Tetrahedron Lett.*, **1975** 3367.
305. N. L. Weinberg, U.S. Patent 4,096,044 (1978); U.S. Patent 4,096,052 (1978).
306. Y. So, J. Y. Becker, and L. L. Miller, *J. Chem. Soc. Commun.*, **1975** 262.
307. G. Bockmair, H. P. Fritz, and H. Gebauer, *Electrochem. Acta*, **23**, 21 (1978).
308. M. I. Usanovich et al., *Elektrokhimiya*, **7**, 915 (1971).
309. N. L. Weinberg, in *Technique of Electroorganic Synthesis*, Part I, N. L. Weinberg, Ed., Techniques of Chemistry Series, Vol 5, A. Weissberger, Ed., Wiley, New York, 1974, pp. 265–271.
310. Y. So and L. L. Miller, *Synthesis*, 468 (1976).
311. Z. Blum, L. Cedheim, and R. Nyberg, *Acta Chem. Scand.*, **26**, 715 (1975).
312. M. A. Stadtherr and D. F. Rudd, *Chem. Eng. Sci.*, **33**, 923 (1978).

Chapter **X**

ECONOMICS OF ELECTROORGANIC SYNTHESIS

K. B. Keating and V. D. Sutlic
E. I. DuPont De Nemours and Company, Wilmington, Delaware

---

1   INTRODUCTION

Purpose of Chapter

The question that is often asked of the industrial researcher when he pro-
poses an electroorganic synthesis is, How much will it cost to produce this
chemical by an electrolytic route? This question is asked, of course, so that
the electrochemical process can be compared to conventional synthesis procedures, the economics of which are much better understood. Until recently,
using sources available in the open literature, one would have been hard
pressed to answer this question. This lack of available information has been
recently filled [1], however, and the answer to this question, so important for
the commercial application of this technology and therefore ultimately to the
health of the science itself, is now within reach. Certainly reasonable estimates can be made to guide research efforts and, given the complexity of an
actual plant design and its economic assessment, these estimates are all that
can be generally deduced with any hope of being broadly applicable.

It is the purpose of this chapter to enable the researcher to make a rough
economic evaluation of an electrolytic approach to a given desired product. It
is the authors' opinion that the detail presented in the development of the
capital cost of an electrochemical cell is justified, not only because there is a
dearth of such material in the open literature, but because it gives the reader
a perspective on items of cost in an electrochemical plant. Thus, with a modi-
cum of experimental data on the process in hand and supported by a professional in plant design and economic assessment, a very legitimate *estimate* of
capital costs and electric costs (for synthesis only) can be made by the researcher.

There are many sources of information on the prospects for the costs of
electric power in the United States and the world and this Chapter does not
deal with this extremely complex matter [2–4].

Background

Electrochemical processes are becoming increasingly important to the
modern technologist and process engineer. One immediately thinks of the
synthetic process introduced by Monsanto, the electrochemical synthesis of
adiponitrile [5, 6]. In addition, an electrochemical process for producing

aminoquinidine bicarbonate is in industrial practice in India [7]. Nalco produces tetramethyl lead electrochemically [8], and the Kolbe reaction is used in the USSR for the synthesis of dimethylsebacate from sodium monomethyl adipate [9]. The 3M Company, in plants in Europe, manufactures perfluorooctanoic acid, electrolytically, and BASF is manufacturing Δ-3,5-dihydrophthalic acid by the electrolytic reduction of phthalic acid [10].

Fitzjohn [11] has reviewed the economics of the significant industrial electrochemical processes vis-à-vis processes of the traditional kind and has concluded that electroorganic processes can be quite competitive with conventional technology. Certainly interest in electrochemical processes is increasing in both industrial and academic laboratories. Discoveries with significant commercial import can certainly be expected in the coming years. An additional stimulus is the recently published bibliography of electroorganic synthesis prepared by Swann and Alkire [12].

Electrochemical synthesis offers many potential advantages over conventional processes. One of the more important of these is the prospect of operation at much lower temperatures, not far from ambient, which promises less by-product formation and corresponding less yield loss and lower capital and operating costs devoted to environmental control. Also, since electricity is a reagent in electrochemical processes and since electricity costs are rising more slowly than material and labor costs, an advantage can be expected here. Electrochemical processes can be very selective and for some products may be the only practical route.

If we are to compare the economics of an electrochemical synthesis versus a competing chemical synthesis, we must know the costs of electrochemical cells and the supporting hardware. To do this, we chose the route outlined in Fig. 10.1. The design and layout of individual cells, cell banks, and battery limits for the plant are determined by the production estimate, basic parameters for the system, rectification economics, and general design parameters.

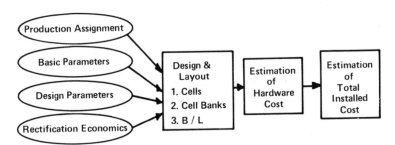

Fig. 10.1. Steps in estimation of costs of electrochemical synthesis processing. Reprinted with permission from the American Institute of Chemical Engineers.

From the layout, one can proceed to an estimation of hardware cost and thence to an estimation of the total installed cost. For this study, an actual cell design was selected, detailed drawings were generated, and professional estimation of individual parts was solicited. It is found in the appropriate section that there is good agreement of our final high spot estimation of battery limit costs with the sparse literature on this subject. In this chapter we do not compare the costs of any specific electrochemical process with a chemical alternate. We try to point out how costs for an electrochemical synthesis can be developed for research guidance purposes.

2   CONSIDERATIONS IN ELECTROCHEMICAL CELL DESIGN

Selection of Cell Configuration

Most electrochemical cells, regardless of their configuration, have several common components: anode, cathode, separator, and cell body. These components are seen schematically in Fig. 10.2, which shows a sample inorganic reaction, the formation of caustic soda and chlorine, in a two-compartment cell. Chloride ions are oxidized in the anode compartment, and sodium ions migrate through the membrane toward the cathode. Chlorine gas leaves as a product from the anolyte compartment. Water is electrolyzed to hydroxyl ions and hydrogen gas in the catholyte compartment. Thus the solution in the catholyte becomes alkaline and caustic soda is produced.

A variety of cell configurations have been developed to varying degrees, such as flat-plate, packed-bed, fluidized-bed, and pumped-slurry electrode cells. We consider only cells suitable for flow-through operation, which have well-known advantages over batch-process types of cells, which are not dealt with here.

The flat-plate electrode cell is the most highly developed and widely known configuration. It is used to produce a variety of products, including organic and inorganic compounds. By comparison, other cell configurations have seen limited use, basically because of insufficient hardware development. A flat-plate cell can be fabricated in the tank arrangement or the plate-and-frame arrangement. As its name implies, the tank configuration is simply a group of plate electrodes arranged in a tank or vessel. An example of such a cell is the Hooker type-S series cell for caustic soda and chlorine production [13a].

The plate-and-frame configuration consists of the basic cell components arranged in individually framed sections and assembled as a cell bank in a filter-press arrangement. Monsanto uses such a configuration in the adiponitrile process.

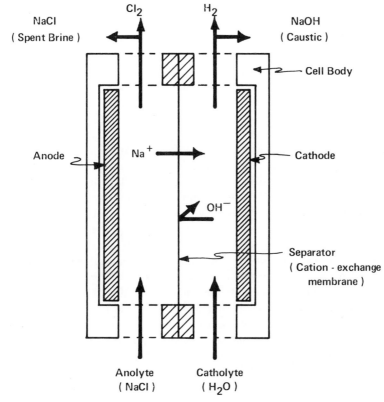

Fig. 10.2. Schematic of two-compartment cell. Reprinted with permission from the American Institute of Chemical Engineers.

In most electroorganic syntheses, a membrane is necessary to protect the products of reaction, and the plate-and-frame design is best suited for the incorporation of a membrane. It is for this reason, despite the fact that a tank-type disk-stack cell without a membrane can be fabricated at less cost [13b], that this chapter focuses on the plate-and-frame configuration. This cell can be applied most readily to the potential electrochemical synthesis of a compound with the least amount of hardware development work. Thus it should be considered as the first alternative in most situations. We refer to this plate-and-frame cell as "conventional geometry" throughout this chapter.

Fundamental Design Factors

An electrolytic facility comprises four distinct component groups: cell feed preparation equipment, electrolytic cells, electrical power supplies, and

product recovery equipment. The technology and costs for cell feed preparation and product recovery are well established and need not be considered
further here. Instead, attention is given to the electrochemical cells and associated power equipment.

The basic design goal of any process is to obtain a large amount of product
(high yield) of high quality for a minimum input of energy and investment.
To maximize the performance of a cell for a given reaction, the following
three basic parameters affecting process economics must be optimized: electrical current efficiency, cell voltage, and current density.

Electrical current efficiency is defined as the fraction of current passed
that is effective in producing the required product. It depends on a large
number of factors, such as mass transport and the kinetics of competing processes [14]. If overall cell voltage is kept to a minimum, power costs may be
reduced significantly for an electrochemical facility. The total cell voltage is
the sum of the reversible potential of the reactions, the overvoltages of the
reactions, and the voltage drop due to the sum of ohmic losses throughout
the cell. Typically, these voltage components are reduced when the temperature is increased. For any given surface area requirement, the ability to increase the current density will reduce the number of cells needed, thereby reducing capital costs. From another perspective, the productivity of a given
number of cells becomes greater as the current density is increased. However,
a level of current density is reached beyond which the system becomes mass
transfer limited. This limit depends on such factors as the geometry of the
electrode surfaces, temperature, and fluid velocity [14]. In practice, too high
a current density often leads to unwanted side reactions (see Chapter I for additional details).

Optimization of the Plate-and-Frame Cell Design

To optimize the foregoing parameters, certain design criteria must be applied to the particular reaction and hardware in question. In our study we
assumed that the reaction was optimized in the laboratory, thereby fixing
such parameters as temperature and current density. Thus the emphasis was
placed on criteria relating directly to the design of the cell hardware, which
fall into one of the following categories.

Overall Fluid Distribution

Ordinarily, flow is arranged in a parallel scheme (Fig. 10.3). There are a
number of reasons that a parallel flow arrangement is usually advantageous.
First, the net heat built up within the cell bank is kept to a minimum, since
each electrolyte stream flows immediately into a product header. The series
flow arrangement allows temperature to increase steadily as the electrolyte
flows through the cell bank. With the parallel flow arrangement, an internal

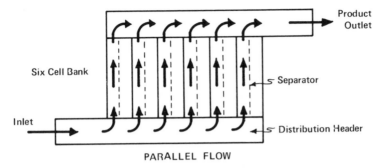

PARALLEL FLOW

Fig. 10.3. Schematic flow in a plate-and-frame cell bank. Reprinted with permission of American Institute of Chemical Engineers.

cooling system (coils, jackets) is generally not required, since no significant temperature difference exists in the cell bank. If temperature control is necessary, an external heat exchanger located on the inlet side of the cell bank is sufficient.

The second reason involves the release of a gaseous product into the outlet stream. In the series flow arrangement, excess gas bubbles accumulate, increasing the resistivity of the electrolyte and raising power requirements. If gas becomes trapped within the cells, hot spots capable of ruining components could form.

Finally, for a given number of cells, the overall pressure drop across a cell bank arranged for parallel flow is much lower than that for a cell bank that has been piped for series flow. Therefore, the capital investment in pumps and the pumping costs are minimized.

To minimize pumping losses, careful consideration must also be given to sizing hydraulic components such as piping and flow channels in a cell. This is especially important when liquid/gas product mixtures are involved, as in a caustic chlorine cell, since losses due to friction in this case are normally higher than those for single phase liquid flow, assuming equivalent pipe sizes.

Fluid Distribution in a Cell

An electrolyte stream must be distributed uniformly within a cell to eliminate stagnant zones. Also, sufficient fluid velocity past the electrodes must be attained to enhance mass transfer. Any significant shortcoming in these areas could cause concentration polarization—a condition in which reactants are converted faster than they are delivered to an electrode surface. This condition can lead to unwanted side reactions, low current efficiency, and high electrode overvoltages. Hot spots may also develop because of poor fluid distribution. These situations can be alleviated by proper design of distribution ports in a cell, by turbulence promoting inserts in the electrolyte compartments to enhance agitation, and by sufficiently large feed systems to circulate the electrolyte streams in the cell at adequate velocities [15].

The formation of a solid phase product imposes special requirements on a fluid distribution system. Flow channels must be modified to allow easy removal of the product from the cell. This can be accomplished by increasing the electrode-to-diaphragm spacing, enlarging the outlet flow ports in a cell compartment, and using gravity to help sweep the product from within a cell by flowing electrolyte from top to bottom.

Electrical Utilization

The first electrical consideration in cell bank design is whether to connect the individual cells in series or in parallel. Full-size commercial cells normally operate at high currents (>500 A) and low voltages (usually 3 to 5 V). In general this leads to a series intercell electrical connection.

Eliminating power loss due to ohmic resistance in the cell is one of the most important factors within a designer's control to improve system economics. The components contributing to total ohmic loss are the individual electrolytes, the membrane, electrodes, and internal and external bussing. Membranes and electrodes are discussed in a later section.

To minimize ohmic losses, all electrical connections must be kept as short as possible. Therefore, internally made intercell connections are preferable to the longer, external cell-to-cell connections.

Contact resistance can also contribute to power losses. When joints are not properly prepared or tightened, the power losses through them may generate enough heat to overburden any cooling system servicing that particular cell bank, not to mention increased power costs.

A final source of ohmic loss is through the electrolyte itself. This loss is directly proportional to the resistivity of the electrolyte, the width of the electrolyte compartment, and the current density. The width of the electrolyte path should obviously be kept as short as possible, consistent with good fluid flow and fluid distribution.

Electrode Surface Area to Cell Volume Ratio

This ratio in the plate-and-frame cell may be increased by decreasing the width of the electrolyte compartment and using electrodes with active surface areas higher than their projected planar areas. These include expanded sheet electrodes and those made from pleated sheet. Owing to their configuration, these electrodes provide better agitation of the electrolyte, thus improving the release of gases. However, their use may be disadvantageous when a solid phase product is formed, since the product may cling to surfaces not directly in the path of the electrolyte stream. Commercial-size expanded sheet electrodes can provide a 70 to 80% increase (50% sheet expansion) in surface area over solid, flat sheet electrodes of equivalent perimeter. However, the voltage drop within an expanded sheet electrode increases in proportion to the voidage (open space), since the cross-sectional area is reduced. This can be compensated for in a design by increasing the electrode's thickness.

Critical Component Selection

Electrode material selection is usually based on the results of laboratory studies on a particular electrochemical reaction. However, besides being an efficient electrocatalyst for the reaction, the electrode material must also have adequate electrical and thermal conductivity, good mechanical properties (strength, machinability), corrosion resistance, and low cost [16, 17] (see Chapter IV for details).

A separator normally divides the two electrolyte streams within a cell. This component is ordinarily necessary to minimize intermixing of the anolyte and catholyte, which can lead to side reactions and associated reduction in current efficiency. The prevention of such intermixing may also be dictated by safety considerations.

Generally, the newer ion-exchange materials are far superior to the older diaphragms commonly made of ceramic or asbestos. These ion-exchange membranes usually possess higher mechanical strength, improved selectivity to specific ion groups (anions or cations), and greater overall stability in a spectrum of solvents over wider temperature ranges.

The particular reaction and operating conditions normally determine which membrane material is chosen. Several factors affect selection. A material with the lowest possible resistance, compatible with its function, is desirable. Other factors include resistance to plugging, dimensional stability, useful life, and cost.

Most plate-and-frame cells employ a membrane mounted between two thin frames. The frames give the necessary support and often act as sealing surfaces when sandwiched between two cells. Normally, the frames are cemented or heat sealed to the membrane, forming an integral unit.

Although membranes are available in a variety of sizes, bulging due to differential pressures limits their maximum size if they are used alone. Therefore, most commercial-size cells use an insert such as plastic spacers or plastic netting to limit membrane bulging and maintain the desired electrode-to-membrane spacing [14]. The membrane can also be internally stiffened.

Materials of Construction

Most commercial electrolytic cells employ plastics such as polypropylene and poly(vinyl chloride) (PVC) for cell bodies, spacers, and membrane frames. Fiber reinforced plastics are also used. These lightweight materials are easily machined. Most are available in resin form for molding. Some types of rubber may also be useful.

Physical properties, useful life, and costs also affect selection. One important consideration is the coefficient of thermal expansion of the material, which should be as low as possible. When dissimilar materials are assembled to form an integral unit (for example, metal electrodes and plastic frames), the different coefficients of thermal expansion may lead to severe component distortion, leakage of electrolyte, and eventual failure. Differential expansion must be allowed for in the design.

Gasketing materials commonly used in electrochemical cells are elastomers such as Neoprene and Viton fluoroelastomer. They are available in sheet or molding compound form.

Assembly and Maintenance

One of the most attractive features of the plate-and-frame cell design is ease of maintenance. However, as is mentioned above, an inherent problem with this design is potential fluid leakage at the multiple sealing surfaces. Therefore, a good design will minimize the number of sealing surfaces.

Normally multiple-cell frames are clamped together in a filter-press unit to form a cell bank. These usually have movable end plates to provide the clamping force necessary to compress the gaskets. The end plates can be manually or automatically actuated.

Operating Conditions

The maximum cell operating pressure is usually determined by the seal design, as this, rather than the frame material strength, is often the weakest link. Material properties (strength, hardness), gasket sealing area, and gasket clamping force determine the maximum pressure a seal will hold.

The maximum operating temperature of a cell depends on the properties of the component materials and on the cell design itself. The maximum working temperature of the component material will establish a limiting value; however, because of differential expansion, the amount of clearance designed into a cell may further limit this maximum operating temperature.

The maximum current density for cell operation can also be related to the maximum operating temperature of the cell (assuming the electrodes will permit the current levels). Normally an energy balance must be carried out to determine the net heat produced within a cell. If the net heat produced causes a temperature rise above the maximum allowable, an appropriate internal or external cooling system must be provided.

3   CELL DESIGN AND COST ESTIMATION

Preliminary Design Parameters

The design parameters established for this study are as follows: plate-and-frame cell ("conventional geometry"), 4000 ft^2 active area; current density,

200 A/ft^2; current efficiency, 90%; cell voltage, 5 V; membrane/electrode spacing, 0.04 to 0.12 in.; membrane, cation-exchange material; operating temperature, 45 to 85°C; flow, 1.0 to 10.0 ft/sec past the electrodes; cell pressure <50 lb/in.^{2}; pressure differential across membrane, <5 lb/in.^{2}

Selected Design Concept

From among a number of conceptual designs considered, the cell shown in Fig. 10.4 was selected. This is the expanded metal flat-plate design. In this design, each cell body contains an anode and a cathode. A series electrical connection is made within the confines of the body by a number of intercell connectors (four are illustrated) that link the two electrodes along a vertical edge. The membrane assemblies in this design are sandwiched between every pair of cell bodies. Both sides of the membrane frame act as seating surfaces for the two gaskets per cell required by this design. With the chosen design,

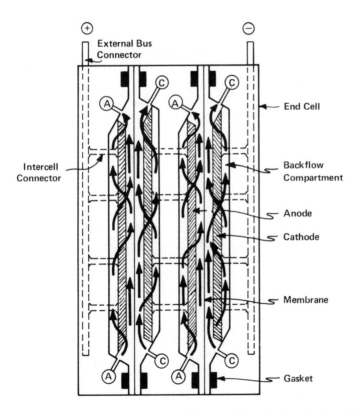

Fig. 10.4. Cross section of selected cell concept with expanded metal electrode. Reprinted with permission from the American Institute of Chemical Engineers.

sealing of fluid passages is minimized, and lower total sealing force per cell is necessary, since less gasketing material is used, allowing lighter duty filter-press equipment. This design also provides good fluid distribution within an electrolyte compartment since the fluid is introduced across the whole width of the chamber. Machining the multiple fluid ports poses no special problem. This also simplifies assembly and disassembly.

Preliminary engineering drawings of this cell concept were completed to evaluate fabrication techniques and estimate costs. For design purposes the electrode membrane spacing was set at 0.12 in. and the electrolyte velocity at 3.3 ft/sec.

CELL SIZE

One of the primary design considerations is to establish the size of each cell, based on active electrode area. Size availability of critical materials (electrodes, membranes) and ease of handling were considered.

Availability of ion-exchange materials is a size-limiting factor. Du Pont, manufacturer of Nafion® perfluorosulfonic acid ion-exchange membranes, produces sheets up to 48 in. wide. Other manufacturers are Ionics, Asahi Chemical, Asahi Glass, Teijin, and Tokuyama Soda. A cell with a maximum electrode size of 18 in. × 40 in. (5 ft²) allows for maximum membrane interchangeability for various applications.

We chose a cell size of 5 ft² as representing a practical limit for handling, and based our calculation on this electrode size.

NUMBER OF CELLS

To make up the required 4000 ft² of active electrode surface area, 444 individual cells utilizing expanded sheet electrodes with a 50% voidage (active surface area of approximately 9 ft²) were required.

ELECTRODE DESIGN

The electrode materials chosen were a precious metal oxide (titanium substrate) anode and a mild steel (C 1030) cathode. Steel was acceptable, since it is cathodically protected against corrosion in this application.

At a current density of 200 A/ft², the required electrode thicknesses, based on a maximum potential drop of 250 mV through each, are 0.125 in. for the anode and 0.060 in. for the cathode (expanded sheet electrodes).

The flow paths in a two-cell bank with expanded metal electrodes are shown in Fig. 10.4. A bank of any number of cells can be assembled in a similar fashion, with two half cells at the ends. The expanded electrodes require cutouts approximately ⅛-in. deep behind each sheet to allow fluid circulation on both sides.

ELECTRICAL DESIGN

Four equidistant intercell connectors along the vertical edge of an electrode join every anode and cathode through the cell body, also shown in Fig.

10.4. This provides a series cell-to-cell electrical connection within a cell bank. The connectors are designed to give good mechanical rigidity between the electrodes. (A rubber insert seals each electrolyte compartment when compressed by two jam nuts.) To provide sufficient corrosion protection for the copper connectors, a 5-mil tantalum coating is applied to their lateral surfaces.

CELL BODY DESIGN

The material specified for the cell bodies is CPVC [chlorinated poly(vinyl chloride)], which is compatible with the electrolytes up to approximately 85°C. Two alternate methods for fabricating the cell bodies are injection molding, with some machining, and complete machining from plastic sheet stock. According to our estimation, the fully machined parts are more economical in our sample case by approximately 17%.

Based on a maximum fluid velocity of 3.3 ft/sec and a compartment cross section measuring 0.12 in. $\times$ 18 in., a nominal flow rate of 22 gal/min is required through each cell. Fourteen fluid distribution ports are needed in each compartment to deliver the electrolyte at 3.3 ft/sec with minimum maldistribution.

MEMBRANE ASSEMBLY DESIGN

The ion-exchange membrane chosen is made of reinforced Nafion perfluorosulfonic acid (XR-400 series), 10 mils thick. This membrane was chosen over the nonreinforced grade because it has greater dimensional stability and is less prone to bulge under a hydrostatic pressure difference.

The thickness of each membrane frame determines the electrode-to-membrane spacing, since the electrodes are mounted flush with the outer surfaces of a cell body. The frames are made from ⅛-in. CPVC sheet with the appropriate opening in the center blanked out.

The outer dimensions of the membrane frames are 24 in. $\times$ 47 in. Since Nafion® perfluorosulfonic acid is made in sheet sizes up to 48 in. $\times$ 96 in., four assemblies can be made at the same time by blanking out four openings in two 4 ft $\times$ 8 ft sheets, applying adhesive to them, and clamping the sections together with the membrane in between. The four assemblies may then be cut from the larger single piece.

Since there is a 4 to 8% increase in area when the membrane is wetted, it must be assembled in that state.

MISCELLANEOUS COMPONENTS

Vexar® plastic netting, with a strand thickness of 60 mils (60-PDS-49), is used on both sides of the membrane to maintain uniform membrane-to-electrode spacing. The vertical edges of the netting are epoxy bonded to the inside edges of the membrane frame assembly. The sheets must be installed so that the strands closest to the electrodes run in the direction of electrolyte flow, thus creating flow paths.

One Viton® fluoroelastomer gasket is used to seal each cell body and membrane frame assembly. It is cemented in place in a machined groove provided on both sides of the cell body. The gasket dimensions allow for a 20 to 25% compression, enough to provide an effective seal around each electrolyte compartment.

CELL OPERATING CONDITIONS

Based on the various cell components of our design, the maximum cell operating temperature range is 77 to 82°C, slightly below the maximum body material working temperature of 85°C. The established operating range is based on the difference in thermal coefficients of the plastic and metallic parts.

The maximum cell operating pressure, when all seals are compressed the required 20 to 25%, is 50 lb/in.^{2}. Thus the inlet pressure at the cells should be below this value to allow for a margin of safety.

Cell Costs

Estimated costs for the cells are based on second quarter 1974 prices. Costs in later years must be adjusted from this base. The cell costs are shown in Table 10.1. A detailed breakdown of these costs is given in Table 10.2. These figures represent costs based on final design drawings.

Since the cell bodies represent approximately 25% of the total cell cost in our case, alternate fabrication techniques should be considered for higher quantities (e.g., 800 to 1000 cells). One such technique is molding with follow-up machining, and another is plastic sheet thermal forming.

Table 10.1.   Estimated Cell Costs[a]

Item	Expanded Sheet Electrodes (444) (thousands of dollars)
Cell bodies	91.8
Membrane assemblies	101.4
Electrodes	140.8
Misc. components and cell assembly	71.1
Totals	405.1
$/ft^2 (4000 ft^2 total)	101.3

[a]Cost information from second quarter 1974.

Table 10.2. Cell Cost Details (second quarter 1974); 4000 ft², total electrode surface area

Item	Material	Method of fabrication	$/Unit	Quantity	Expanded Sheet Electrodes (9 ft²/cell)	
					Cost (thousands of dollars)	a
1. Cell bodies						
(with 5% extras)	CPVC	Machine	197.0	466	91.8	22.7
2. Membrane assemblies						
(with 5% extras):						
Plastic frames (with blanking)	CPVC	Blank and cut	8.0	932	7.5	
Blanking dies		—	—	—	20.0	
Plastic netting (with cutting)	Vexar®	Cut	0.6	932	0.6	
Membranes	Nafion®	Cut	119.0	466	55.5	
Assembly	—	—	6.0	466	2.8	
Undeveloped design	—	—	—	—	15.0	
					101.4	25.0
3. Electrodes						
Anodes	Metal oxide/Ti	Cut	178.8	444	133.1[b]	
Cathodes	Steel	Cut	13.8	444	7.7[c]	
					140.8	34.8
4. Miscellaneous						
Cu connectors	Cu	Machine	—	—	3.4	
Body supports	Al	Cast	—	—		
Gaskets	Viton®	Mold	60.0	932	56.0	
Gasket molding dies	—	—	—	—	5.0	
Cell Assembly	—	—	15.0	444	6.7	
					71.1	17.5
5. Totals					405.1	100.0%
6. $/Unit					912.4	
7. $/ft²					101.3	

[a] % of total costs. [b] $299.8/unit; $33.3/ft². [c] $17.3/unit; $1.9/ft².

4   BANK DESIGN AND COST ESTIMATION

Cell Bank Size

The most economical bank is determined by the cell bank cost and rectification equipment cost. The optimum bank size can be determined by examining the sum of the costs as a function of bank size.

Total cell bank costs comprise cells, filter-press hardware, and pumps, piping, and valves. Cell costs have been established as $405.1 thousand for expanded sheet electrodes. This component remains constant, regardless of bank size. The sum of the remaining elements provides a net reduction in total bank costs of less than 2% over the range of 20 to 40 cells/bank. Therefore, the total cell bank cost is considered constant.

A specification of 20 cells/bank was chosen because it is at about this value that total rectification costs versus bank size (number of cells per bank) levels off rapidly. Thus there is no incentive to go to larger size cell banks. Smaller banks may even be advantageous from a production standpoint, since the temporary loss of one unit would not reduce production as much. To simplify our analysis, we therefore selected a size of 20 cells/bank.

For our application, rectification equipment economics favor a series connection [19]. This is also valid when several banks are connected together. Some cost savings are possible with single rectifier/multiple bank systems; however, the cost of more extensive bussing then needs to be considered. Also, the power regulation and monitoring scheme becomes more complex.

All rectification costs given are for constant current, automatic voltage regulation, the preferred control mode.

Cell Bank Hardware

Bank (or filter-press) hardware is based on 20 cells/bank. Also, two-end cells are employed. These contain the appropriate bus hardware, which links the intercell connectors to an exterior bus terminal.

Electrolytes are fed and products are recovered by manifolds mounted on the exterior of the filter-press frame. The hydraulic connections are made with flexible hoses and quick disconnect fittings to each cell. Electrolyte is fed to each cell compartment from the bottom, while the product leaves from the top.

The electrolyte feed manifolds and associated connecting hardware were sized so that, at the maximum bank flow rate of 440 gal/min [20 cells × 22 gal/(min)(cell)], maximum flow maldistribution is no greater than ±8.5% between cells. This ensures accurate metering of the electrolyte to each cell within a bank.

Table 10.3 gives the important parameters for the expanded sheet electrode cell bank in this study.

Table 10.3. Parameters for Expanded Sheet Electrode Cell Banks

Item	Expanded Sheet Electrodes
Total active area	4000ft^2
Active area/cell	9ft^2
Number of cells	444
Cells/bank	20
Number of banks	22
Electrode/membrane spacing	0.12 in.
Compartment cross sect. (ft²)	0.30
Compartment volume (ft³)	(Includes backflow comp.)
	0.10
Area/volume (ft²/ft³)	180^a
Electrolyte velocity (ft/s)	1.7
Electrolyte flow rate (cell, gal/min)	22
Electrolyte flow rate (bank, gal/min)	440
Voltage/cell (V)	5
Voltage/bank (V)	100
Current density (A/ft²)	200
Current/cell (A)	1800
Current/bank (A)	1800
Rectifier rating	200 kVa^b
Total power consumption (all banks)	3960 kVa
Electrical connection in bank	Series
Flow in bank	Parallel

[a] Based on 0.050 ft³ compartment volume.
[b] Rectifier is 2000 A at 100 V bank requirement, however, is 180 kVa (kilovolt amps) each.

Cell Bank Costs

The cost for fabricating and testing the 22 expanded sheet electrode cell banks was estimated to be $621.8 thousand.

These costs include cells, filter-press hardware, assembly, and cell bank testing for leaks and electrical shorts. The various cost elements are detailed in Table 10.4.

Note that no provision has been made for the heat generated by the reaction and by resistive losses in the cell. If these become excessive, cooling will be required, and an additional cost will be incurred.

As the active electrode surface area increases, rectification costs may be only one of the several factors (such as fluid distribution and controls) to be

Table 10.4.   Cell Bank Costs^a (4000 ft^2, Total Electrode Area)

	Expanded Sheet Electrodes (9 ft^2/cell)			
Item	Cost/Bank, (thousand $)	No. of Banks	Total Cost, (thousand $)	%
1. Cells	18.4	22	405.1	65.1
2. End cells (two per bank)	0.5	22	11.0	1.8
3. Frames (filter press)	1.4	22	30.8	5.0
4. Manifold pipes, misc. hardware, press. test manifolds	1.9	22	41.8	6.7
5. Assembly and testing	2.8	22	61.6	9.9
6. Subtotals	25.0	—	550.3	—
7. Contingency	3.3	22	71.5	11.5
8. Totals:	28.3	—	621.8	100
9. $/ft^2	—	—	155	—

^{a}Estimated cost with a second quarter, 1974 basis.

considered in the optimal arrangement of cell banks. Such cases should be optimized on an individual basis.

5   BATTERY LIMIT (B/L) COSTS

Basis for Total Capital Investment

The capital cost data discussed here include only the direct manufacturing investment. Our battery limits include electrochemical equipment except liquid feed preparation and product separation equipment, heat exchangers, power substations, general facilities, and buildings. The items included in the specified B/L and comprised in Fig. 10.5 are complete cell bank units, rectifiers and DC bussing, pumps, and piping to feed each electrochemical cell bank, cell bank instrumentation (for power, flow, pressure, and temperature but not for compositional analysis), and safety interlock control loop. Included also are such common facilities as feed and product headers, drain and rinse lines, and AC distribution system (from substations to rectifiers), as well as such control room items as motor control centers (MCC) and instrument control centers (ICC). The cell banks, rectifiers, and pumps were grouped together as a complete cell module. The space requirement for the appropriate number of modules and the general facility area is estimated to be 4900 ft.^{2} These figures include the service areas between the modules.

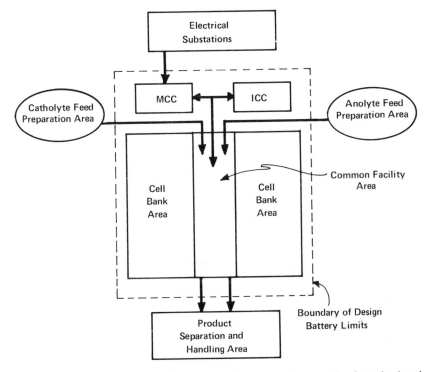

Fig. 10.5. Battery limits for estimating purposes. Reprinted with permission from the American Institute of Chemical Engineers.

Facilities serving all modules consist of fluid handling, power distribution, and control sections. The fluid handling section comprises a waste tank and pump, used in conjunction with the drain line to handle rinse water from a cell bank(s) during flushing operations. The power distribution system consists of the cable trays carrying the motor and rectifier power supply feeds. The system is rated at 6000 kVa, of which 4000 kVa feeds the rectifiers and 2000 kVa serves the power demands of the pump motors, electric valves, and miscellaneous tie-ins.

The MCC and ICC areas consist of the starters, panels, meters, and power transformers necessary to feed and monitor the cell bank modules. Three 2000-kVa substations feed power to this area. The safety or emergency control loop automatically shuts down any module if there is a loss of fluid flow or power to the cell bank, or if a preset temperature is exceeded.

A surge tank and a cooler are generally required for an electrolytic plant. The capacity requirements (and the materials of construction) for these items can be quite variable. They are not included in the B/L for this study.

Overall Costs

Total installed cost for the specified B/L items is estimated to be $2.36 million ($589/ft^2).

Table 10.5 gives the complete cost breakdown for the above estimates and Fig. 10.6 shows the relative contribution to this cost of major items. Figure 10.7 is a plot of the cost per square foot of electrode surface area (installed) as a function of year forecasted. It can be seen that it will cost approximately $1200/ft^2 to install an electrochemical plant using expanded metal electrodes in 1981.

The cost of items not included in the battery limits (that is, general facilities, buildings) may be estimated from available investment figures.

It should be noted that special circumstances, such as an extremely corrosive reaction environment (which would require exotic materials of construction) and a particularly severe safety hazard (toxicity, explosion, etc.), could markedly affect the cost of items on the B/L and substantially alter the costs quoted here. Such considerations, of course, are also shared with conventional processes.

Cost Analysis

The cost of the rectification equipment, including DC bussing, comprises 25% of the total installed module cost. The indirect cost allowance of 40% includes overhead, engineering design, and construction charges.

Comparison with Literature Cost

Danly has stated [20] that a 9-ft^2 cell could range in cost from $250 to $1000. Using a ratio of 1.78 to adjust these figures to mid-1974 dollars, we obtain a range of $50 to $200/ft^2. Adjustment of these figures to a battery limit as we define it (our factor is 5.7) leads to an installed cost range of $285 to $1140/ft^2. This range should be compared to our value of $589/ft^2 installed.

6   ESTIMATES

It can easily be shown that the investment calculations derived from the data developed in this chapter will predict a relationship between cost (1974) and capacity (million lb/year) as shown by the dashed curves (Keating-Sutlic) in Fig. 10.8. The calculations are for 90% current efficiency, a reaction involving 2 F, and a product molecular weight of 100. Also in Fig. 10.8, which is derived from Fitzjohn's work, are the rules of thumb of Nohe [21], Fitzjohn's envelope, and published data on caustic-chlorine manufacture [22]. Nohe's guidelines (1978), for capacities between 100 and 5000 metric tons/year ranged from 1500 to 6000 DM per metric ton capacity.

Table 10.5. Total Battery Limit Costs (4000 ft^2, total)[a]

Item	Expanded Sheet Electrodes (thousand $)
Module Costs	
Cell bank	28.3
Rectifier and bussing	11.1
Pumps, press. reg.	3.9
Subtotal	43.3
Installation	9.0
Installed equip. cost	52.3
Module piping (pipe, valves, labor)	2.4
Instrumentation and electrical	9.5
NET	64.2
Indirect costs	25.7
Gross—per module	89.9
Total—all modules (22)	1977.8
$/ft^2	495
Common Facility Costs	
Feed headers	59.0
Product headers	66.0
Drain header	12.5
Power distribution system	40.0
Motor control centers and instrument control centers	87.0
Waste tank and pump	7.0
NET	271.5
Gross	380.1
Grand total	2357.9
$/ft^2 (electrode surface area)	589
Comparison	68.5%

[a]Estimated cost with a second quarter, 1974 basis.

It can be seen that calculations based on the considerations in this chapter will reflect the very narrowly defined battery limits chosen for this development. This will not only make such estimates somewhat low, but will result in a power dependence of the cost versus capacity, which is close to unity. Above capacities of 500 million lb/year, our estimation procedure will give high results because extrapolation of our envelope may not take into account

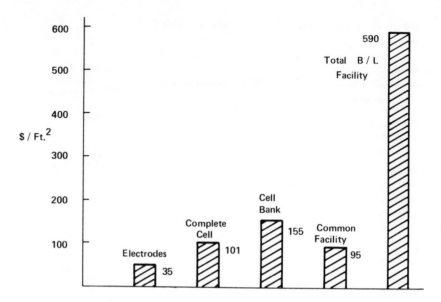

Fig. 10.6. Relative costs of B/L facility elements.

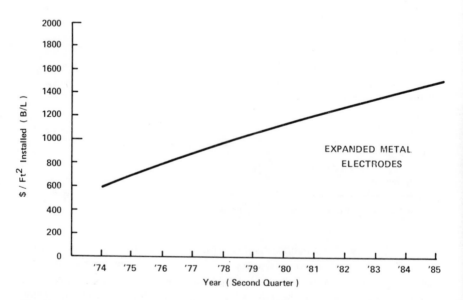

Fig. 10.7. Installed cost of electrochemical facility (B/L) as influenced by actual or estimated inflation.

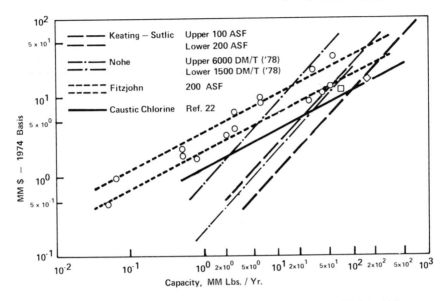

Fig. 10.8. Direct investment—electrochemical synthesis. After Fitzjohn [11].

economies of scale available inside this B/L. Product quantities of this scale, however, are exceptional. Below capacities of 5 million lb/year, this technique is unacceptably low because the error introduced by excluding areas such as liquid feed preparation and product separation becomes relatively significant.

Power cost calculations can be made by simply estimating the voltage required to run the cell in question and calculating the power necessary. For example, for a material whose molecular weight is 100 (2 F) manufactured in a cell operating at 200 Amps. per sq. ft. and 5 V, 6.1 cents/lb is required if electrical efficiency is 90% and power costs 5 cents/KWh.

7 FUTURE TRENDS

The incentive for future development in the area of electrochemical cell technology is to reduce the capital investment per unit of active electrode surface area and thus to further reduce the mill cost of a product produced electrochemically. This can best be achieved by increasing the surface area/cell volume ratio of the electrochemical cells and then operating the cell in the mass transfer limited mode. Such cells would have to possess sufficient current-carrying capacity to allow large electric currents in a relatively small cell volume, without unduly large current densities. This factor may become especially significant for the electrochemical synthesis of organic products, which often require low current densities.

Many electrochemical devices utilizing a high surface-to-volume ratio have been reported in the recent literature. For example, small-scale devices have been constructed utilizing beds of nonconducting particles [23], porous carbon electrodes [24], packed-bed graphite electrodes [25], and fluidized-bed electrodes [26]. A comparison of several electrochemical reactor types for copper removal has also been made [27], and more recently, a cell utilizing nonconducting fluidized beds and mesh electrodes has been described [28].

Recently, an extremely effective porous electrode cell has been invented [29]. This device has been tested on a pilot-plant scale [30, 31] and on a 100-gal/min scale, and these studies are described in the literature [32]. One unit of a 100-gal/min cell of this type contains 725 ft^2 of surface in a volume of approximately 0.58 ft^3. Such a cell, when fabricated from stainless steel mesh for the cathode, cost approximately $14,000 to fabricate in January 1974. Six such cells would provide 4000 ft^2 of electrode surface for $84,000 versus $447,000 (items 1 through 3 in Table 10.4) for expanded metal electrodes.

These cells, however, have thus far been used for collection of heavy metals in low concentrations, and the currents are therefore relatively low. Thus the cells have no provision for heat removal, and the cell bus bars are not designed for high currents. In addition, high-surface-area anodes containing inherently more expensive materials would be needed for oxidative processes. However, it seems reasonable to assume that there would still be a large financial bonus available (with respect to the conventional flat-plate configuration), even after the required modifications were made.

8 SYNOPSIS

A summary of the major points of this chapter is given below.

1. At present, the conventional geometry, that is, the plate-and-frame design is the most widely used cell for electroorganic synthesis. Its advantages are relative simplicity with regard to assembly and maintenance, compatibility with membrane utilization, and availablity as an article of commerce. In such cells, the cost correlates closely with volume. The plate-and-frame design appears to be at its optimum with regard to design improvement and significant cost reduction. Its costs are now relatively well understood.

2. The direct manufacturing investments, including indirect field costs, for a battery limit electrochemical facility, are (second quarter 1974) $589/ft^2 for expanded electrode cells. Such factors as the corrosivity of the process streams, their toxicity, or the condition of temperature and pressure may affect these costs.

3. The cost of the cells comprises approximately 17% of the installed cost, the cost of cell banks about 27%, and the cost of power equipment 20 to 25%.

4. A way to significantly reduce the required investment for electrochemical cells is to increase the area-to-volume ratio in those cells. Examples of this kind of device are beginning to appear in the literature.

5. Even with cells of conventional geometry, certain syntheses can be very competitive with traditional processes of manufacture.

6. Costs derived from this chapter will be reasonably accurate for a very narrowly defined battery limit around the electrolysis cell. For capacities below 5 MM lb/year, they should be adjusted upward and, in cases where complex separations and solution preparations are involved, they should be adjusted upward throughout the capacity range. The estimates assume a current efficiency and yield of $>90\%$.

Acknowledgments

The authors would like to acknowledge the editorial assistance of Dorothy M. Welsh of the Engineering Research & Development Division and several useful technical discussions with J. L. Fitzjohn of the Engineering Technology Laboratory, Engineering Research & Development Division, and Dr. C. H. Bedingfield, Jr., of the Engineering Service Division, all of E. I. du Pont de Nemours & Co., Inc. In addition, the following Du Pont Engineering Department personnel aided in various phases of cost estimation: H. G. Garner, R. A. Dickson, J. L. Pedicone, and J. J. McHenry. Dr. D. E. Danly, Monsanto Company, made a number of useful recommendations that have been incorporated into this chapter.

References

1. Atlanta 1978 AIChE Meeting, February 26 to March 1, *Electroorganic Synthesis Technology, AIChE Symp. Ser.*, No. 185, **75** (1979).

2. "Overview and Strategy," 1980–1984 Research and Development Program Plans, EPRI Planning Staff, Electric Power Research Institute, Palo Alto, CA, July 1979.

3. *New York Times,* Sunday, April 6, 1980; P. G. Hill *Power Generation,* MIT Press, Cambridge, 1977.

4. *Energy-Global Prospects 1985–2000,* Workshop on Alternative Energy Strategies, McGraw-Hill, New York, 1977.

5. M. M. Baizer, *J. Electrochem. Soc.*, **111**, 215 (1964).

6. J. M. Prescott, *Chem. Eng.* **72** (23), 238 (1965).

7. T. P. Balakrishnan, *Ind. Eng. Chem. Proc. Des. Dev.*, **10**, 495 (1971).

8. L. L. Bott, *Hydrocarbon Process.*, **44**, 115 (1965).

9. A. I. Kaneneva, M. Ya. Fioshin, L. I. Kazakova, and Sh. M. Itenberg, *Neftekhimiya*, **2**, 550 (1962).

10. F. Beck, private communication.

11. J. L. Fitzjohn, *Chem. Eng. Prog.*, **7** (2), 85 (1975).
12. S. Swann and R. C. Alkire, *Bibliography of Electro-Organic Synthesis 1801–1975*, Electrochemical Society, Inc., Princeton, NJ, 1980.
13(a) C. L. Mantell, *Electrochemical Engineering*, McGraw-Hill, New York, 1960.
   (b) D. E. Danly, paper presented at 89th National AIChE Meeting, Portland, Oregon, Aug. 1980.
14. R. B. McMullin, "Engineering Aspects of Scaling-Up Electroorganic Processes," *Electrochem. Technol.*, 106–113 (March–April 1964).
15. D. Danly, *Industrial Electroorganic Chemistry*, M. M. Baizer, Ed., Dekker, New York, 1973, Chap. 28.
16. A. T. Kuhn and P. M. Wright, *Electrodes for Industrial Processes*, A. T. Kuhn, Ed., Elsevier, New York 1971, Chap. 14, p. 525.
17. K. B. Keating, "Electrode Selection for Electrochemical Processes," paper presented at 78th National AIChE Meeting, Salt Lake City, Utah, Aug. 1974.
18. K. B. Keating and V. D. Sutlic, *Electroorganic Synthesis Technology*, *AIChE Symp. Ser.*, No. 185, **75**, 76 (1979).
19. C. L. Dawes, "Electrical Engineering," in *Mark's Standard Handbook of Mechanical Engineering*, T. Baumeister, Ed., McGraw-Hill, New York, 7th edition, Chap. 15, (1967).
20. D. E. Danly, *Hydrocarbon Process*, 159 (1969).
21. H. Nohe, *Electroorganic Synthesis Technology, AIChE Symp. Ser.*, No. 185, **75**, 69 (1979).
22. K. M. Guthrie, *Chem. Eng.*, **77**, 13 (June 15, 1970).
23. Anonymous, *American Machinist*, "Electrochemical Unit Treats Plating Waste," **114**, Feb. 23, 94 (1970).
24. D. N. Bennion and J. Newman, *J. Appl. Electrochem.*, **2**, 113 (1972).
25. A. K. P. Chu, M. Fleischman, and G. J. Hills, *J. Appl. Electrochem.*, **4**, 323 (1974).
26. M. Fleischman, J. W. Oldfield, and L. Tennakoon, *J. Appl. Electrochem.*, **1**, 103 (1971).
27. A. T. Kuhn and R. W. Houghton, *Electrochim. Acta*, **19**, 733 (1974).
28. C. L. Lopez-Cacicedo, *Trans. Inst. Met. Finishers*, **53** (2), 74 (1975).
29. J. M. Williams (E. I. du Pont de Nemours & Co.), Apparatus for Electrochemical Processing, U.S. Patent 3,859,195 (1975).
30. K. B. Keating and J. M. Williams, 80th National AIChE Meeting, Boston, MA, Sept. 1975.
31. K. B. Keating and J. M. Williams, *Resour. Recovery Conserv.*, **2**, 39 (1976).
32. J. M. Williams and M. C. Olson, 82nd National AIChE Meeting, Atlantic City, NJ, Sept. 1976.

Chapter XI

ERRATA FOR PART I AND PART II

N. L. Weinberg

Electrosynthesis Company,
East Amherst, New York

PART I, CHAPTER IV: ELECTROCHEMICAL OXIDATION OF ORGANIC COMPOUNDS

Page 533; Additional References

[491] A. P. DeBottens, *Z. Elektrochem.*, **8**, 673 (1902).

[492] I. A. Athanasiu, *Bul. Chim. Pura Apl.*, **31**, 75 (1929). *Chem. Ber.*, **64**, 252 (1931).

[493] C. E. Nabuco De Arajuo Jr., *Chem. Zentr.*, **I**, 3394 (1934); **II**, 2748 (1934).

[494] E. Müller, *Z. Elektrochem.*, **7**, 509 (1901).

[495] E. Müller and Friedberger, *Chem. Ber.*, **35**, 2652 (1902).

[496] A. Kramli, *Arch. Biol. Hung.*, **17**, 337 (1947) [*Chem. Abstr.*, **43**, 8908 (1949)].

[497] F. Fichter and E. Elkind, *Chem. Ber.*, **49**, 239 (1916).

[498] J. P. Billon, *J. Electroanal. Chem.*, **1**, 486 (1959/1960).

[499] K. Koyama, T. Susuki and S. Tsutsumi, *Tetrahedron*, **23**, 2675 (1967).

[500] A. H. Maki and D. H. Geske, *J. Chem. Phys.*, **30**, 1356 (1959).

[501] H. Schmidt and J. Noack, *Z. Anorg. Allgem. Chem.*, **296**, 262 (1958).

[502] J. Vedel and B. Tremillon, *J. Electroanal. Chem.*, **1**, 241 (1960).

[503] K. Uneyama and S. Torii, *Tetrahedron Lett.*, **1971**, 329.

[504] M. Y. Fioshin, I. A. Avrutskaya, A. I. Borisov, and L. A. Chupina, *Elektrokhimiya*, **7**, 397 (1971) [*Chem. Abstr.*, **75**, 14178z (1971)].

[505] L. L. Miller, V. R. Koch, M. E. Larscheid, and J. F. Wolf, *Tetrahedron Lett.*, **1971**, 1389.

[506] A. Cooper and C. L. Mantell, *Ind. Eng. Chem. Proc. Design Dev.*, **5**, 238 (1966).

[507] A. Goosen, H. A. H. Lane, and J. Roudnick, *J. S. Afr. Chem. Inst.* **23**, 200 (1970).

[508] R. Woods, *J. Phys. Chem.*, **75**, 354 (1971).

[509] A. Laurent and R. Tardirel, *Compt. Rend.*, **271**, 324 (1970).

[510] S. A. Shevelev, V. A. Kokorekina, L. G. Feoktistov, and A. A. Fainzil'berg, *Sov. Electrochem.*, **6**, 1215 (1970).

PART II, CHAPTER VIII: ELECTROCHEMICAL REDUCTION OF ORGANIC COMPOUNDS

P. 93.   Second to last arrow should have "+2H$^+$."
P. 106.   Delete *"trans"* in last line of table.
P. 107.   Last table entry should be

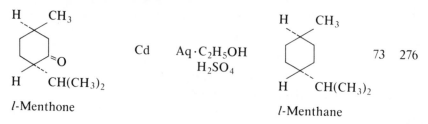

l-Menthone   *l*-Menthane

P. 108.   Fourth table entry should be

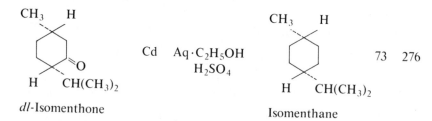

dl-Isomenthone   Isomenthane

P. 110.   Product in second entry is

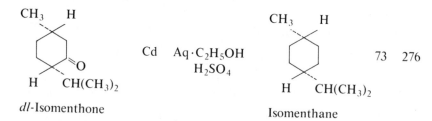

Starting material in third entry is

P. 111.   Second entry, product should be

P. 113. Third starting material entry is "Androsta-1,4-diene-17β-ol-3-one."

P. 118. Delete part of next to last and last entry. Delete

$$R = -OH, -OCH_3 \qquad R = OH \qquad 203$$
$$R = OCH_3$$

P. 119. (a) First entry: "R = OCH_3."

(b) Starting material is "$C_6H_5-\overset{\overset{O}{\|}}{C}-CH_2-\overset{\overset{O}{\|}}{C}-C_6H_5$."
Product in (b) is

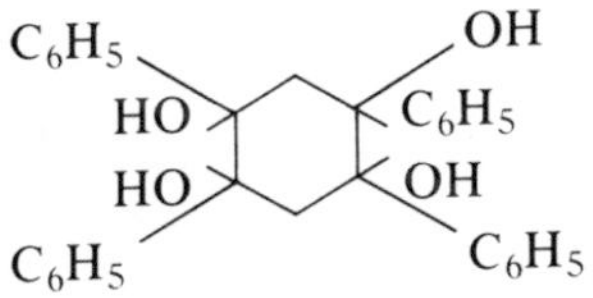

P. 120. Second entry, under Aq. C_2H_5OH, add "R = OCH_3."

P. 121. Fifth product entry line should be "corresponding pinacolone."

P. 125. Fourth reaction product is "2,5-Dihydrophthalic Acid."

P. 126. Seventh reaction product in table should be "m-Carboxybenzyl alcohol."

P. 128. Fifth starting material is

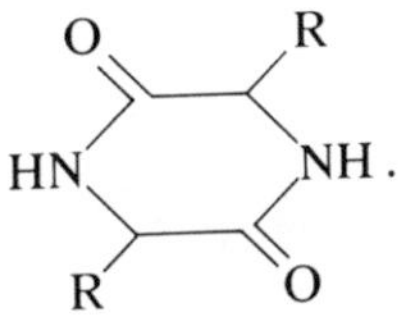

P. 131. Fourth to eighth starting materials and product entries have "N-H" bonds.

P. 135. Starting at second entry:

$R = CH_3, n = 2$ 33

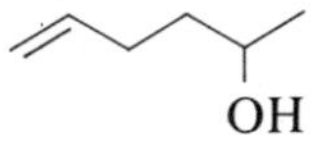

$R = CH_3, n = 3$ 66

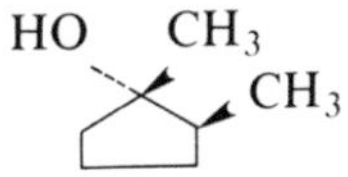

$R = CH_3, n = 4$

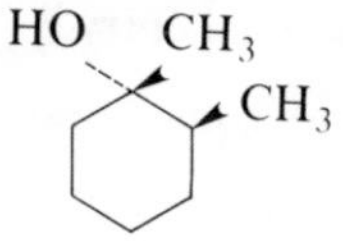

50

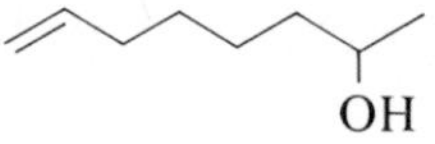

10

$R = CH_3, n = 5$

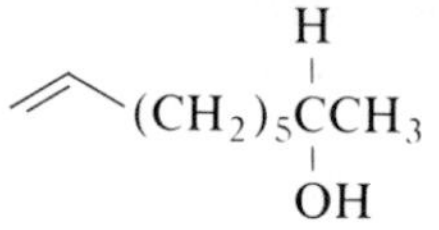

25

$R = C_2H_5, n = 9$ No product

P. 141.   Last equation is

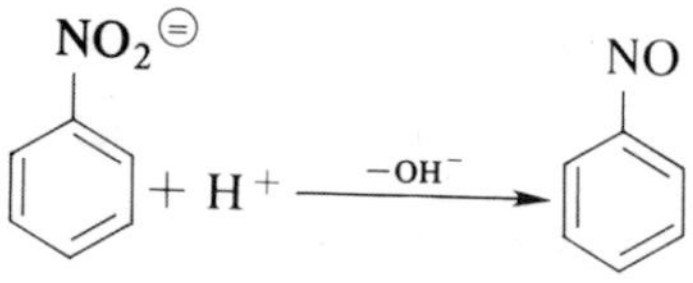

P. 144.   Last line, replace "(849%)" by "(84%)."

P. 145.   Delete fifth product line: "t-Bu-NO$_2$."

P. 155.   Delete third starting material entry formula.

P. 162.   Last line, reaction medium is Aq. HOAc, HCl and product is

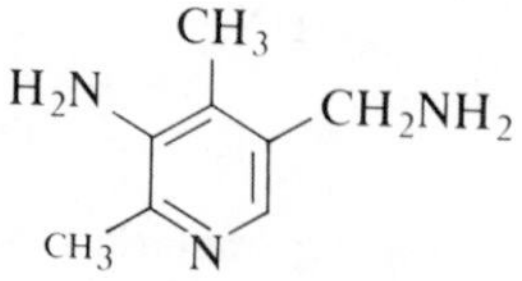

P. 167.   (a)   The next to last entry is "3-(2-Nitrovinyl)-phenanthrene, and
                product 3-(2-Aminoethyl)-phenanthrene, 50% yield."
          (b)   The last entry is "2-(2-Nitrovinyl) phenanthrene, and product
                2-(2-Aminoethyl) phenanthrene, 35–40% yield."
          (c)   A further entry is 9-(2-Nitrovinyl) phenanthrene, and product
                9-(2-Aminoethyl) phenanthrene, 57% yield."

P. 168.   Next to last product entry should be "Tetrahydroquinoline."

P. 175. Last sentence, second last paragraph should be "This is supported by the finding that even the α, ω-dihalide with $n = 1$ affords high yields of cyclopropane (408)."

P. 180. Starting material "$ClCH_2-CH_2-CH_2-CN$" should be "$Cl-CH_2-CH_2-CN$."

P. 184. Line 3, product should be "

"

P. 185. Last product shown should be paracyclophane:

P. 187. The third reaction product entry should be

P. 189. The fourth starting material entry is

This entry is misplaced and should be placed on page 179 as the second entry.

P. 192. (a) Delete second paragraph and substitute the following: "This electrochemical hydrodimerization of activated olefins is a general one and has been successfully applied to activated diolefins to obtain cyclic products."

(b) Move third equation to top of page 196.

P. 199. (a) Equation (2), middle formula is "$R_2\overline{C}-CR_2CR_2-\overline{C}R_2$."

(b) Equation (3), middle formula is "$R_2\dot{C}-CR_2CR_2-\overline{C}R_2$."

(c) Second last line, substitute "H^+" for "K."

P. 204. Seventh reaction product is "2,3-dimethylbutane."

P. 212. Third entry, products should be

$$\overset{\displaystyle O}{\overset{\|}{H_5C_6C}}\text{-}CH_2CH_2CN \quad \text{and} \quad \begin{array}{l} H_5C_6\text{-}CH\text{-}CH_2CN \\ \quad\quad\ \ | \\ H_5C_6\text{-}CH\text{-}CH_2CN \end{array} \quad (87\% \text{ of total})$$

P. 213. First product is "1-Octene."

P. 214. Formulas in last entry should be

$$\begin{array}{ccccccc} & CH_3 & & OH & \\ & | & & | & \\ CH_3\text{-} & C & - & C & \text{-}C{\equiv}CH \\ & | & & | & \\ & CH_3 & & CH_3 & \end{array} \quad \text{and} \quad \begin{array}{ccccccc} & CH_3 & & OH & \\ & | & & | & \\ CH_3\text{-} & C & - & C & \text{-}CH=CH_2 \\ & | & & | & \\ & CH_3 & & CH_3 & \end{array}$$

P. 219. First line, "nonalternant."

P. 226. (a) Product in first entry is

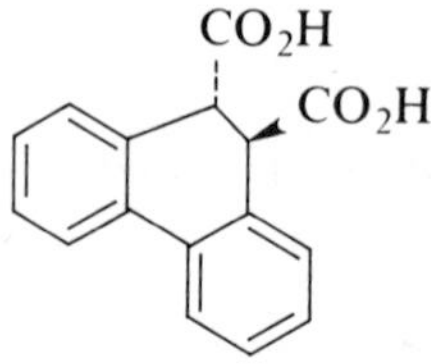
CO2H
CO2H

(b) Starting material in second entry is "anthracene."

P. 227. Product of second equation is

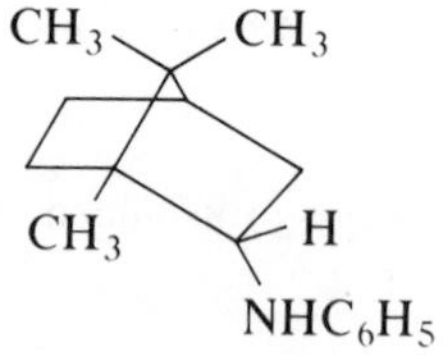
CH3
CH3
CH3
H
NHC6H5

P. 229. Line 5 should be ". . . for example, the semicarbazone of benzal-."

P. 233. Fourth product is "Hydrocinnamaldehyde."

P. 234. Product of last entry is

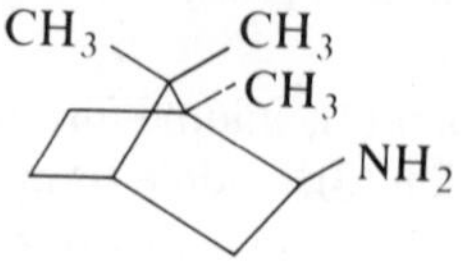
CH3
CH3
CH3
NH2

P. 236. (a) Title is "Semicarbazones."
(b) Next to last line, delete "SNHCl."

P. 238.   Third last line, first word, "detail."

P. 239.   Last line should be "... to afford 1.6 g (100%) of ethylamine hydrochloride."

P. 241.   (a)   Second entry, product is

$$CH_3 \underset{N}{\overset{N}{\diamond}} \begin{matrix} NH_2 \\ CH_2NH_2 \end{matrix}$$

(b)   Sixth starting material is

$$\begin{matrix} NC & & CN & CN \\ \diagdown & & \mid & \diagup \\ C & = & C - C & - \overset{+}{N}(CH_3)_4 \\ \diagup & & & \diagdown \\ NC & & & CN \end{matrix}$$

PART II, CHAPTER IX: PREPARATIVE ELECTROLYSES OF SYNTHETIC AND NATURALLY OCCURRING *N*-HETEROCYCLIC COMPOUNDS

P. 273.   Top equations, replace "MeCOEt" by "$CH_3COCH = CH_2$."

P. 335.   (a)   Entry 75. Starting material is "Pyridoxal oxime." This name should be entered as the third from the last line on page 334, and the entire entry given as 75 on page 335, should be deleted.

PART II, CHAPTER X: ELECTROLYTIC SYNTHESIS AND REACTIONS OF ORGANOMETALLIC COMPOUNDS

P. 443.   The equation is "$RHgCl + e \longrightarrow RHg + Cl^-$."

PART II, CHAPTER XI: ELECTROCHEMICAL SYNTHESIS OF POLYMERS

Pp. 608–611.   Replace monomer "MMA" by "AN" from fourth entry on page 608 to second entry on page 611.

Pp. 611–612.   Replace monomer "MMA" by "Styrenes" from third entry of page 611 to third entry on page 613.

P. 635.   At top, replace "I-BUTYLVINYL ETHER" by "ISOBUTYL VINYL ETHER."

P. 646.   Delete tenth and eleventh lines from the bottom of the page.

INDEX